FROM
MODAL LOGIC
TO
DEDUCTIVE
DATABASES

FROM
MODAL LOGIC
TO
DEDUCTIVE DATABASES

Introducing a Logic Based Approach to Artificial Intelligence

EDITOR
André Thayse

AUTHORS
Paul Gochet, Eric Grégoire, Pascal Gribomont,
Guy Hulin, Alain Pirotte, Dominique Roelants, Dominique Snyers,
André Thayse, Marc Vauclair, Pierre Wolper
with the collaboration of
Philippe Delsarte

JOHN WILEY & SONS
Chichester · New York · Brisbane · Toronto · Singapore

Copyright © 1989 John Wiley & Sons Ltd,
                    Baffins Lane, Chichester,
                    West Sussex PO19 1UD, England

All rights reserved.

No part of this book may be reproduced by any means,
or transmitted, or translated into a machine language
without the written permission of the publisher.

Other Wiley Editorial Offices

John Wiley & Sons, Inc., 605 Third Avenue,
New York, NY 10158-0012, USA

Jacaranda Wiley Ltd, G.P.O. Box 859, Brisbane,
Queensland 4001, Australia

John Wiley & Sons (Canada) Ltd, 22 Worcester Road,
Rexdale, Ontario M9W 1L1, Canada

John Wiley & Sons (SEA) Pte Ltd, 37 Jalan Pemimpin   05-04,
Block B, Union Industrial Building, Singapore 2057

Library of Congress Cataloging in Publication Data:

From modal logic to deductive databases : introducing a logic based
  approach to artificial intelligence / editor, André Thayse ;
  authors, Paul Gochet ... [et al.] with the collaboration of Philippe
  Delsarte.
    p. cm.
  Bibliography: p.
  Includes index.
  ISBN 0 471 92345 1
    1. Artificial intelligence.   2. Logic programming.   I. Thayse,
André, 1940-   .   II. Gochet, Paul.
  Q335.F76 1989
  006.3—dc20
                                                              89-9035
                                                                 CIP

British Library Cataloguing in Publication Data:

From modal logic to deductive databases :
    introducing a logic based approach to
    artificial intelligence.
    1. Artificial intelligence
    I. Thayse, André, *1940–*
    II. Gochet, Paul
    006.3

    ISBN 0 471 92345 1

Printed and bound in Great Britain by Courier International, Tiptree, Essex

List of Contributors

André Thayse   *Philips Research Laboratory, Brussels*
Paul Gochet   *University of Liège*
Eric Grégoire   *University of Louvain, Louvain-la-Neuve*
Pascal Gribomont   *Philips Research Laboratory, Brussels*
Guy Hulin   *Philips Research Laboratory, Brussels*
Alain Pirotte   *Philips Research Laboratory, Brussels*
Dominique Roelants *Philips Research Laboratory, Brussels*
Dominique Snyers   *Philips Research Laboratory, Brussels*
Marc Vauclair   *Philips Research Laboratory, Brussels*
Pierre Wolper   *University of Liège*

with the collaboration of
Philippe Delsarte   *Philips Research Laboratory, Brussels*

Foreword

This book aims at carrying on the study, begun in [Thayse 88], of the concepts and methods of artificial intelligence by taking logic as a guideline.

Knowledge representation is one of the main topics in artificial intelligence. If we want to write a program that is efficient in a given context, we need to provide it with a thorough knowledge of this context, which therefore should be represented in a formalism accessible to the program. The problem of representation lies mainly in determining the most adequate formalisms for representing knowledge and the most efficient methods for handling these formalisms. Knowledge representation thus means encoding domain-specific knowledge into suitable data structures. It is important to be aware that the problem of knowledge representation is strongly connected to the problem of knowledge processing and reasoning. Knowledge representation includes a passive component: a book, a figure, or a memory contains some information. Artificial intelligence systems not only must be able to represent this information, but also must be able to use it in performing their task. One thing that all these systems require is the ability of 'drawing inferences'. To draw an inference means that a new fact is believed to be true on the basis of some initial information. Knowledge representation comprises thus also an active component; knowledge that is represented but cannot be processed is in fact useless for artificial intelligence.

By reviewing the knowledge representation techniques that have been used by humans we will be aware of the importance of the *language*. Knowledge is generally expressed in words and therefore is transmitted by means of a language. Language is a means of knowledge representation and of communication. There essentially are two different types of languages, namely, *natural languages* and *formal languages*. Natural languages such as English and French arose and developed naturally; they did not evolve under the control of any theory. In their current form, such languages have a strong expressive power. They may be employed to analyse highly complex situations and very subtle forms of reasoning. It is the richness of their semantic component and their close relationship with the practical aspects of the contexts in which they are used that give natural languages their great expressive power and their value as a tool for subtle reasoning. Therefore, natural languages are the most adequate tools for representing and processing philosophical, linguistic, symbolic and, more generally, non-formalized knowledge. The qualifier 'natural' is used in opposition with formal languages which are 'artificially' constructed to achieve a particular task. Scientific knowledge is generally represented in terms of formal languages that have been constructed for that purpose. The separation of languages into two well-defined classes (natural and artificial or formal) is somewhat arbitrary. In fact, we can see that certain languages, such as Euclidean geometry for example, present some hybrid characteristics. Euclidean geometry mainly uses the support of a natural language, supplemented with a language of geometric figures. Analytical geometry, which was developed later on, is a translation of the axiomatized theory of Euclid into the formal language of algebra.

Summarizing, any type of knowledge, whether of scientific nature or of purely human nature, is represented and processed by means of a natural language or by means of a formal language. The problem of knowledge representation in artificial intelligence can therefore be considered as the problem of translating natural and formal languages into a formalism that can be understood by a computer, that is, into a programming language. It is at that very point that a particular formal language, namely, the language of formal logic, comes in.

By definition, *formal logic* is concerned with the study of correct forms of reasoning. The framework of formal logic is made up of various tools allowing the formalization and the validation of reasoning: formal inference systems and semantic deductive systems. These reasoning techniques are developed on the top of different logic languages which can be used directly for representing knowledge. The main advantages of these languages are the rigorous syntax and semantics they are provided with. In particular, the semantics plays a reference role by stating an unambiguous correspondence between the knowledge that is specific to the domain of discourse, and the representation of this knowledge in the language. It also justifies the inferences we can draw from the represented knowledge. The framework of logic can therefore be used as such for representing knowledge and for reasoning about it. Knowledge is represented in a declarative way by means of formal expressions of the logic language. Reasoning is defined as being an operation which states the validity or the consistency of a logical assertion. Automatic reasoning in artificial intelligence is achieved by using some particular deductive rules, such as resolution, which can easily be implemented. This gives rise to logic programming techniques which can directly be used for representing knowledge and for reasoning. In view of these arguments, it is not surprising that logic has played a prominent role in artificial intelligence and that most of the existing knowledge representation formalisms are more or less directly based on formal logic.

Our general approach to the topics just outlined (knowledge representation, natural and formal languages, logic languages) is as follows. The problem of knowledge representation in artificial intelligence can be viewed as a two-fold translation problem: translation from natural or formal languages, in which the knowledge is originally represented, into a logic language, followed by translation from this last language into a formalism that can be understood by a computer, that is, into a programming language. Let us see how the different chapters of this book are organized so as to meet this approach to artificial intelligence.

The **first chapter** sets forth the most distinguishing characteristics of *natural* and *formal languages* and describes the formalism of *modal logic*.

We first explain why researchers consider that both natural and formal languages have some principles in common. This will lead us to lay down the foundations of a theory for the translation of a natural or formal language into a particular formal language, namely, the language of logic. The ultimate goal of this process is to allow the manipulation of natural and formal languages by use of computing science techniques, such as those provided by the Prolog language. The logic formalism however does not result in a unique language but in a collection of languages some of which (the *propositional* and the *predicate* languages) have already been studied in [Thayse 88]. These classical logic languages deal with the formalization of valid (that is, absolutely correct) forms of reasoning.

The deductive systems of logic allow us to formalize reasoning in terms of a rigorous proof of a theorem. But on the other hand, artificial intelligence is often concerned with the formalization of less strict forms of reasoning which are in everyday use. A good deal of what we know about the world is 'almost always true' and much commonsense reasoning is only plausible, in the sense that its conclusions and even the facts and rules on which it is based are only approximately true, incomplete or time dependent. The foundations for the representation of revisable reasoning by means of some types of non-classical logics such as *logics of belief and knowledge* have been examined in [Thayse 88]. These logics belong to a class of logics called modal logic. We present in the first chapter a systematic statement of syntax, semantics and axiomatic systems of *modal logic* and we show how different logics used in artificial intelligence such as epistemic, temporal and dynamic logics can be obtained from particular interpretations of modal logic.

Finally we examine three examples of application of logic in the domain of knowledge representation and of reasoning about this knowledge. We introduce the 'wise man problem' which McCarthy has used as a test of models of knowledge, the 'knights and knaves' problem, and the axiomatization of set theory by a collection of clauses, setting down a foundation for the expression of most theorems of mathematics in a form acceptable to a resolution based automated theorem prover.

The **second chapter** deals with the representation of natural language by means of a particular logic language, that is, *intensional logic*.

The formal representation of a natural language has already been considered in [Thayse 88] through Chomsky's phrase structure grammars. These grammars are formed with a set of syntactic rules, also called productions, that are approximations of the syntactic rules of the natural language grammars. It has been shown how the formal rewriting operation on which the syntactic rules of Chomsky's grammars are based, can be interpreted as the first-order resolution rule of logic. More generally the phrase structure grammar formalism can be translated into the Horn clause language which in turn leads to the Prolog logic programming language. The translation of natural language into logic language is thus performed by means of the syntax. In this second chapter a different approach is introduced: *the translation by means of the semantics.*

The method of constructing the syntax of a formal language has been well known since the work of Carnap. Similarly, the method of defining the semantics of a formal language is well established. Tarski showed that the semantic notions of truth, of logical consequence and of validity may be defined rigorously for a formal language such as propositional logic or predicate logic. The semantics of possible worlds, introduced by Kripke, allowed these concepts to be extended to modal logic. The inventors of formal syntax and semantics were the first to ask whether the methods which they had invented for the study of the syntax and semantics of artificial languages could be applied to natural languages.

Between 1968 and 1970, Montague, an American logician from Tarski's school, wrote three articles, aiming at showing that no important theoretical differences exist between natural languages and the formal language of logic. For that purpose, Montague introduced his *intensional logic* and showed how a well-defined 'fragment of English' can be translated in this logic. In the second chapter we wish to acquaint the reader with the fundamentals of the formal language of intensional logic developed by Montague for representing natural languages. Its most salient feature is that it allows techniques of mathematical logic to be applied to the semantics of natural languages. In fact, the translation of a natural language into intensional logic allows us not only to verify whether a sentence is true but also to understand the meaning of this sentence, independently of its truth conditions. Generally, we do not know the reference (or extension) of the components of a sentence but only their understanding (or intension). That is why we are able to understand a sentence without knowing whether it is true. So extensional logic has the possibility of determining whether a sentence is true while intensional logic goes a step further and tries to understand the sentence.

The **third chapter** introduces the concept of *categorial grammar*, by taking as a typical example the grammar defined by Montague for describing a fragment of English language.

After having developed a formal language (intensional logic) as close as possible to natural language, we want to bridge the remaining gap between both. The syntactic rules of the fragment of English are formulated in the basic papers by Montague in terms of recursive definitions specifying how complex sentences are to be formed out of simpler ones. These syntactic rules constitute a *categorial grammar*. Previously, researchers in the domain of formal representation of natural languages tried to develop a notation to restrain natural language to an existing formal language such as predicate calculus. The spirit of Montague's contribution is exactly the opposite: it models logical calculations according to the syntax of natural language, with the aim of formalizing valid reasoning processes which resist being so restraint. Montague did this in defining an adequate categorial grammar, called the *PTQ* grammar, that describes the syntactic rules for a fragment of English. In his approach, Montague does not interpret the words of the vocabulary and the *PTQ* syntactic rules. Instead, he proceeds indirectly by translating each natural language basic expression into the intensional logic language that has been presented in the second chapter. The semantics of the intensional logic language provides us (indirectly) with an interpretation of the fragment of English.

Montague does not ignore the obvious differences between natural and formal languages, and acknowledges that both evolved differently and have different functions, but he denies that these practical differences indicate a theoretical difference. Montague's general approach, towards formulating a syntax and a semantics for natural language which have the same rigour as those of formal languages, is explained in the second and third chapters. In particular, the third chapter describes the *PTQ* grammar and the translation of the fragment of English into the intensional logic language.

The **fourth chapter** is devoted to *temporal logic* and its applications in artificial intelligence.

In the first chapter we have developed the formalism of modal logic and have shown how it can be interpreted in terms of the logics of knowledge and belief, of temporal logic and of dynamic logic. The first chapter contains only a description of the syntax, semantics, axiomatic systems and inference rules of these logics. Stated otherwise, it provides some elementary building blocks for potential constructions that could be realized by means of each of these logics; it does not explicitly develop examples of applications for them. The purpose of the fourth chapter is to present one of these logics in a more detailed manner. We present the specific theory of temporal logic and some of its most typical applications (just as the theory and applications of the logics of belief and knowledge has been presented in chapter 4 of [Thayse 88]). Temporal logic has been studied as a branch of logic for several decades. It was developed as a logical framework to formalize reasoning about time, temporal relations such as 'before' or 'after' and related concepts such as tenses. As said above, a temporal logic can be viewed as a special case of modal logic and, therefore, its origins can be traced to the same source. The theory of temporal logic is an integral concern of philosophical inquiry, and questions of the nature of time and of temporal concepts have preoccupied philosophers since the inception of the subject.

In its *linear* version, temporal logic allows us to describe *sequences of states* and to reason about them. We can interpret these states in different ways. In a pure temporal interpretation, the sequence of states represents the evolution of the state of the world for a given interval of time. In another interpretation we consider, it is suggested that temporal logic can be a useful tool to formalize reasoning about the execution sequence of programs and especially of concurrent programs. In that approach, the sequence of states a machine goes through during a computation is viewed as the temporal sequence of worlds described by temporal logic. This interpretation allows us to use temporal logic for stating and proving properties about programs and especially about concurrent programs. As the program cannot be represented explicitly in terms of temporal logic formulas, it has to be encoded in a set of statements that basically represent the allowable transitions in each state. It is sometimes interesting, as for example in the context of the verification and specification of non-deterministic programs, to consider *branching-time*, that is, 'non-linear', temporal logic, where a given time instant can have several futures. We also give a syntax and an axiomatic system for this type of logic and present some of its most significant applications.

The **fifth chapter** deals with logic and *revisable reasoning* in artificial intelligence.

The deductive systems of classical logics provides a formalization of valid reasoning in terms of an unfailing proof of a theorem. Artificial intelligence is mainly concerned with the formalization of a less restrictive type of reasoning, like that used in everyday life. A good deal of our knowledge about the world is 'almost always true'. Commonsense reasoning is quite often no better than plausible, and might be subject to revision, because the facts and rules on which it is based are uncertain, incomplete or time sensitive. In the fifth chapter, we continue the presentation of logical approaches to the formalization of revisable reasoning that was undertaken in chapter 4 of [Thayse 88]. Logics designed to formalize revisable reasoning are called *non-monotonic logics*.

Many domains of artificial intelligence can benefit from a formal theory of revisable reasoning. As a first example, we can mention the formalization of certain aspects of our elementary perception, like vision, which involve revisable reasoning because they operate with the help of iterated approximations and corrections. Similarly, high-level and specialized human tasks like diagnosis can involve revisable forms of reasoning. The formalization of reasoning about plans and actions, the representation of prototypal information and hierarchical knowledge that contain exceptions, and a correct reasoning about these forms of knowledge may require formal theories of revisable reasoning. Such formal tools can also be useful in the development of speech acts theories. Indeed, the communication process is often based upon implicit information and conventions. Lastly, a correct treatment of incomplete, negative or implicit information both in databases (classical or deductive ones) and in logic programming systems requires the formalization of forms of revisable reasoning. In this chapter, we present non-monotonic logics that can be applied in the domains just mentioned. The fundamental principle upon which these logics are based is as follows. Reasoning can be non-monotonic when it is built from pieces of incomplete or time-dependent information and thus requires specific 'augmenting conventions' about the unknown knowledge. These conventions often are *minimization conventions* in that they often circumscribe and restrict the positive *implicit* information that incomplete knowledge may involve. The non-monotonic logics we are interested in are based upon such augmenting conventions. We successively study the *closed-world assumption, negation as failure, predicate completion* and *circumscription*. These techniques are developed in the framework of classical logic; they are best studied from a semantic point of view.

The **sixth** chapter describes the semantics of relational databases and of some of their extensions in terms of classical logic.

Two approaches, the *model-theoretic* and *proof-theoretic* approaches, are presented and compared. The model-theoretic approach derives from the traditional view of *relational databases*. There, models are associated with the database. A formula is evaluated to true in the database if it is true in all associated models. Recently, when extending relational databases with deduction rules, the proof-theoretic approach has been developed. In that approach, the database is represented by a first-order theory. A formula is then evaluated to true in the database if it is a theorem of the associated theory. The semantics of a relational database can precisely be described with a single model, which is the database itself, roughly speaking. Moreover, the theory constructed in the proof-theoretic approach has exactly that model as its unique model.

The distinction between the two approaches becomes more fruitful when considering *deductive databases*. A deductive database is a relational database plus a set of rules defining new relations. These rules may be recursive. In deductive databases with recursion but without negation, it is still possible to choose a single model that characterizes exactly the information in the database. Moreover, there exists a theory that has that model as its unique model. For deductive databases without recursion, an efficient query evaluation algorithm can easily be constructed from the theory. For the more interesting case where recursion is allowed, constructing such a theory becomes much more complex and hence query evaluation has to be based on the model-theoretic approach.

In deductive databases with both negation and recursion, it is in general necessary to consider a *set* of models, each of which represents a world possibly modelled by the database. For some special cases (one is presented: the stratified databases) the selection of a unique natural model is still possible and a complete query evaluation algorithm can be developed.

Finally in relational databases with *incomplete information*, the model-theoretic approach has to consider a set of models among which it is impossible to choose. Incomplete information is also easily represented in the proof-theoretic approach. The constructed theory is, however, not complete. The query evaluation algorithms derived from both approaches are sound but not complete.

In this foreword, frequent reference is made to the book *From Standard Logic to Logic Programming* [Thayse 88], which deals with classical logics and their applications to artificial intelligence, and introduces some non-standard logics in connection with revisable reasoning. It should, however, be stressed that the present book is self-contained and can be read independently of this former volume. The reading presupposes some acquaintance with classical logic, but, in the concern to achieve completeness, two subsections in the first chapter have been devoted to summarizing the main concepts of classical logics that are described in detail in [Thayse 88]. The fact remains that the first volume deals with more elementary subjects than the present one and that it may therefore be considered as a first part of a comprehensive work describing how logic can be used in the context of artificial intelligence. The connections between the chapters of both volumes can be schematized as in the figure given at the end of this foreword.

The authorship of the different chapters of this book is as follows.

1. Languages and logics:
   Dominique Snyers and André Thayse.

2. Intensional logic and natural language:
   Paul Gochet and André Thayse.

3. Montague's semantics:
   Paul Gochet and André Thayse.

4. Temporal logic:
   Pascal Gribomont and Pierre Wolper.

5. Formalization of revisable reasoning:
   Eric Grégoire.

6. Logic and databases:
   Guy Hulin, Alain Pirotte, Dominique Roelants and Marc Vauclair.

A third volume is planned; it will be devoted to the use of the various logic languages and formal grammars described in the first two volumes, for some of the most significant application areas of artificial intelligence and computing science. The following subjects will be considered.

1. Natural language processing.

2. Logic and grammars in speech understanding.

3. Knowledge base construction for expert systems.

4. Revisable reasoning and truth maintenance systems.

5. A logic for requirements engineering.

6. Logic and learning.

7. Meta-level programming in Prolog.

André Thayse
Brussels, December 1988

xvi

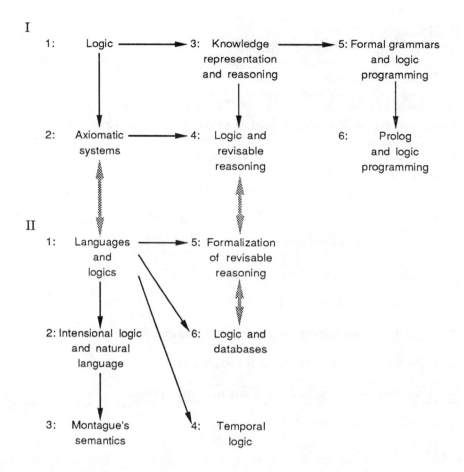

Acknowledgements

Pierre Dupont, Daniel Dzierzgowski, Christine Froidevaux, Yves Kamp, Jean Pirnay, Pierre-Yves Schobbens, Michel Sintzoff and Philippe Smets performed a critical reading of parts of the manuscript.

Jay Darien and Matt Noesen participated in the translation of the text from French to English.

The work of Eric Grégoire was supported by the Belgian governmental organization: Institut pour la Recherche Scientifique dans l'Industrie et l'Agriculture, under Grant Number 4856.

The work of Guy Hulin, Alain Pirotte, Dominique Roelants and Marc Vauclair was supported by the European Community Esprit project under Grant ESTEAM 316.

Anne-Marie De Ceuster and Edith Moës participated in the typing of the text.

Contents

1 Languages and logics   **1**
   1.1 Logic processing of languages 1
     1.1.1 Introduction 1
     1.1.2 Natural languages and formal languages 1
     1.1.3 Properties of natural languages 3
     1.1.4 Properties of formal languages 4
     1.1.5 Computer manipulation of natural languages 6
     1.1.6 Computer manipulation of formal languages 7
     1.1.7 Conclusion 8
   1.2 Modal logic 9
     1.2.1 Introduction 9
     1.2.2 Propositional logic 10
     1.2.3 Syntax and semantics of modal logic 13
     1.2.4 Examples and applications 16
     1.2.5 Binary relations and formula schemata 19
     1.2.6 Valuations and tautologies 20
     1.2.7 Logics 21
     1.2.8 Axiomatic systems 23
     1.2.9 Multimodal languages 24
     1.2.10 Temporal logic 25
     1.2.11 Dynamic logic 30
     1.2.12 Logics of knowledge and of belief 32
     1.2.13 First-order predicate logic 32
     1.2.14 Model for a first-order logic 35
     1.2.15 Predicate modal logic 36
     1.2.16 Model and semantics for a modal predicate logic 40
     1.2.17 First-order dynamic logic 42
   1.3 Examples and applications 43
     1.3.1 Introduction 43
     1.3.2 Modal logic and reasoning about knowledge 44
     1.3.3 Predicate logic and formalization of reasoning 46
     1.3.4 Mathematics and predicate logic 51

2 Intensional logic and natural language   **55**
   2.1 Logic processing of natural languages 55
     2.1.1 Introduction 55
     2.1.2 Formal semantics and natural language 55

		2.1.3	Davidson's recursive semantics	57
		2.1.4	Montague's innovations	59
		2.1.5	Montague's approach	60
	2.2	An extensional type-theoretic logic		62
		2.2.1	Introduction	62
		2.2.2	Syntax and semantics of a propositional language	62
		2.2.3	Absolute truth and truth relative to a model	65
		2.2.4	Syntax and semantics of a first-order language	68
		2.2.5	Example	70
		2.2.6	Function-theoretic definition of semantics	71
		2.2.7	A type-theoretic version of the language $\mathcal{L}_1$	75
		2.2.8	A higher-order type-theoretic language	79
		2.2.9	Lambda calculus and the language $\mathcal{L}_1$	82
		2.2.10	Lambda calculus and the language $\mathcal{L}_t$	84
		2.2.11	The multimodal language $\mathcal{L}_m$	88
		2.2.12	Example	91
	2.3	An intensional type-theoretic logic		93
		2.3.1	Introduction	93
		2.3.2	Sense and reference	94
		2.3.3	Compositionality in non-referential context	95
		2.3.4	Intension and extension	96
		2.3.5	Example	97
		2.3.6	Formal definition of intension	98
		2.3.7	Syntax of intensional logic	100
		2.3.8	Model-theoretic interpretation of intensional logic	102
		2.3.9	Semantics of intensional logic	104
		2.3.10	Summary of intensional logic	105
		2.3.11	The adverb 'necessarily'	110
		2.3.12	Extensional and intensional adjectives	110
		2.3.13	Uniformization of the semantic treatment	112
3	**Montague's semantics**			**115**
	3.1	Formal representation of natural languages		115
		3.1.1	Introduction	115
		3.1.2	Montague's programme	116
		3.1.3	Natural and formal languages	117
		3.1.4	Conditions for the suitability of a semantics	119
		3.1.5	A formal semantics for natural language	121
		3.1.6	A categorial grammar	122
		3.1.7	A recursive semantics	123
	3.2	Montague's grammar		125

	3.2.1	Introduction	125
	3.2.2	Montague's categorial grammar	126
	3.2.3	Syntax of the *PTQ* categorial grammar	129
	3.2.4	Examples	131
	3.2.5	Introduction to the translation formalism	137
	3.2.6	Translation of basic expressions	138
	3.2.7	Translation of the rules of functional application	140
	3.2.8	Adequacy criterion for a logic of natural language	141
	3.2.9	Translation rules for determiners	143
	3.2.10	Additional syntactic and translation rules	144
	3.2.11	Examples of extensional sentence translations	146
	3.2.12	Examples of intensional sentence translations	148
	3.2.13	Meaning postulates	150
	3.2.14	Examples of meaning postulates	152
	3.2.15	Translation of sentences with modalities	153
3.3	Conclusion and alternative formalisms	155
	3.3.1	Difficulties inherent in Montague's semantics	155
	3.3.2	Theory of discourse representations	156
	3.3.3	Situation Semantics	158
	3.3.4	Boolean semantics	160
	3.3.5	Generalized phrase structure grammars	162

4 Temporal logic 165
4.1	Propositional linear time temporal logic	165
	4.1.1 Introduction	165
	4.1.2 Syntax of linear time temporal logic	165
	4.1.3 Semantics of linear time temporal logic	166
	4.1.4 Examples	168
	4.1.5 Axiomatization	169
4.2	Temporal logic and finite automata	171
	4.2.1 Finite automata on infinite words	172
	4.2.2 ω-Regular languages	175
	4.2.3 From temporal logic to Büchi automata	179
	4.2.4 The Büchi automaton corresponding to a formula	182
	4.2.5 Building the local automaton	184
	4.2.6 Building the eventuality automaton	185
	4.2.7 Composing the two automata	187
	4.2.8 The automaton on the alphabet 2^P	189
	4.2.9 Algorithmic considerations	189
	4.2.10 Extended Temporal Logic	190
	4.2.11 Definition of extended temporal logic	192

	4.2.12 Automata on finite words 195
	4.2.13 Conclusion 196
4.3 Predicate linear temporal logic 196
	4.3.1 Introduction 196
	4.3.2 Syntax 197
	4.3.3 Semantics 198
	4.3.4 Axiomatic system 200
	4.3.5 First-order temporal theories 202
4.4 Applications of temporal logic 202
	4.4.1 Introduction 202
	4.4.2 An introductory example 203
	4.4.3 Program modelling 207
	4.4.4 Example: Peterson's algorithm 208
	4.4.5 The temporal theory of a program 210
	4.4.6 Program properties 214
	4.4.7 Invariance properties for Peterson's algorithm 215
	4.4.8 Fairness of Peterson's algorithm 216
	4.4.9 Verification using propositional temporal logic 218
	4.4.10 Modelling programs with finite automata 219
	4.4.11 The automata representing the processes 220
	4.4.12 The automata representing the variables 222
	4.4.13 Modelling the fairness hypothesis 224
	4.4.14 Specification and verification 225
4.5 Propositional branching-time temporal logic 227
	4.5.1 Introduction 227
	4.5.2 Syntax 228
	4.5.3 Semantics 228
	4.5.4 Computation tree logic 230
	4.5.5 Application to program verification 231

5 Formalization of revisable reasoning   235
5.1 Introduction 235
	5.1.1 Revisable reasoning in artificial intelligence 235
	5.1.2 Forms of revisable reasoning 236
	5.1.3 Main properties of non-monotonic logics 237
	5.1.4 Reiter's, McDermott's, and Moore's logics 237
	5.1.5 Minimization based approaches 238
5.2 Logic framework 240
	5.2.1 Logic language 240
	5.2.2 Model-theoretic and proof-theoretic considerations . . . 241
	5.2.3 Clausal forms 242

Table of Contents xxiii

	5.2.4	Different sublanguages	243
	5.2.5	Canonical Herbrand models	246
5.3	The closed-world assumption		247
	5.3.1	Introduction	247
	5.3.2	Formal definition	248
	5.3.3	Properties and limitations	248
	5.3.4	Additional conventions	250
	5.3.5	Restricted closed-world assumption	251
5.4	Predicate completion		252
	5.4.1	Introduction	252
	5.4.2	Formal definition	254
	5.4.3	Example	255
	5.4.4	Properties and drawbacks of predicate completion	256
	5.4.5	Closed-world assumption and predicate completion	257
5.5	Specific application domains		257
	5.5.1	Logic programming	257
	5.5.2	Deductive databases	260
5.6	Circumscription		262
	5.6.1	Introduction	262
	5.6.2	Minimal models	262
	5.6.3	Formal definition	263
	5.6.4	Example	264
	5.6.5	Properties	266
5.7	The preferential-model approach		268
	5.7.1	Introduction	268
	5.7.2	Preferred models	269
	5.7.3	Application to non-monotonic logics	270
	5.7.4	Possible generalizations	272
	5.7.5	The rationalities of revisable reasoning	273
5.8	Conclusion		277

6 Logic and databases   279

6.1	Introduction		279
6.2	Relational model		281
	6.2.1	Importance of the relational model	281
	6.2.2	Fundamental relational concepts	282
	6.2.3	Relational algebra	286
	6.2.4	Relational language	288
	6.2.5	Well-formedness of queries	291
	6.2.6	From relational language to relational algebra	292
6.3	Logical approach to databases		294

	6.3.1	Definition	294
	6.3.2	Model-theoretic approach	295
	6.3.3	Proof-theoretic approach	296
	6.3.4	Correspondence between approaches	298
	6.3.5	Summary	300
6.4	Definite deductive databases	301
	6.4.1	Introduction	301
	6.4.2	Deductive databases	302
	6.4.3	Model-theoretic approach	305
	6.4.4	Proof-theoretic approach with $\mathcal{T}_B^{CWA}$	306
	6.4.5	Hierarchical databases	308
	6.4.6	Recursive rules in definite databases	312
6.5	Negation in deductive databases	317
	6.5.1	Introduction	317
	6.5.2	The closed-world assumption and deductive databases	319
	6.5.3	The generalized closed-world assumption	322
	6.5.4	Stratified databases	326
6.6	Incomplete information	332
	6.6.1	Introduction	332
	6.6.2	Syntax	334
	6.6.3	Model-theoretic approach	336
	6.6.4	Query evaluation in the model-theoretic approach	339
	6.6.5	Proof-theoretic approach	342
	6.6.6	Query evaluation in the proof-theoretic approach	344
	6.6.7	Correspondence between approaches	348
	6.6.8	Alternative method of query evaluation	348
6.7	Conclusion	349

Bibliography 351

Index 370

Chapter 1

Languages and logics

1.1 Logic processing of languages

1.1.1 Introduction

There are essentially two recognized types of languages, namely, natural languages and formal languages. Natural languages such as French and English began and developed naturally, that is, without the control of any theory. Theories of natural languages, such as grammars, were established *a priori*, that is, after the language had already matured. Formal languages, such as mathematics and logic, were generally developed from a theory, established *a posteriori*, which provided the basis for these languages. The goal of this section is to set forth the most distinguishing characteristics of these two classes of languages and to present the reasons that led researchers to consider the possibility of the existence of principles common to both natural and formal languages. This will lead us to lay the foundations of a theory the goal of which is to permit the translation (or the expression) of a natural or a formal language into a particular formal language, namely, the language of logic. The ultimate goal of this process is to allow the manipulation of natural and formal languages using computing science techniques, such as the Prolog language. We know that logic programming languages such as Prolog are derived from the formal language of logic: Prolog executes statements which are sentences of a particularly elementary logic language, namely, the language of Horn clauses. The relation that exists between logic and Prolog has been studied in chapters 5 and 6 of [Thayse 88]. Prolog is, in some way, logic in action; it 'executes' logical formulas. Stated otherwise, logic programming is the programming counterpart of logic.

1.1.2 Natural languages and formal languages

A language is considered to be a set of sentences. Usually that set is infinite. We distinguish two classes of languages, namely, the *natural languages* such as French and English and the *formal languages* such as mathematics and logic. The English language can be defined as the set (theoretically infinite) of all English sentences. These sentences are naturally consistent with practical human experience which is automatically organized while at the same time organizing the language itself [John and Bennett 82]. An English sentence is a finite sequence of English words; we know that the set of these words is finite. An English sentence can therefore be regarded as a finite sequence of elements taken from a given finite set. Nevertheless, not all the combinations of words are permitted; it is necessary that these combinations be correct (with respect to a syntax) and have a sense (with respect to a semantics). These syntax and semantics constitute a sort of theory of the English language: that which permits definition of all English sentences and thus of the English language. In the case of a natural language such as English, the formation of sentences of the language preceded the formalization of that language by means of a theory or a grammar. Such a language is for that reason qualified as natural which is to say non-artificial or non-constructed. The qualifier 'natural' opposes the languages in question to formal languages which are constructed and therefore artificial. A formal language such as logic consists of a set of *sentences*, often called *formulas* or *well-formed expressions*, which we can obtain by application of the laws of logic. We have qualified this 'artificial language', meaning that the language was formed by means of rules and axioms of formation. An extreme point of view may lead us to say that in fact it is the pure imagination of the theoretician which the rules and axioms of a new artificial language come from. The qualifier 'formal' refers specifically to the fact that the sentences of these languages consist of a list of symbols (logical or mathematical) subject to diverse interpretations. Contrary to natural languages where a word in a sentence possesses a well-determined meaning, a symbol in a formal sentence does not carry any intrinsic meaning. A number, a symbol can represent anything. They are able to hold whatever meaning we choose in any given application. This meaning must be rigorously specified by the methods of interpretation of formal systems; the methods in question constitute the semantics of the formal language.

Thus, we are considering two types of languages, which differ significantly from each other by their origin and by their application area. First of all we are going to attempt to identify the most important properties of these two types of languages. This will lead us to examine one of the questions which founded artificial intelligence, namely, 'to what extent can the natural languages be represented (translated) by means of formal languages ?' More specifically, we will be interested in the utilization of computer languages aiming at representing and manipulating natural languages.

Remark. The partition of the set of languages into two well-defined classes is somewhat arbitrary. In effect, we can conceive some languages, Euclidean geometry, for example, which presents some hybrid characteristics. Historically, Euclidean geometry was first developed by use of the support of a natural language supplemented by a language of (geometric) figures; thereafter, it formed itself into a completely axiomatized theory, within a natural language and a language of geometric figures. Analytical geometry is a translation of the axiomatized theory of Euclid into a formal algebraic language. Since the introduction of analytical geometry by Descartes, the two formalisms (algebraic and natural languages) are used jointly. Neither of the two languages has ever supplanted the other. The use of the theory of real functions offers an example of a rather different evolution. This theory was developed initially with the aid of natural language. Such an approach requires the use of increasingly bulky 'literary' constructions, and one has passed on progressively to formalization by means of an algebraic notation (language). The switch to a formal language permitted the theory of real functions to attain a state of development which the usage of natural language might never have. In this case, formal language progressively took the place of natural language.

1.1.3 Properties of natural languages

Language is the function that expresses thoughts and communication between people. This function is carried out by way of vocal signals (speech) and possibly by written signs (writing) which make up the natural language. Language, inasmuch as it serves as a means of expression for thought, seems to be an essential element in the make-up of a human being. With respect to our world, language allows us to designate actual things (and to reason about them) and to create meaning. Contrary to what certain formal linguistic theories would lead one to believe, natural language was not founded on a priori rational truth (in which case the ideal form of the language would be that of a formal language), but it was developed and organized from human experience, in the same process as that by which human experience was organizing itself. In their current form, natural languages such as French and English have a great expressive power. Such languages may be employed to analyse highly complex situations and to reason very subtly. It is the richness of their semantic component and their close relationship with the practical aspects of the contexts in which they are used which give natural languages their great expressive power and their value as a tool for subtle reasoning. In what follows herein, we will see how difficult it is to formalize the semantic component of a natural language, i.e., the constituent of the language by which sentences have or acquire their meanings [Flores and Winograd 79]. On the other hand, the syntax of a natural language may easily be modelled by a formal language similar to those found in logic and mathematics. Another unique property of natural languages is polysemy, i.e., the possibility that a word or a sentence have several meanings, several values. For example, the word 'pair' can first be considered as a substantive; it is then used in phrase structures as: 'arranged in pairs', 'carriage and pair', 'the happy pair'. It can, however, also be interpreted as a transitive verb; typical phrase structures are in this case: 'two vases that pair', 'to pair off with someone'. The polysemic character of a language obviously tends to increase the richness of its semantic component. Moreover, this trait makes formalization difficult if not impossible. The polysemy of natural languages is often considered to be a property acquired recently [Atlan 86]. The earliest forms of natural language would have been similar to formal languages while polysemy would be the result of a progressive enrichment.

In sum, natural languages are distinguished by the following properties:

- Development by progressive enrichment before any attempts at formation of a theory.

- Importance of expressive character due greatly to the richness of the semantic component (polysemy).

- Difficulty or impossibility of a complete formalization.

1.1.4 Properties of formal languages

Natural languages differ greatly from formal languages such as mathematics and logic. The definition (e.g. axiomatic) of a theory of a given formal language preceded the formation of sentences (or formulas) of this language. The processes of generation and development of a formal language are reversed with respect to those of a natural language. Consequently, the words and the sentences of a formal language are perfectly defined (a word keeps the same meaning regardless of context or use). In addition, the meaning of symbols is determined exclusively by the syntax, without reference to any semantic content; a function and a formula can designate anything at all. Only the operators and relations which allow us to write a formula, such as equality, inequality, membership, non-membership, logical connectives $\vee, \wedge, \neg$, etc., and algebraic operators $+, \times, \sqrt{}$, etc., have specific meaning. Formal languages are, therefore, necessarily devoid of any semantic component save that of their operators and relations. Once again, this differs from the polysemic nature of natural languages. It is thanks to this lack of special meanings that formal languages may be used to model a theory of a mechanical, electrical, linguistical or other nature which then assumes the status of the semantic component of such a language. That is to say that, during the conception of formal languages from which all ambiguity of meaning is eliminated, it is as if this reduction to

1.1. LOGIC PROCESSING OF LANGUAGES

unique meaning must manifest itself as an elimination of the 'world of meanings' in the process of constructing the formulas, as far as the abstract level of these constructions is concerned [Atlan 86]. It is only by way of an additional step that meaning is attached to the formulas; this step later allows us the possibility of applying a true/false criterion to each formula. The world of meanings, that is, the semantic component, only exists in the theory which one attempts to express through the formal language. For example, a semantic component often associated with the formal language of conic theory is the motion of celestial bodies; (finite) linear systems of all sorts are possible semantic components of matrix theory. Our primary interest in this book is that of assigning the semantic component of natural languages to (formal) computer languages. One of the principal goals of artificial intelligence is the manipulation of natural languages with computing science methods. Such a manipulation is based on the assignment of the semantic component of natural language (or a subset thereof) to certain logic languages such as first-order languages or modal languages.

One cannot avoid mentioning the importance of numbers in formal languages. In a numbering system as well as in a calculation system, numbers always have the potential to refer to a certain 'content' which will then belong to the semantic component of the language: that of possible objects, when these are countable or measurable. The association of a meaning with a number or calculation is not always obvious, however. It is worth remembering that in physics, when one completes a calculation and then seeks to interpret it, one only keeps the positive numbers among the results. The negative or imaginary solutions to equations which are supposed to describe reality are the most often rejected because they do not correspond with 'physical reality'.

In sum, formal languages are characterized by the following properties:

- Development from a pre-established theory.

- Minimal semantic component.

- Possibility of increasing the semantic component according to the theory to formalize.

- Syntax which produces unambiguous sentences.

- Importance of the role of numbers.

- Complete formalization and therefore the potential for computer implementation.

1.1.5 Computer manipulation of natural languages

A fundamental goal of artificial intelligence is the manipulation of natural languages using the tools of computing science. In this perspective, the various computer languages play a primary role: they often form the necessary link between natural languages and their manipulation by machine. We must remember first of all that a programming language such as Prolog is a tool for writing computer programs. The definition of Prolog originates from formal logic: Prolog executes statements which are none other than sentences from a particularly elementary logic language, namely, the language of Horn clauses. Moreover, a paradoxical thesis of formal semantics states that there is no essential difference between natural languages and formal (computer) languages which would prevent the former from acting as guides in the analysis of syntactic and semantic phenomena [Montague 73], [Montague 74a], [Montague 74b], [Montague 74c], [Montague 74d], [Nef 84]. This thesis does not state a possible identification between artificial languages such as first-order logic or modal logic, and natural languages such as French or English. We have indeed shown in the preceding subsections the contrast which exists between the (syntactic and semantic) poverty of formal languages and the richness of natural languages. One can admit that natural languages are much more complex than formal logic languages, but nevertheless try to represent fragments (of increasing size) of these natural languages by equipping computer systems with more and more expressive components (by passing from propositional logic to predicate logic and modal logic).

A second fundamental difference between natural language and logic language, which should be taken into account in the representation of the former by the latter, is that which results from the central place occupied by the concept of truth in logic. The role of logical analysis is above all to determine if a sentence (formula) of a language is valid (true in all of its interpretations), absurd (false in all of its interpretations) or simply consistent (true in at least one of its interpretations). A good number of sentences in a natural language are non-declarative, vague or indeterminate statements such as 'did Paul understand ?', 'Jack is young', or 'I may come' to which it seems at first glance difficult to assign truth values (true or false). A way to bypass the difficulty resulting from the importance given or not given to the concept of the truth of a sentence is to introduce the notions of *proposition* and of *expressed proposition* [Nef 84].

A *proposition* is the content of a sentence, i.e., the set of situations (possible worlds) in which the sentence is true. The concept of proposition takes into account the fact that an important function of natural language is to refer to objects and situations, that is, to interpret each sentence of the language as a fragment of reality. The truth value of a proposition depends, therefore,

1.1. LOGIC PROCESSING OF LANGUAGES

not only on the relations between the words of the language and the objects in the world but also on the state of the world and on the knowledge about that state. The truth value of the sentence 'Paul runs' depends not only on the person denoted by 'Paul' and the meaning of the verb 'to run', but also on the moment when this sentence is expressed. Paul probably runs now and then, but he certainly does not always run. The truth value of the sentence 'Peter thinks that Paul runs' depends as well on Peter's knowledge: Peter is probably wrong now and again. If we translate natural language, necessarily interpreted as reality, into logic language, the latter automatically takes on this interpretation. We must still guarantee the complete formalization of this interpretation with a formal semantics like that of possible worlds [Kripke 63]. As a result, the truth analysis of the sentences (formulas) of this logic language allows us to reason about the reality that it represents.

Let us now look at the difference between proposition and expressed proposition. It is not always necessary that an agent be able to say whether a sentence that he expresses is true or false with respect to a certain state of reality; a language should not be reduced to the sole function of referring to a real world or to a possible world. As to the formalization of a natural language by way of a logic language, it is necessary that the latter be sufficiently expressive to reflect all of the subtleties of the former. In other words, these logic languages must be able to reflect, in the formalization, all of the extra-referential functions of the language, which are often much more complex than the referential functions. By 'expressed proposition' we thus mean the specific sense of this proposition, independently of any truth analysis.

1.1.6 Computer manipulation of formal languages

The difficulties (evoked in the preceding subsection) inherent in the representation of natural languages by means of a logic language do not occur in the process of translating a formal language into logic. This is primarily due to the fact that predicate logic was constructed with the representation and study of mathematical languages in mind. Like the sentences of logic, the mathematical statements do not depend upon the context in which they are stated. One of the crucial problems in the representation of a natural language by logic, namely, the problem related to referentiation, disappears in this case. One of the historical goals of the representation of the mathematical discourse in logical form was certainly to be able to benefit from the rigour and precision of logic language in the foundation and development of certain parts of mathematics. For example, Gödel had this perspective when he formalized the theory of sets by way of a finite number of first-order logic formulas [Gödel 40]. The current motivations for logic representation result mainly from the connection of such a representation with automatic proof theory. A recent paper proposed the reformulation of the logic axioms of set theory in the language of clauses [Boyer et al. 86]. These clauses can be used directly in theorem proving methods based on the resolution rule. The authors of that paper illustrate the expressive power of the language of clauses by using it for expressing certain conjectures of the theory of numbers such as the Goldbach conjecture and the conjecture stating that there are infinitely many pairs of twin primes. They also propose a proof by resolution of the theorem stating that the composition of two homomorphisms is a homomorphism. Numerous research works have been devoted to the question of representing elementary geometries in logic language [Coelho and Pereira 86], [Chou and Schelter 86], [Chou 87], [Bundy 83], [Wu Wen-tsun 86]. Most of these works aim towards a reformulation of Euclidean geometry into a language of Horn clauses. A logic programming language such as Prolog can then be used to produce a mechanical proof of the theorems of geometry.

As a conclusion, let us quote the editor's preface to 'Basic principles of mechanical theorem proving in elementary geometries' by Wu Wen-tsun.

Geometry is one of the oldest mathematical topics to inspire mankind to pursue the question 'how do we reason ?'. It is therefore particularly satisfying when progress is made in automating geometric reasoning. The papers by the Chinese mathematician Wu Wen-tsun not only lay the foundation for efficient theorem proving in geometry but immediately establish a remarkable standard against which to judge future geometric deduction engines. Among the theorems proved by Wu's method are Simon's theorem, the nine point circle theorem, Pascal's theorem, Feuerbach's theorem, Morley's trisector theorem and Thébault's conjecture which was open over forty years and was proved only in 1983. [...] The implementation of Wu's method may well constitute not just a contribution to mechanical theorem proving but to geometry itself [Moore 86].

1.1.7 Conclusion

One of the main problems of artificial intelligence and of 'compositional linguistics' is that of the representation and manipulation of natural and formal languages with the tools of logic. Indeed, natural and formal languages make up the favoured vehicle for representing knowledge and for reasoning about that knowledge. The formal languages of logic allow us, thanks to the logic programming languages, to automatize the deductive processes which occur in the formalization of reasoning. The translation of natural and formal languages into logic formalism constitutes therefore a possible approach to the automatic processing of knowledge.

Section 1.2 contains an introduction to modal logics which are used to formalize reasoning; it provides an extension of the material given in chapters 1 and 4 of [Thayse 88]. Moreover, in order to make this work as self-contained as is reasonably possible, Section 1.2 includes a summary of the most important concepts that appear in propositional logic and in first-order predicate logic.

1.2 Modal logic

1.2.1 Introduction

Classical propositional and predicate logics play an important role in artificial intelligence and, more generally, in the field of computing applications; their fundamentals are explained in chapters 1 and 2 of [Thayse 88]. By definition, these logics are concerned with the study of correct forms of reasoning.

A main advantage of the logic languages is their rigorous syntax. The semantics they are provided with is also a strong point. Such a semantics plays a reference role by stating an unambiguous correspondence between the world of discourse and its representation in the language. It also allows us to justify the inferences we can draw from the represented knowledge.

It has been pointed out in chapter 4 of [Thayse 88] that not all types of knowledge and of reasoning can be formalized within classical logic. Certain non-classical logics, such as modal logics, are better suited for dealing with beliefs and time assertions. Many extensions and modifications of classical logic appear in the literature. These may take the form of rejection of certain axioms of classical logic, the intuitionistic logic is an example, or of addition of logical operators or connectives, as is being done in modal logic. The term 'non-standard logic' is a generic term which is used to refer to any logic other than classical propositional or predicate logic.

Non-standard logics can be divided into two groups: those that compete with classical logic and those that generalize it. In the first group we find many-valued logics [Davio et al. 78], [Rescher 69], partial logic, free logic, intuitionistic logic [Gabbay and Guenther 83] and fuzzy logic [Zadeh 81]. In the second group we find modal logics that specialize in temporal logic, dynamic logic and others [Gabbay and Guenther 84]. Logics that extend classical logic are characterized by theorems that generalize those of classical logic. Important philosophical considerations and attitudes towards the foundations of mathematics and logic have motivated the introduction and study of non-standard logics. The relevance of non-standard logics to artificial intelligence has been a topic of debate since the paper by McCarthy and Hayes [McCarthy and Hayes 69]. The employment of non-standard logics in artificial intelligence is not too surprising. Let us cite [Turner 84]:

In many respects the tasks of the philosophical logician and the artificial intelligence worker are quite similar. Both are concerned with the formalization of certain aspects of reasoning which is in everyday employment. It is true that the philosophical logician has been traditionally concerned with those aspects of reasoning which are of some philosophical significance. Hence the development of modal logics of necessity, time, knowledge and belief. But such issues are also of central concern to the artificial intelligence worker interested in knowledge representation.

Syntax, semantics and axiomatic systems of classical propositional and predicate logics are recalled in Subsections 1.2.2 and 1.2.13 (that constitute a summary of chapters 1 and 2 of [Thayse 88]). The other subsections of this section are devoted to the definition of syntax, semantics and axiomatic systems of the most frequently used modal logics. We follow rather closely the introduction to modal logic given in the book by Goldblatt [Goldblatt 87].

1.2.2 Propositional logic

Propositional logic deals with statements (declarative sentences) that are either true or false. The *propositional vocabulary* consists of a set of *propositions* that are the elementary sentences (also called *atomic formulas* or *atoms*) of the language. The vocabulary of propositional logic gives rise to combinations, obtained by assembling logical connectives $\neg, \vee, \wedge, \supset, \equiv$ (also called *logical constants*) and propositions in strings. The purpose of the construction rules (that constitute the *syntax*) is to allow the specification of particular combinations, called *formulas*. The *language* of propositional logic is the set of formulas.

SYNTAX

- **Base**: every proposition is a formula.

- **Induction**: if X and Y are formulas, then $\neg X$, $(X \wedge Y)$, $(X \vee Y)$, $(X \supset Y)$ and $(X \equiv Y)$ are formulas.

- **Closure**: all formulas are obtained by applying the base and induction rules a finite number of times.

The study of both natural and formal languages comprises, among others, a *syntax*, which allows the distinction of well-formed sentences, and a *semantics*, the purpose of which is the assignment of a meaning to well-formed sentences. This classification applies also to propositional logic.

We have mentioned that a proposition is either true or false. This leads to the introduction of the *semantic domain* {**T, F**}. The *interpretation* of a formula consists of the assignment of a *truth value* (**T** or **F**) to this formula. (The value **T** is also represented by 1 and the value **F** by 0.)

1.2. MODAL LOGIC

The semantics, that is, the set of interpretation rules of the language, should be *compositional*: the meaning of a formula must be a function of the meanings of its components. More specifically, the truth value assigned to a formula will depend only on the structure of this formula and on the truth values assigned to the propositions contained in it. In other words, the connectives (logical constants) of propositional logic are *truth-functional*; the semantics of these connectives is given below.

SEMANTICS

- Any proposition is interpreted as true (**T**) or false (**F**).

- If X and Y are formulas, then

    $\neg X$ is true if X is false,

    $(X \wedge Y)$ is true if both X and Y are true,

    $(X \vee Y)$ is true if either X or Y is true,

    $(X \supset Y)$ is true if either X is false or Y is true,

    $(X \equiv Y)$ is true if either X and Y are both true or else X and Y are both false.

An *interpretation function*, or *interpretation*, is a function which assigns a truth value to each proposition. The domain of this function is the set of propositions; the semantic rules determine an extension of the interpretation to the whole set of formulas. This extension is still called an *interpretation*. A formula is said to be *consistent* if it can be interpreted as the value **T** for at least one of its interpretations. A non-consistent formula is said to be *inconsistent*. A formula is said to be *valid* when it is always interpreted as **T**, for any interpretation of the propositions it contains. Valid formulas of propositional logic are also called *tautologies*. If A is a formula, the notation

$$\models A$$

means that A is a tautology. More generally, if E is a set of formulas, the notation

$$E \models A$$

means that all the interpretations which make true all formulas of E, also make true the formula A. In this case, A is said to be a *logical consequence* of E.

Let us recall that there is an older, more traditional way of characterizing validity and logical consequence for logic languages in terms of a *derivation*, or *proof*, and *inference rules*. This may be accomplished either through an *axiomatic system* or, in most contemporary logic texts, through a *natural deduction system*.

The notion of axiomatic system is very old. Such a system consists of a set of *axioms* (or *axiom schemata*), which are statements considered as valid, and a set of *inference rules* , which are machineries for producing new valid statements from previously obtained statements. In this framework, valid statements produced by the system are called *theorems* . A *derivation* , or *proof* , of a theorem is the ordered list of axioms, inference rules and previously obtained theorems which were needed to produce that theorem. For a given axiomatic system, the derivability, or provability, of a formula A is written

$$\vdash A.$$

More generally, if E is a set of formulas, the expression

$$E \vdash A$$

means that A is derivable from E, that is, A is derivable by the system if the elements of E are considered as additional axioms. In this case, the elements of E are viewed as *hypotheses*. The general subject of axiomatic systems is examined in chapter 2 of [Thayse 88]. The most vital requirement of an axiomatic system is *soundness* : every theorem must be a valid formula. The converse property is called *completeness* ; it will be requested too. An axiomatic system is *sound* if it enjoys the property

$$\vdash A \Rightarrow \vDash A,$$

for every formula A. Conversely, a system is *complete* if it enjoys the property

$$\vDash A \Rightarrow \vdash A,$$

for every formula A. Thus, in a sound and complete system, the provability symbol $\vdash$ is strictly equivalent to the logical consequence symbol $\vDash$.

Let us point out that a sound axiomatic system is syntactically consistent; it cannot generate both a formula and its negation. Indeed, within such a system, only tautologies are derivable.

The correspondence between the semantic definitions given earlier and the deductive definitions relative to an axiomatic system is summarized in the table of Figure 1.1.

1.2. MODAL LOGIC

As an example, let us mention the sound and complete axiomatic system made up of the three axiom schemata (A1–A3) and of the *Modus Ponens* inference rule (*MP*) given below.

(A1) $(X \supset (Y \supset X))$,
(A2) $((X \supset (Y \supset Z)) \supset ((X \supset Y) \supset (X \supset Z)))$,
(A3) $((\neg X \supset \neg Y) \supset ((\neg X \supset Y) \supset X))$.

When we substitute arbitrary formulas for the symbols X, Y and Z in each of these schemata, we obtain a tautology.

(*MP*)   If X and $(X \supset Y)$ are theorems, then Y is a theorem.

SEMANTIC DEFINITIONS	AXIOMATIC DEFINITIONS
A formula is valid if it is true with respect to every possible interpretation.	A formula is a theorem if there is a proof of it from the axiom schemata.
A formula is inconsistent if it is false with respect to every possible interpretation.	A formula is inconsistent if its negation is a theorem.
A formula A is the logical consequence of a set E of formulas if all the interpretations in which all the formulas of E are true are interpretations in which A is true.	A formula A is provable from a set E of formulas if there is a proof of it from the axiom schemata and from the formulas of the set E.
Two formulas are logically equivalent if they are true for exactly the same interpretations.	Two formulas are logically equivalent if each is provable from the other.

Figure 1.1 : Semantic and axiomatic definitions

1.2.3   Syntax and semantics of modal logic

The class of *propositional modal logics* to be considered here has, besides the usual propositional connectives, the modal operators $\Box$ and $\Diamond$ that are respectively called *universal modal operator* and *existential modal operator*. These operators act on propositional logic formulas in order to modify their meaning.

SYNTAX

- All formation rules of propositional logic are also formation rules of propositional modal logic.

- If A is a formula, then $\Box A$ and $\Diamond A$ are formulas.

The existential modal operator $\Diamond$ has the meaning of a negated universal modal operator; it can be defined by

$$\Diamond A =_{def} \neg\Box\neg A. \tag{1.1}$$

Just as the logical connectives can be 'read' in natural language ($\neg$: not, $\wedge$: and, $\vee$: or, $\supset$: if ... then, $\equiv$: if and only if), it is customary to assign a 'natural interpretation' to the modal operators $\Box$ and $\Diamond$. The various readings of $\Box A$ and of $\Diamond A$ must satisfy relation (1.1). Clearly there is no limit to the number of readings, and philosophers have proposed a great variety of them. Some interesting readings are gathered in the table of Figure 1.2.

READING OF $\Box A$	READING OF $\Diamond A$
It is necessarily true that A.	It is possibly true that A.
It will always be true that A.	It will sometimes be true that A.
It ought to be that A.	It can be that A.
It is known that A.	The opposite of A is not known.
It is believed that A.	The opposite of A is not believed.
After every terminating execution of the program, A is true.	There is an execution of the program that terminates with A true.

Figure 1.2 : Possible readings of $\Box A$ and $\Diamond A$

Unlike the classical connectives $\neg, \vee, \wedge, \supset, \equiv$, the modal operators do not admit a truth-functional interpretation. Instead, their interpretation enlists the notions of frame and model. To make the interpretations (or readings) of the modal operators more precise, we need to introduce these notions.

A *frame* is a pair $\mathcal{F} = (W, R)$, where W is a non-empty set and R is a binary relation on W, i.e., a subset of $W \times W$, called *accessibility relation*. The elements of W are called *points*.

Let P be a set of atomic formulas (or propositions). The set of all formulas generated from P will be denoted $\mathcal{L}(P)$, or simply $\mathcal{L}$. A *P-model* on a frame (W, R) is a triple $\mathcal{M} = (W, R, V)$, with V a mapping from P to 2^W (the set of subsets of W). Thus V is a function assigning to each atomic formula $p \in P$ a subset $V(p)$ of W. Informally, $V(p)$ is to be thought of as the set of points

1.2. MODAL LOGIC

$w \in W$ at which p is 'true'. Generally, we drop the prefix P in discussing models where the context is clear; we shall then simply speak of *model* $\mathcal{M}$.

The relation 'formula A is true at point w in model $\mathcal{M}$', denoted

$$\mathcal{M} \vDash_w A,$$

is defined recursively on the formation of formulas A as follows.

SEMANTICS

$\mathcal{M} \nvDash_w \mathbf{F}$
$\mathcal{M} \vDash_w p$ if $w \in V(p)$
$\mathcal{M} \vDash_w A_1 \supset A_2$ if $\mathcal{M} \vDash_w A_1$ implies $\mathcal{M} \vDash_w A_2$
$\mathcal{M} \vDash_w \Box A$ if wRt implies $\mathcal{M} \vDash_t A$ for all $t \in W$

The last rule expresses that a formula $\Box A$ is true at a point w of a model (W, R, V) if the formula A is true at all points t of the set W which are accessible from the point w, i.e., for which wRt. These definitions constitute the basic semantic relations.

The semantic evaluation of the formulas $\mathbf{T}$, $\neg A$, $\Diamond A$, $(A_1 \vee A_2)$, $(A_1 \wedge A_2)$ and $(A_1 \equiv A_2)$ is obtained from these basic semantic relations and from the following definitions:

$\neg A \quad =_{def} \quad A \supset \mathbf{F}$
$\mathbf{T} \quad =_{def} \quad \neg \mathbf{F}$
$A_1 \vee A_2 \quad =_{def} \quad \neg A_1 \supset A_2$
$A_1 \wedge A_2 \quad =_{def} \quad \neg(A_1 \supset \neg A_2)$
$A_1 \equiv A_2 \quad =_{def} \quad (A_1 \supset A_2) \wedge (A_2 \supset A_1)$
$\Diamond A \quad =_{def} \quad \neg \Box \neg A$

For example, using the definitions of $\neg A$ and of $\Diamond A$ we can prove the following semantic rules:

$\mathcal{M} \vDash_w \neg A$ if $\mathcal{M} \nvDash_w A$,

$\mathcal{M} \vDash_w \Diamond A$ if there exists a point $t \in W$ such that wRt and $\mathcal{M} \vDash_t A$.

From this truth analysis of modal formulas, we can introduce the notions of *truth in a model*, *truth in a frame* and *validity* of a modal formula.

- A formula A belonging to $\mathcal{L}$ is said to be *true in a model* $\mathcal{M} = (W, R, V)$ if A is true at all points of the model, i.e., if $\mathcal{M} \vDash_w A$ for all $w \in W$; this will be denoted $\mathcal{M} \vDash A$.

- A formula A belonging to $\mathcal{L}$ is said to be *true in a frame* $\mathcal{F} = (W, R)$ if A is true in every model $\mathcal{M} = (W, R, V)$, i.e., if $\mathcal{M} \vDash A$ for any model $\mathcal{M} = (W, R, V)$; this will be denoted $\mathcal{F} \vDash A$.

- A formula A belonging to $\mathcal{L}$ is said to be *valid* if A is true in every frame, i.e., if $\mathcal{F} \vDash A$ for any frame $\mathcal{F}$; this will be denoted $\vDash A$.

1.2.4 Examples and applications

The following formulas are true in every model and hence in every frame:

$$\Box(A \supset B) \supset (\Box A \supset \Box B),$$
$$\Box(A \supset B) \supset (\Diamond A \supset \Diamond B),$$
$$\Box(A \land B) \equiv (\Box A \land \Box B),$$
$$\Diamond(A \lor B) \equiv (\Diamond A \lor \Diamond B),$$
$$\Box(A \supset \Diamond(B \supset C)) \supset \Diamond(B \supset (\Box A \supset \Diamond C)).$$

We prove the first of these formulas, by the *ab absurdo* method. We try to discover a model in which the formula is false. An implicative formula is false only if its first operand (the antecedent) is true and if its second operand (the consequent) is false. Thus we have to find a model in which

$$\Box(A \supset B) \text{ is } true \text{ and } (\Box A \supset \Box B) \text{ is } false.$$

(For the sake of readability, we sometimes write *true* and *false* instead of **T** and **F**, respectively.) The same reasoning applied to the consequent of the initial formula yields the constraints

$$\Box A \text{ is } true \text{ and } \Box B \text{ is } false.$$

The condition $\Box A$ *true* implies that A is true at every point. This condition, together with $\Box(A \supset B)$ *true*, forces $\Box B$ to be *true*. Consequently, the second constraint, i.e., $\Box B$ *false*, can no longer be satisfied; hence the formula is proved.

The following formulas do not have the property of being true in all frames:

$$\Box A \supset A,$$
$$\Box A \supset \Box\Box A,$$
$$\Box(A \lor B) \supset (\Box A \lor \Box B).$$

We have indicated in Figure 1.2 several 'possible readings' for the modal formulas $\Box A$ and $\Diamond A$. These different readings give rise to various applications of modal logic.

Before considering some of these applications, let us comment on the terminology of *possible-world semantics*. It is Kripke who introduced the technique of possible worlds as a uniform semantics for various modal systems [Kripke 71]. The basic idea behind this approach is that a modal formula will

1.2. MODAL LOGIC

be evaluated within a set of possible worlds which is provided with a certain accessibility relation. More precisely, the truth value of a modal formula will depend on the considered possible world. Roughly speaking, $\Box p$ is true in a given world if and only if p is true in all possible worlds that are accessible from that world. In a dual way $\Diamond p$ is true in a given world if and only if p is true in at least one possible world accessible from that world. Thus, frame $\mathcal{F}$ is a pair (W, R), where W is the set of possible worlds, sometimes called the *universe*, and R is a binary relation on W called *accessibility relation*. Let s and t be two worlds belonging to W; if world t is accessible from world s, we write sRt (as above).

ALETHIC LOGIC

Possibility and *necessity* are called *alethic modalities* or *modes of truth*. A system having modal operators for 'it is possible that' and 'it is necessary that' is called an *alethic logic*. For 'necessarily true' we will use the symbol $\Box$, so that $\Box A$ can be read 'it is necessarily true that A'. Its dual will be $\Diamond$, so that $\Diamond A$ is read 'it is possibly true that A'. In Kripke's terminology 'possibly true' means 'true in at least one possible world' and 'necessarily true' means 'true in all possible worlds'. In this context, the accessibility relation R is assumed to be reflexive, i.e., to satisfy sRs for all $s \in W$.

It is often useful to single out one of the worlds of the universe as the 'actual world'. A formula A that is necessarily true in the 'actual world' will be true in all the worlds that are accessible from the actual world and, therefore, in particular, in the actual world itself; hence the formula

$$\Box A \supset A$$

is true in the considered frame. If a formula A is true in the actual world, it will be true in at least one world accessible from the actual world; hence the formula

$$A \supset \Diamond A$$

is true in the considered frame.

Different notions of necessity can be entertained. Thus logical necessity may be contrasted with physical necessity, the former taking $\Box A$ to mean 'A is a consequence of the laws of logic' while the latter taking $\Box A$ to mean 'A is a consequence of the laws of physics'. For example in our world, $\Box(t > -273)$ is true under the physical reading, with t denoting the absolute temperature. On the other hand it is logically possible that $(t > -273)$ be false.

Temporal logic

In temporal logic, the same sentence may have a different truth value at different times: a sentence true at some time in the past may not be true now, and a sentence true now may not remain so in the future. Here the members of the universe W are taken to be the moments of time. If sRt means 't is after s' then $\Box A$ means 'A will be true at all future times', while $\Diamond A$ means 'A will be true at some future time'. In a dual way, if sRt means that t is before s, then $\Box A$ means 'A has always been true in the past', while $\Diamond A$ means 'A was true at some past time'. Natural time frames (W, R) are given by taking W as one of the number sets $\mathbf{N}$ (natural numbers), $\mathbf{Z}$ (integer numbers), $\mathbf{Q}$ (rational numbers) or $\mathbf{R}$ (real numbers) and R as one of the relations $<, \leq, >, \geq$ [Rescher and Urquhart 71], [Goldblatt 87], [Turner 84].

Dynamic logic

Dynamic logic is based on the idea of associating a modal operator with each command of a programming language. In this context, W is to be regarded as the set of possible states of a computation process, with sRt meaning that there is an execution of the program that starts in state s and terminates in state t. A non-deterministic program may admit more than one possible 'outcome' t when started in s. Then $\Diamond A$ means 'every terminating execution of the program brings about A true' while $\Diamond A$ means that 'there is some execution of the program that terminates with A true'.

Logics of knowledge and of belief

Another domain of application of modal logic is concerned with the formalization of belief and knowledge. Various logic systems proposed in the literature make use of formal languages including the modal operators of 'belief' and of 'knowledge". In the context of belief and of knowledge logics, the operator $\Box$ reads 'is believed' and 'is known', respectively. Its dual, the operator $\Diamond$, reads 'the inverse is not believed' and 'the inverse is not known', respectively. Logic languages dealing with knowledge and belief are studied in chapter 4 of [Thayse 88]. We shall examine a typical example of introspective reasoning based on such a logic in Subsection 1.3.2.

1.2. MODAL LOGIC

1.2.5 Binary relations and formula schemata

To start with, let us give a list of interesting properties that a binary relation R can hold.

1. Reflexive : $\forall s\,(sRs)$
2. Symmetric : $\forall s \forall t\,(sRt \supset tRs)$
3. Serial : $\forall s \exists t\,(sRt)$
4. Transitive : $\forall s \forall t \forall u\,(sRt \wedge tRu \supset sRu)$
5. Euclidean : $\forall s \forall t \forall u\,(sRt \wedge sRu \supset tRu)$
6. Partially functional : $\forall s \forall t \forall u\,(sRt \wedge sRu \supset t = u)$
7. Functional : $\forall s \exists ! t\,(sRt)$
8. Weakly dense : $\forall s \forall t\,(sRt \supset \exists u\,(sRu \wedge uRt))$
9. Weakly connected : $\forall s \forall t \forall u\,(sRt \wedge sRu \supset tRu \vee t = u \vee uRt)$
10. Weakly directed : $\forall s \forall t \forall u(sRt \wedge sRu \supset \exists v\,(tRv \wedge uRv))$

(The notation $\exists ! t$ means 'there is exactly one t'.) Corresponding to this list is a list of formula schemata:

1. $\Box A \supset A$ (T)
2. $A \supset \Box \Diamond A$ (B)
3. $\Box A \supset \Diamond A$ (D)
4. $\Box A \supset \Box\Box A$ (4)
5. $\Diamond A \supset \Box \Diamond A$ (5)
6. $\Diamond A \supset \Box A$
7. $\Diamond A \equiv \Box A$
8. $\Box\Box A \supset \Box A$
9. $\Box(A \wedge \Box B \supset B) \vee \Box(B \wedge \Box B \supset A)$ (L)
10. $\Diamond \Box A \supset \Box \Diamond A$

Some of these formula schemata are taken as axiom schemata in axiomatic systems. The historical names for some well-known schemata are indicated within parentheses in the list above.

The following theorem states in a formal way the correspondence between particular binary relations and the associated schemata. A proof of this theorem can be found in [Goldblatt 87].

Theorem 1.1. *Let $\mathcal{F} = (W, R)$ be a frame; the relation R satisfies one of the properties 1–10 if and only if the corresponding schema is true in $\mathcal{F}$.*

Theorem 1.1 is fundamental; its far-reaching effect explains the great success that the possible-world semantics enjoyed upon its introduction by Kripke [Kripke 63]. Frames are easy to deal with and many modal schemata were shown to have their models characterized by simple properties of the relation

R. Note, however, that there are some naturally occurring properties of a binary relation R that do not correspond to the validity of any modal schema. For example, none of the following conditions corresponds to a modal schema.

irreflexivity	:	$\forall s \, \neg(sRs)$
antisymmetry	:	$\forall s \forall t \, (sRt \wedge tRs \supset s = t)$
asymmetry	:	$\forall s \forall t \, (sRt \supset \neg(tRs))$

1.2.6 Valuations and tautologies

In this subsection and in the next one we introduce a group of concepts that will be used in the definition of the classical axiomatic systems of modal logic. These two subsections may be omitted by the reader who is only interested in an intuitive understanding of these axiomatic systems.

The set $Sf(A)$ of all subformulas of a given formula $A \in \mathcal{L}(P)$ is defined recursively as follows.

- If p is an atom, $Sf(p) = \{p\}$.
- $Sf(\mathbf{F}) = \{\mathbf{F}\}$.
- $Sf(A_1 \supset A_2) = \{A_1 \supset A_2\} \cup Sf(A_1) \cup Sf(A_2)$.
- $Sf(\Box A) = \{\Box A\} \cup Sf(A)$.

Given a P-model $\mathcal{M} = (W, R, V)$, an element s of W and a set P of propositions, the *valuation* function V_s is defined as a mapping from P to $\{true, false\}$, such that, for $p \in P$, we have

$$V_s(p) = \text{true if } s \in V(p),$$
$$= \text{false if } s \notin V(p).$$

Using the semantic rules for logical connectives (§ 1.2.2), we can extend this function V_s to the set of formulas constructed from propositions of P and from logical connectives only. A model on a frame gives rise to a collection $\{V_s : s \in W\}$ of valuations of P, while, conversely, such a collection defines the model in which the function V is given by

$$V(p) = \{s \in W : V_s(p) = true\}.$$

A formula A is said to be *quasi-atomic* if either it is atomic ($A \in P$), or else it begins with the symbol $\Box$, i.e., $A = \Box B$ where B is any formula.

Let P^q be the set of all quasi-atomic formulas of P; any formula A can be obtained from members of $P^q \cup \{\mathbf{F}\}$ and the connective $\supset$. Hence, by using the semantic rule relative to this connective (i.e., $X \supset Y$ is true if and only if X is false or Y is true), any valuation for quasi-atomic formulas,

1.2. MODAL LOGIC

$$V_s : P^q \to \{true, false\}, \quad s \in W,$$

extends uniquely to a valuation of all formulas of the modal language $\mathcal{L}$ generated by P, that is, to a function

$$V_s : \mathcal{L} \to \{true, false\}.$$

A formula A is a *tautology* if $V(A) = true$ for every valuation V of its quasi-atomic subformulas.

It can be shown that any (modal) tautology is a substitution instance of a tautology of propositional logic, i.e., a tautology free of the modal operator $\Box$. For example, the tautology

$$(\Box(p \supset \Box q) \supset \Box \neg r) \supset (\neg \Box \neg r \supset \neg \Box(p \supset \Box q))$$

is obtained from the propositional tautology

$$(p \supset q) \supset (\neg q \supset \neg p)$$

by means of the uniform substitution

p is rewritten as $\Box(p \supset \Box q)$, q is rewritten as $\Box \neg r$.

1.2.7 Logics

Given a language $\mathcal{L}(P)$ based on a set P of atoms, a *logic* (or *theory*) Λ is defined to be any subset of formulas generated from P and satisfying the following conditions:

- Λ includes all tautologies;

- if A and $A \supset B$ are elements of Λ, then B is also an element of Λ (i.e., Λ is closed under the *Modus Ponens* rule (§ 1.2.2));

- If A is an element of Λ, then A' is an element of Λ whenever A' is the result of a *uniform substitution* of formulas for atomic formulas in A (i.e., Λ is closed under uniform substitution (§ 1.2.4 of [Thayse 88])).

The set $\mathcal{L}(P)$ of formulas generated from the set P of propositions, and the set of all tautologies in $\mathcal{L}(P)$ constitute two examples of logics.

The elements of a logic are called its *theorems*. We write $\vdash_\Lambda A$ to mean that A is a theorem of the logic Λ; thus we have the equivalence

$$\vdash_\Lambda A \iff A \in \Lambda.$$

Let $\mathcal{C}$ be a frame or a model; the logic Λ is said to be *sound with respect* to $\mathcal{C}$ if every theorem A of Λ is a true formula in $\mathcal{C}$, i.e., if we have

$$\vdash_\Lambda A \Rightarrow \mathcal{C} \models A.$$

A logic Λ is said to be *complete with respect to* $\mathcal{C}$ if every true formula A in $\mathcal{C}$ is a theorem of Λ, i.e., if we have

$$\mathcal{C} \models A \Rightarrow \vdash_\Lambda A.$$

The logic Λ is said to be *determined by* $\mathcal{C}$ if it is both sound and complete with respect to $\mathcal{C}$, i.e., if we have

$$\vdash_\Lambda A \iff \mathcal{C} \models A.$$

A logic Λ is said to be *normal* if it contains the formula schema

$$K \ : \ \Box(A \supset B) \supset (\Box A \supset \Box B),$$

and if it is provided with the modal *inference necessitation rule*, i.e., if we have

$$\vdash_\Lambda A \Rightarrow \vdash_\Lambda \Box A.$$

This rule states that if A is a theorem of the logic Λ, then $\Box A$ is a theorem of Λ; we write this:

$$A \vdash_\Lambda \Box A.$$

If $\{\Lambda_i \mid i \in I\}$ is a collection of normal logics, then

$$\bigcap \{\Lambda_i \mid i \in I\}$$

is a normal logic. In particular, the logic denoted K and defined by

$$K = \bigcap \{\Lambda_i \mid \Lambda_i \text{ is a normal logic}\}$$

is the smallest normal logic. The following theorem, a proof of which can be found in [Goldblatt 87], illustrates the importance of this 'minimal logic'.

Theorem 1.2. *A formula A is a theorem of the logic K if and only if A is valid (i.e., true in all frames).*

1.2. Modal logic

The following results hold for any normal logic Λ:

$\vdash_\Lambda A \supset B \Rightarrow \vdash_\Lambda \Box A \supset \Box B$ and $\vdash_\Lambda \Diamond A \supset \Diamond B$,
$\vdash_\Lambda A \equiv B \Rightarrow \vdash_\Lambda \Box A \equiv \Box B$ and $\vdash_\Lambda \Diamond A \equiv \Diamond B$,
$\vdash_\Lambda \Box A \wedge \Box B \equiv \Box(A \wedge B)$,
$\vdash_\Lambda \Diamond(A \vee B) \equiv \Diamond A \vee \Diamond B$,
$\vdash_\Lambda \Box A \vee \Box B \supset \Box(A \vee B)$,
$\vdash_\Lambda \Diamond(A \wedge B) \supset \Diamond A \wedge \Diamond B$.

1.2.8 Axiomatic systems

An *axiomatic system* for a normal logic is made up of the following three components:

- An axiomatic system of propositional logic.
- The axiom schema denoted $K : \Box(A \supset B) \supset (\Box A \supset \Box B)$.
- The modal inference rule of necessitation.

It has become customary to represent the smallest normal logic Λ containing the formula schemata $\Sigma_1, ..., \Sigma_n$ by the notation

$$\Lambda = K\Sigma_1 \cdots \Sigma_n.$$

This logic is defined as follows:

$$\Lambda = \bigcap \{\Lambda_i \mid \Lambda_i \text{ is normal and } \Sigma_1 \cup \cdots \cup \Sigma_n \subset \Lambda_i\}.$$

Classical names for some schemata are (§1.2.5):

$D : \Box A \supset \Diamond A$
$T : \Box A \supset A$
$4 : \Box A \supset \Box\Box A$
$B : A \supset \Box\Diamond A$
$5 : \Diamond A \supset \Box\Diamond A$
$L : \Box((A \wedge \Box A) \supset B) \vee \Box((B \wedge \Box B) \supset A)$
$W : \Box(\Box A \supset A) \supset \Box A$

Names of some frequently used logics are:

$$S4 \; : \; KT4$$

$$S5 \; : \; KT45$$

Theorem 1.3. *A formula A is a theorem of the logic KT if and only if A is true in every frame where R is reflexive.*

Theorem 1.4. *A formula A is a theorem of the logic S4 if and only if A is true in every frame where R is reflexive and transitive.*

Theorem 1.5. *A formula A is a theorem of the logic S5 if and only if A is true in every frame where R is reflexive, transitive and Euclidean (this means that R is an equivalence relation).*

Theorems 1.3–1.5 and analogous theorems dealing with proof theory in other logics can be proved by making use of the following lemma, a proof of which is given in [Goldblatt 87].

Lemma 1.6. *If a normal logic Λ contains any one of the schemata 1–10 (of Subsection 1.2.5), then there exists a canonical relation R^Λ for Λ, which satisfies the corresponding property (of Subsection 1.2.5).*

As shown in chapter 4 of [Thayse 88], the logics $KT, S4$ and $S5$ are useful to the formalization of reasoning about belief and knowledge. Interpreting the modal operator $\square$ as belief, the axiom schema K asserts that whenever B is implied by A, an agent who believes A will also believe B. As expected, normal modal logics assert consequential closure when the modal operator is interpreted as belief. Similarly, the distribution axiom schema can be interpreted as follows: if an agent believes that B is implied by A, then if he believes A he will also believe B. Some axiom schemata can be added to K to get more elaborate modal systems that better characterize epistemic operators. Details on that subject can be found in chapter 4 of [Thayse 88].

1.2.9 Multimodal languages

There exist logic languages with more than one modal operator; they make use of a collection of symbols $\{[i] \mid i \in I\}$ each of which corresponds to a universal operator. The dual existential operator is denoted $<i>$ and is defined as $\neg[i]\neg$; the logics making use of more than one modal operator are called *multimodal logics*.

SYNTAX

- All formation rules of propositional logic are also formation rules of multimodal logic.

1.2. MODAL LOGIC

- If A is a formula, then $[i]A$ and $<i>A$ are formulas.

A *frame* $\mathcal{F}$ for multimodal language is defined as follows:

$$\mathcal{F} = (W, \{R_i \mid i \in I\});$$

it comprises a non-empty set W (the universe) and a collection of binary relations $R_i \subset W \times W$, for each $i \in I$. A model $\mathcal{M} = (\mathcal{F}, V)$ on a frame $\mathcal{F}$ is defined as previously by means of a function V from the set P of atoms of $\mathcal{L}$ to the set 2^W. The semantics of a multimodal language is defined as in Subsection 1.2.3, except for the last rule which is replaced by the following set (with $i \in I$).

SEMANTICS

$\mathcal{M} \models_s [i]A$ if sR_it implies $\mathcal{M} \models_t A$ for each $t \in W$.

The definitions of formula (A) *true in a model* $(\mathcal{M} \models A)$, *true in a frame* $(\mathcal{F} \models A)$ and *valid* $(\models A)$ remain unchanged (§ 1.2.3). The notion of *tautology* is defined as previously (§ 1.2.6), taking all formulas of the form $[i]A$, along with members of P, in the definition of 'quasi-atomic formula'.

A *logic* continues to be a subset of the set of formulas generated from P; a logic Λ is assumed to include all tautologies, and to be closed under *Modus Ponens* and *uniform substitution*. A logic Λ is said to be *normal* if it contains the schemata

$$K_i : [i](A \supset B) \supset ([i]A \supset [i]B)$$

and satisfies the necessitation rule

$$\vdash_\Lambda A \Rightarrow \vdash_\Lambda [i]A,$$

for every $i \in I$. The smallest normal logic is generally denoted K_I.

1.2.10 Temporal logic

To form a tensed language from propositional logic, we need to add two types of operators: future tense operators and past tense operators. We shall consider a multimodal language with two universal operators, $[F]$ and $[P]$, meaning *at all future times* and *at all past times* respectively. The corresponding existential operators, $<F>$ and $<P>$, will mean *sometimes in the future* and *sometimes in the past* respectively. According to the preceding subsection, a frame for this tensed language has the form (W, R_F, R_P). If $\mathcal{M}$ is a model, we have

$$\mathcal{M} \models_s [F]A \text{ if } sR_Ft \text{ implies } \mathcal{M} \models_t A \text{ for each } t \in W,$$
$$\mathcal{M} \models_s [P]A \text{ if } sR_Pt \text{ implies } \mathcal{M} \models_t A \text{ for each } t \in W.$$

We interpret the properties sR_Ft and sR_Pt as 't is in the future of s' and 't is in the past of s' respectively.

Given a frame $\mathcal{F} = (W, R_F, R_P)$, we can state the following propositions:

- $\mathcal{F} \models A \supset [P]\!<\!F\!>\!A \iff \forall s \forall t\, (sR_Pt \Rightarrow tR_Fs)$.
- $\mathcal{F} \models A \supset [F]\!<\!P\!>\!A \iff \forall s \forall t\, (sR_Ft \Rightarrow sR_Pt)$.
- If a normal logic Λ contains the schema

$$A \supset [P]\!<\!F\!>\!A,$$

then the canonical relations R_P^Λ and R_F^Λ have the property

$$sR_P^\Lambda t \Rightarrow tR_F^\Lambda s.$$

- If a normal logic Λ contains the schema

$$A \supset [F]\!<\!P\!>\!A,$$

then the canonical relations R_P^Λ and R_F^Λ have the property

$$tR_F^\Lambda s \Rightarrow sR_P^\Lambda t.$$

The two schemata above state formally that:

- If a formula A is true (now), then there exists some past time in which A will become true in the future.
- If a formula A is true (now), then there exists some future time in which A was true in the past.

Any temporal logic (also called tense logic) should at least contain these two schemata. In the frames for such a logic, the accessibility relations R_F and R_P can be defined from each other, so we may as well take one relation as primitive, and use frames $\mathcal{F} = (W, R)$, where $R \subset W \times W$ and

$$\mathcal{M} \models_s [F]A \iff (sRt \Rightarrow \mathcal{M} \models_t A),$$
$$\mathcal{M} \models_s [P]A \iff (tRs \Rightarrow \mathcal{M} \models_t A).$$

A *normal temporal logic* is obtained by imposing no restriction on the relation R. An axiomatic system for a normal temporal logic is made up of the following three components:

1.2. Modal logic

- An axiomatic system of propositional logic.

- The axiom schemata

$$K_F : [F](A \supset B) \supset ([F]A \supset [F]B),$$
$$K_P : [P](A \supset B) \supset ([P]A \supset [P]B),$$
$$C_F : A \supset [F]<P>A,$$
$$C_P : A \supset [P]<F>A.$$

- The inference rules of necessitation

$$A \vdash [F]A,$$
$$A \vdash [P]A.$$

A normal temporal logic is therefore a normal logic in the multimodal language of operators $[F]$ and $[P]$. The smallest normal temporal logic is known as K_t in the literature [Rescher and Urquhart 71]. All the other temporal logics are extensions of minimal temporal logic, and are obtained by imposing further constraints on the temporal accessibility relation.

A relation R is said to be *left linear* if

$$\forall t \forall s \forall u \, ((tRu \wedge sRu) \supset tRs \vee t = s \vee sRt).$$

This property means that if t precedes u and if s precedes u, then either t precedes s, either t is identical to s, or s precedes t. The concept of 'branching-time' is obtained by imposing the conditions of left linearity and of transitivity (item 4 of lists of Subsection 1.2.5) to R. Considering a sequence of events, say E_1, E_2, E_4, we may view it as one among various physically possible alternatives, including also e.g. E_1, E_3, E_5. We can represent all these possible sequences as in the tree-like diagram of Figure 1.3. A node in such a branching structure is to be thought of in temporal contexts as an event, while a branch is to be thought of as a possible world history. Here E_1, E_2 and E_3 may be called *branching events* for obvious reasons. Branching-time thus insists on there being only one past but allows the future to remain open. To axiomatize this logic we need to adjoin three schemata to the axioms of K_t. The transitivity property imposed to the relation R involves the introduction of the following two schemata:

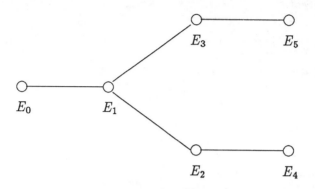

Figure 1.3 : Sequences in a branching-time structure

$[F]A \supset [F][F]A,$

$[P]A \supset [P][P]A.$

The left linear property corresponds to the following axiom schema:

$$([P](A \vee B) \wedge [P](A \vee [P]B) \wedge [P]([P]A \vee B)) \supset ([P]A \vee [P]B).$$

The transformation of this schema in terms of the existential operator $<P>$ provides us with the equivalent axiom schema:

$$(<P>A \wedge <P>B) \supset (<P>(A \wedge B) \vee <P>(A \wedge <P>B) \vee <P>(B \wedge <P>A)).$$

This schema indicates that if events A and B occur in the past, then either A and B occur at the same time in the past, either B occurs in the past of A, or A occurs in the past of B.

If we adjoin the formulas we have just discussed to K_t as axioms, we obtain an axiomatic system for branching-time logic, which is called K_b. The logic K_b is complete with respect to all branching-time frames [Rescher and Urquhart 71]; this means that, for any formula A, we have

$\vdash_{K_b} A$ if and only if A is true in all branching-time frames.

The standard picture of time is that of a linear series. Such a conception is assumed in a great portion of physics. The absolute time of Newtonian physics is a one-dimensional linear continuum; even in relativistic physics the ordering

1.2. MODAL LOGIC

of 'local' time series is linear. To capture this conception of time we must rule out branching in the future as well as branching in the past. This is achieved by insisting that R be *linear* (or connected):

$$\forall t \forall s \, (sRt \lor s = t \lor tRs).$$

It is obvious that linearity is equivalent to the conjunction of left linearity (see above) and of right linearity given as

$$\forall t \forall s \forall u \, ((uRs \land uRt) \supset tRs \lor s = t \lor sRt),$$

to which corresponds the schema

$$([F](A \lor B) \land [F](A \lor [F]B) \land [F]([F]A \lor B)) \supset ([F]A \lor [F]B).$$

We are now confronted with the problem of axiomatizing the tense logic, denoted K_l, corresponding to the class of linear tense frames. To axiomatize K_l it suffices to add to K_b the corresponding axiom for right linearity [Rescher and Urquhart 71].

The basic linear tense logic K_l leaves unanswered many fundamental questions about the nature of time. Is there a first or a last moment in time or is the time infinite in both directions? We might for instance wish to hold that there is a first moment to conform to the primeval atom theory of cosmogony, while if we choose the steady-state universe model we shall take the second option given above. Is there a moment between any two distinct moments? Is time continuous like the real numbers? Different answers to these various questions give rise to different temporal logics.

If we wish to express the property that time has no beginning and no end we can make use of the following two conditions:

$$\forall s \exists t \, (sRt) \quad \text{and} \quad \forall s \exists t \, (tRs).$$

The tense logic corresponding to this extension is called K_s. Its axiomatization is obtained from the addition of the following two schemata to the axiom schemata of the K_l logic :

$$[F]A \supset <F>A,$$
$$[P]A \supset <P>A.$$

Is there a moment between any two moments? A positive answer leads to a view of time in which the time line has the structure of the rational numbers; it forces the time to be 'dense', a property that we can represent as follows:

$$\forall s \forall t \exists u \, (sRt \supset (sRu \land uRt)).$$

Density in time can be represented by adding the following two axiom schemata to those of the K_l logic:

$$[F][F]A \supset [F]A,$$
$$[P][P]A \supset [P]A.$$

The resulting logic is called K_{ld}. The connection of the first of these schemata with density of time can be seen by looking at the equivalent schema $<F>A \supset <F><F>A$. If $<F>A$ is true at time t, there is a moment t' in the future of t at which A is true; but if time is dense, there is a moment t'' between t and t' at which $<F>A$ is true, hence $<F><F>A$ is true at time t. The same argument shows that the second schema is valid in dense time series.

In contrast to the condition of density, we can impose a discrete ordering of time, so that the ordering of temporal instants is isomorphic to that of the integers or of a subset of the integers. To express the assumption of discreteness we have to adjoin the axiom schema

$$\Box([F]A \supset A) \supset ([F]A \supset [P]A),$$

where $\Box A$ is defined as $[F]A \wedge [P]A \wedge A$.

The purpose of the present subsection was to introduce the main concepts (syntax, semantics, axiomatic systems) relative to some usual temporal logics. All these logics are treated with great care in [Rescher and Urquhart 71]. Chapter 4 of this book is devoted to a more detailed investigation of temporal logics and of some of their applications to artificial intelligence and computing science.

1.2.11 Dynamic logic

The language of dynamic logic is a multimodal language which is based on the idea of associating with each command α of a programming language a modal operator $[\alpha]$, with the formula $[\alpha]A$ being read as 'after every terminating execution of command α, formula A is true'. The dual formula $<\alpha>A$ means 'there is an execution of α that terminates with A true'. An interesting theory emerges about the ways in which properties of complex programs can be expressed by the modal operators of their constituent programs ([Pratt 76], [Goldblatt 87]). These constituent programs are generated from a well-defined set Π of 'atomic' programs that plays a similar role for programs as the role of P does for formulas of propositional logic.

SYNTAX

- All formation rules of multimodal propositional logic are also formation rules of dynamic logic; the modal operators of the language are $[\alpha]$ and $<\alpha>$ where α is a program.

- Syntax of programs.

1.2. MODAL LOGIC

- **Base:** every atomic program is a program.
- **Induction:** if α, α_1 and α_2 are programs and if A is a logic formula, then $(\alpha_1; \alpha_2)$, $(\alpha_1 \cup \alpha_2)$, α^* and $A?$ are programs.
- **Closure:** all programs are obtained by applying the base and induction rules a finite number of times.

The operations on programs are interpreted as follows:

Program	Meaning
$\alpha_1; \alpha_2$	do α_1 and then α_2
$\alpha_1 \cup \alpha_2$	do either α_1, or α_2 (non-deterministically)
α^*	repeat α some finite number of times
$A?$	test A: continue if A is true, otherwise 'fail'
$[\alpha]A$	after α, A

These basic operations allow us to define the classical statements of programming languages:

if A **then** α **else** β	$=_{def}$	$(A?; \alpha) \cup (\neg A?; \beta)$
while A **do** α	$=_{def}$	$(A?; \alpha)^*; \neg A?$
repeat α **until** A	$=_{def}$	$\alpha ; (\neg A?; \alpha)^*$

A model $\mathcal{M}$ for a language of dynamic logic is a triple

$$\mathcal{M} = (W, \{R_\alpha \mid \alpha \in \Pi\}, V)$$

where R_α is a binary relation on W and V is a mapping from P into 2^W.

SEMANTICS

$$\mathcal{M} \models_s [\alpha]A \text{ if } sR_\alpha t \text{ implies } \mathcal{M} \models_t A.$$

The binary relations R_α have to reflect the intended meanings of programs α. A model $\mathcal{M}$ is *standard* if its accessibility relation R satisfies the following conditions:

$$\begin{aligned} R_{\alpha;\beta} &= \{(s,t) \mid \exists u(sR_\alpha u \wedge uR_\beta t)\} \\ R_{\alpha \cup \beta} &= R_\alpha \cup R_\beta \\ R_{\alpha^*} &= (R_\alpha)^* \quad \text{(the reflexive and transitive closure of } R_\alpha) \\ R_{A?} &= \{(s,s) \mid \mathcal{M} \models_s A\} \end{aligned}$$

Consider the smallest normal logic that contains the following axiom schemata.

AXIOMS

$$\begin{aligned} {[}\alpha;\beta]A &\equiv [\alpha][\beta]A \\ [\alpha \cup \beta]A &\equiv [\alpha]A \wedge [\beta]A \\ [A?]B &\equiv (A \supset B) \\ [\alpha^*]A &\supset (A \wedge [\alpha][\alpha^*]A) \\ [\alpha^*](A \supset [\alpha]A) &\supset (A \supset [\alpha^*]A) \end{aligned}$$

This logic is known to be determined by the class of standard models [Goldblatt 87] (see Subsection 1.2.7 for the definition of 'determined logic').

1.2.12 Logics of knowledge and of belief

The languages that deal with the knowledge or belief of several agents are clearly multimodal languages. In such a context, formula $[i]A$ means 'agent i knows A', or 'agent i believes A'. These logics have been studied in chapter 4 of [Thayse 88]. We shall examine a typical example of introspective reasoning based on these logics in Subsection 1.3.2.

1.2.13 First-order predicate logic

We now recall the basic concepts of predicate logic; this logic has been explained in detail in chapters 1 and 2 of [Thayse 88].

As in propositional logic, we have first to define the atoms of predicate logic; they are made up of four types of symbols:

- the *variables* (denoted x, y, z, etc.);
- the *individual constants* (denoted a, b, c, etc.);
- the *function constants* or *function names* (denoted f, g, h, etc.);
- the *predicate constants* or *predicate names* (denoted P, Q, R, etc.).

Individual constants, function constants and predicate constants are also called *non-logical constants*. These basic symbols are used in the definition of the following four concepts.

- A *term* is a variable or a functional form.

- A *functional form* is the application of a function constant to an adequate number of terms. If f is a function constant of arity n and if $t_1, \ldots, t_n$ are terms, then the corresponding form is denoted $f(t_1, \ldots, t_n)$. If $n = 0$, the form is written f instead of $f(\ )$; this is nothing but an individual constant.

- A *predicate form* is the application of a predicate constant to an adequate number of terms. If P is a predicate constant of arity m and if $t_1, \ldots, t_m$ are terms, then the corresponding form is denoted $P(t_1, \ldots, t_m)$. If $m = 0$, the form is written P instead of $P(\ )$.

- An *atom* is a predicate form or an equality, that is, an expression such as $(s = t)$, where s and t are terms.

1.2. MODAL LOGIC

Note that the names 'functional form' and 'predicate form' are often replaced by 'function' and 'predicate', especially in the applications.

The formulas of the logic language are obtained recursively from the atoms by means of a collection of syntactic rules. These syntactic rules use the *connectives* $\neg, \vee, \wedge, \supset, \equiv$ of propositional logic plus two symbols $\forall$ and $\exists$ that are called *universal quantifier* and *existential quantifier* respectively. The logical connectives and the quantifiers are also called *logical constants*. If x is a variable, then $\forall x$ is read 'for all x' and $\exists x$ is read 'for at least one x'.

SYNTAX

- **Base:** an atom is a formula.

- **Induction:** if X and Y are formulas and if x is a variable, then $\neg X, (X \wedge Y), (X \vee Y), (X \supset Y), (X \equiv Y), \forall x X$ and $\exists x X$ are formulas.

- **Closure:** all formulas are obtained by applying the base and induction rules a finite number of times.

In predicate calculus as in propositional calculus, formulas can be interpreted, i.e., they can be mapped on *truth values*. However, components of a formula of predicate calculus are not only subformulas but also terms. It is therefore necessary to give an interpretation to terms. Intuitively, a term refers to an *object*. Thus, an interpretation should involve a set of objects, which will be called the *domain of the interpretation*.

More precisely, an *interpretation* I is a triple (S, I_c, I_v) with the following properties.

- S is a non-empty set, called the domain of the interpretation.

- I_c is a function which maps each function constant f of arity n to a function $I_c(f)$ from S^n into S, and which maps each predicate constant P of arity m to a function $I_c(P)$ from S^m into $\{\mathbf{T}, \mathbf{F}\}$. (By convention, a function from S^0 into a given set is an element of this set.)

- I_v is a function which maps each variable to an element of S.

Before giving the interpretation rules for both universal and existential quantifications let us introduce a useful notation. If S_I is the domain of a given interpretation I, if x is a variable and if d is an element of S_I, then $I_{x/d}$ denotes the interpretation J such that $S_J = S_I, J_c = I_c, J_v(x) = d$ and $J_v(y) = I_v(y)$, for each variable y distinct from x.

Interpretation rules corresponding to the interpretation $I = (S, I_c, I_v)$ can now be given. These rules associate a truth value $I(A)$ with each formula A and associate an element $I(t)$ of S with each term t.

SEMANTICS

- If x is a variable, then $I(x) =_{def} I_v(x)$.

- If f is a function constant of arity n and if $t_1, \ldots, t_n$ are terms, then $I(f(t_1, \ldots, t_n)) =_{def} I_c(f)(I(t_1), \ldots, I(t_n))$.

- If P is a predicate constant of arity m and if $t_1, \ldots, t_m$ are terms, then $I(P(t_1, \ldots, t_m)) =_{def} I_c(P)(I(t_1), \ldots, I(t_m))$.

- If s and t are terms, then $I(s = t)$ is **T** if $I(s) = I(t)$.

- If A and B are formulas, then $\neg A, (A \wedge B), (A \vee B), (A \supset B)$ and $(A \equiv B)$ are interpreted as in propositional calculus.

- If A is a formula and if x is a variable, then $I(\forall x A)$ is **T** if $I_{x/d}(A)$ is **T** for every element d of S.

- If A is a formula and if x is a variable, then $I(\exists x A)$ is **T** if $I_{x/d}(A)$ is **T** for at least one element d of S.

A formula A of predicate calculus is said to be *true for an interpretation I* if $I(A) = \mathbf{T}$.

As in propositional calculus, formulas can be classified into three categories: *valid formulas* are true for all interpretations, *inconsistent formulas* are false for all interpretations and *contingent formulas* are true for some interpretations and false for some others. In propositional calculus these categories are recursive; this is no longer the case in predicate calculus. No general algorithm can decide whether a formula of predicate calculus is valid, contingent or inconsistent. A formula is said to be *consistent* if it is not inconsistent, i.e., if it is true for at least one interpretation.

The concepts of *tautology, logical consequence, axiomatic system, inference rule, theorem,* and *proof,* that have been introduced in the context of propositional logic are defined in the same way for predicate logic.

The axiomatic system presented in Subsection 1.2.2 can be adapted to the framework of predicate calculus without equality. The axiom schemata ($A1 - A3$) and the *Modus Ponens* rule (MP) are kept unchanged. Two additional schemata, ($A4$) and ($A5$), and an additional inference rule, called the *generalization rule* and denoted (G), are introduced to deal with quantification.

We need some preliminary definition and notation. The variable x is *free for the term t in the formula Q* if neither x nor any variable of t occurs bound in Q. In this case, $Q_{x/t}$ denotes the formula obtained by uniformly replacing x by t in Q. The five axiom schemata and the two inference rules are listed below.

1.2. MODAL LOGIC

AXIOMATIC SYSTEM

- (A1) $(P \supset (Q \supset P))$,
- (A2) $((P \supset (Q \supset R)) \supset ((P \supset Q) \supset (P \supset R)))$,
- (A3) $((\neg P \supset \neg Q) \supset ((\neg P \supset Q) \supset P))$,
- (A4) $\forall x (P \supset Q) \supset (P \supset \forall x Q)$
  (x does not occur in P and is not bound in Q),
- (A5) $(\forall x Q \supset Q_{x/t})$
  (x is free for t in Q).

- (MP) If X and $(X \supset Y)$ are theorems,
  then Y is a theorem.
- (G) If P is a theorem and if x is not bound in P,
  then $\forall x\, P$ is a theorem.

The existential quantification is not introduced explicitly; the formula $\exists x A$ is considered as an abbreviation for $\neg \forall x \neg A$.

1.2.14 Model for a first-order logic

The semantics that has been defined in the preceding subsection is based on the notion of *interpretation*. It is also possible to define a semantics from the notion of *model*; the two types of definitions are equivalent. As the semantics of modal logic is generally defined from model theory, it is useful to develop this approach also in the case of a first-order logic without equality. Let us remember that a *logic* (or *theory*) relative to a language is a set of formulas of this language. Logics based on predicate calculus are called *first-order logics* (or *first-order theories*).

The interpretation rules of Subsection 1.2.13 define a semantics for the predicate language; these rules are based on an interpretation I that is defined as a triple (S, I_c, I_v) where S is a non-empty set of individuals (the domain of interpretation), I_c is a function (valuation) that assigns values to predicate and function constants, and I_v is a function (of assignment) that assigns an element of S to each variable.

A *model* $\mathcal{M} = (S, V)$ for a first-order logic $\mathcal{L}$ (without equality) consists of a non-empty domain S (the interpretation domain), together with a function V which assigns to each n-place function symbol a function from S^n to S and to each n-place predicate symbol a function from S^n to $\{\mathbf{T}, \mathbf{F}\}$. The valuation V plays exactly the same role as the function I_c of an interpretation I.

In order to provide a first-order language with a semantics relative to such a model $\mathcal{M} = (S, V)$, we introduce an assignment function which assigns an element of S to each variable. This function is denoted g and corresponds to the function I_v of an interpretation I. The semantics of the formulas of the language makes use of the notation

$$\mathcal{M} \vDash_g A$$

which means 'the formula A is true in the model $\mathcal{M}$, for the assignment g'.

If α is an expression (a term or a formula) of the language $\mathcal{L}$, we shall represent the semantic value of α in the model $\mathcal{M}$ and for the assignment g by the notation $[\![\alpha]\!]^{\mathcal{M},g}$. Thus, for every formula A of $\mathcal{L}$, we have the following equivalent semantic notations:

$\mathcal{M} \vDash_g A \iff [\![A]\!]^{\mathcal{M},g} = 1$,
$I(A) = 1 \iff [\![A]\!]^{\mathcal{M},g} = 1$.

SEMANTICS

- If x is a variable, then $[\![x]\!]^{\mathcal{M},g} =_{def} g(x)$.

- If f is an n-place function constant and if $t_1, ..., t_n$ are terms, then
  $[\![f(t_1, ..., t_n)]\!]^{\mathcal{M},g} =_{def} V(f)([\![t_1]\!]^{\mathcal{M},g}, ..., [\![t_n]\!]^{\mathcal{M},g})$.

- If P is an n-place predicate constant and if $t_1, ..., t_n$ are terms, then
  $[\![P(t_1, ..., t_n)]\!]^{\mathcal{M},g} =_{def} V(P)([\![t_1]\!]^{\mathcal{M},g}, ..., [\![t_n]\!]^{\mathcal{M},g})$.

- $\mathcal{M} \vDash_g P(t_1, \cdots, t_n)$ if and only if $[\![P(t, \ldots, t_n)]\!] = \mathbf{T}$.

- $\mathcal{M} \vDash_g \neg A$ if and only if $\mathcal{M} \not\vDash_g A$.

- $\mathcal{M} \vDash_g A \wedge B$ if and only if $\mathcal{M} \vDash_g A$ and $\mathcal{M} \vDash_g B$. (Similar definitions for the formulas $A \vee B, A \supset B$ and $A \equiv B$.)

- $\mathcal{M} \vDash_g \forall x A$ if and only if $\mathcal{M} \vDash_{g'} A$ for every assignment g' that differs from g only by the value assigned to the variable x. (Dual definition for the formula $\exists x A$.)

A formula A is said to be *valid* if $\mathcal{M} \vDash_g A$ for each model $\mathcal{M}$ and each assignment function g.

1.2.15 Predicate modal logic

Subsections 1.2.3 to 1.2.8 contain a presentation of propositional modal logic. The modal operators $\Box$ and $\Diamond$ act on propositions whose truth value depends not only on the declarative sentences they represent but also, for example, on the time at which these sentences have been stated. Let us now turn to the usual area of quantification theory that includes quantifiers over individuals. The modal operators will now act on predicate forms whose truth value depends on the value taken by the arguments. If these arguments are variables,

1.2. MODAL LOGIC

they will usually be under the scope of universal or existential quantifiers. The variables x range over a certain domain and the quantified expressions $\forall x$ and $\exists x$ respectively mean: 'for all values in the domain' and 'for at least one value in the domain'. A basic question in a modal context is this: over what domain will these variables range ? The universe W of possible worlds (§ 1.2.3) introduces an additional degree of freedom which should be taken into account in the definition of the variables domain. Three natural definitions of the domain are the following:

1. The set Γ_a of all the individuals existing in the actual world.

2. The set Γ_w of all the individuals existing in a given possible world w of the universe W.

3. The set Γ of all the individuals existing in any world ($\Gamma = \cup_{w \in W} \Gamma_w$).

As universal and existential quantifiers over these domains (and thus as candidates for $\exists$ and $\forall$) we shall have the three pairs:

1. $\forall_a x$ means 'for all x in the actual world', and $\exists_a x$ means 'for an x in the actual world'.

2. $\forall_w x$ means 'for all x in the world w', and $\exists_w x$ means 'for an x in the world w'.

3. $\forall^* x$ means 'for all x', and $\exists^* x$ means 'for at least one x'.

Internally, as regards the patterns of their mutual interrelationships, each of these pairs of quantifiers observes all of the standard rules of predicate logic, such as:

$$\forall x A(x) \supset \exists x A(x),$$
$$\neg \forall x A(x) \equiv \exists x \neg A(x).$$

No logical novelty arises with each pair: the only change from case to case is a modification of the domain of quantification. But semantically, the situation as regards the meaning and the interpretation of these quantifiers is quite different. Let us, for example, give a temporal interpretation to the different worlds w of the universe W; the formula

$$\forall x (Man(x) \supset Mortal(x))$$

can be interpreted in three different ways according to the definition taken for the universal quantifier:

1. All now-existing men are mortal.
2. All existing-at-time-w men are mortal.
3. All men are mortal.

The third interpretation can take a more restrictive meaning when considering past-oriented temporal logic or future-oriented temporal logic:

3'. All up-to-now-existing men are mortal.
3''. All from-now-existing men are mortal.

There exists a particular formulation of modal operators that allows their easy integration into the formalism of predicate logic. The expression $T_s(p)$ means 'p is true at world s'. This expression allows us to define the modal operators $\Box$ and $\Diamond$ as follows:

$$\Box p = \forall s T_s(p),$$
$$\Diamond p = \exists s T_s(p).$$

These modalities are immediately adaptable to the context of predicate logic, where predicate forms rather than propositional constants are considered. (Note that in this case we are no longer in the framework of first-order logic.)

A *modal predicate logic* must be able to build each of the following eight formulas (we also give the quantified version of the modal operators):

1. $\Box \forall x A(x) \quad = \quad \forall s T_s(\forall x A(x))$
2. $\forall x \Box A(x) \quad = \quad \forall x \forall s T_s(A(x))$
3. $\exists x \Box A(x) \quad = \quad \exists x \forall s T_s(A(x))$
4. $\Box \exists x A(x) \quad = \quad \forall s T_s(\exists x A(x))$
5. $\Diamond \forall x A(x) \quad = \quad \exists s T_s(\forall x A(x))$
6. $\forall x \Diamond A(x) \quad = \quad \forall x \exists s T_s(A(x))$
7. $\Diamond \exists x A(x) \quad = \quad \exists s T_s(\exists x A(x))$
8. $\exists x \Diamond A(x) \quad = \quad \exists x \exists s T_s(A(x))$

Assume that the following equivalences are satisfied:

$$T_s(\forall x A(x)) \equiv \forall x T_s(A(x)), \tag{1.2}$$
$$T_s(\exists x A(x)) \equiv \exists x T_s(A(x)). \tag{1.3}$$

From these equivalences, we deduce the following relations between formulas 1–8 above:

1.2. MODAL LOGIC

$$\square \forall x A(x) \equiv \forall x \square A(x)$$
$$\diamond \forall x A(x) \supset \forall x \diamond A(x)$$
$$\diamond \exists x A(x) \equiv \exists x \diamond A(x)$$
$$\exists x \square A(x) \supset \square \exists x A(x)$$

The equivalences (1.2), (1.3) are, however, not always satisfied; it depends on the interpretation that is chosen for the quantifiers $\forall$ and $\exists$. Let us, for example, take the interpretation $\forall_a x$, i.e., 'for every x in the actual world'. From a particular interpretation of a predicate $A(x)$ we deduce the following sequence of interpretations:

- $A(x)$: if x is a man, x is mortal.

- $\forall_a x A(x)$: all now-existing men are mortal.

- $T_s(\forall_a x A(x))$: 'all now-existing men are mortal' is true at time (or world) s.

- $T_s(A(x))$: 'if x is a man, x is mortal' is true at time (or world) s.

- $\forall_a x (T_s(A(x)))$: for all the now-existing men, it is true that if x is a man, x is mortal at time (or world) s.

Consequently, the formula $T_s(\forall_a x A(x))$ states that 'at time s all the men are mortal' while the formula $\forall_a x(T_s(A(x)))$ expresses that 'all the now existing men are mortal at time s'. These two formulas have thus different meanings and the equivalences (1.2) and (1.3) do not hold for the interpretations $\forall_a$ and $\exists_a$ of the quantifiers. A formal proof of the incorrectness of formulas (1.2) and (1.3) can be found in [Rescher and Urquhart 71]. Furthermore, this observation raises the question of the validity of the *Barcan formulas*

$$\diamond \exists x A(x) \supset \exists x \diamond A(x), \quad \forall x \square A(x) \supset \square \forall x A(x), \qquad (1.4)$$

which are incorporated in some axiomatic systems of modal predicate logic. The expressions $\diamond \exists x A(x)$ and $\exists x \diamond A(x)$ involved in the first Barcan formula do not have the same meaning. Assume for example that the formula $A(x)$ means 'someone is coming'. In the latter case, the meaning of the formula $\diamond \exists x A(x)$ is 'it is possible that someone is coming'. Both the existence and the coming of someone are problematical. The formula $\exists x \diamond A(x)$ means 'there exists someone that will probably come'. In this last case, the existence of this 'someone' is guaranteed while its coming remains problematical.

The syntax and an axiomatic system for modal predicate logic are defined as follows.

SYNTAX

- All formation rules of classical predicate logic are also formation rules of modal predicate logic.

- If A is a formula, then $\Box A$ and $\Diamond A$ are formulas.

AXIOMATIC SYSTEM

- An axiomatic system of classical predicate logic (§ 1.2.13).

- Some modal axiom schemata (§ 1.2.5), with possibly the Barcan schemata.

- The inference rules of classical predicate logic.

- The modal necessitation inference rule, to which some particular rules are possibly added in order to formalize some particular forms of reasoning. For example, the non-monotonic inference rule, that is, 'one cannot deduce $\neg p$' $\vdash \neg \Diamond p$, can be added to systems that formalize human belief ([Thayse 88], § 4.5.3).

The semantics for modal predicate logic is given in the next subsection. Some interesting applications, concerned with natural language processing, make use of some special kinds of modal logics, for which the terms are only individual constants or variables. A semantics for this restricted class of modal logics is introduced in Chapter 2.

1.2.16 Model and semantics for a modal predicate logic

A model $\mathcal{M}$ for a modal predicate logic is a four-tuple (W, R, S, V) where

- W is a non-empty set of possible worlds;

- R is a binary relation of 'accessibility' on W, i.e., a subset of $W \times W$;

- S is a non-empty domain of individuals;

- V is a function which

  - assigns to each pair consisting of an n-place function constant f and of an element w of W a function $V(f, w)$ from S^n to S;

1.2. MODAL LOGIC

- assigns to each pair consisting of an n-place predicate constant P and of an element w of W a function $V(P,w)$ from S^n to $\{1, 0\}$.

The point is that certain predicates, for example, might be true over different sets of objects in different possible worlds. A more general notion of model would allow the domain of individuals itself to vary from world to world. Such models will be considered in Section 2.2 in the context of the construction of a logic language that will be able to represent a fragment of a natural language.

The interpretation of a modal predicate logic language in such a model differs from the interpretation of a classical predicate logic language in that the universe W of possible worlds plays a key role. The interpretation, however, is still given relative to an assignment function g that assigns elements of S to the variables. The semantics of the formulas of a modal predicate language is described by means of the notation

$$\mathcal{M} \models_{w,g} A$$

with the meaning: 'A is true in the model $\mathcal{M}$, at the world w and for the assignment function g'.

If α is an expression of the language, we shall represent the semantic value of α in the model $\mathcal{M}$, at the world w and for the assignment function g by the notation $[\![\alpha]\!]^{\mathcal{M},w,g}$. We have thus the following equivalence that holds for any formula A of the language:

$$\mathcal{M} \models_{w,g} A \iff ([\![A]\!]^{\mathcal{M},w,g} = 1).$$

If c is a function or predicate constant we have

$$[\![c]\!]^{\mathcal{M},w,g} = V(w,c).$$

The interpretation of the formulas of the modal predicate language is then given by the semantic rules of the classical predicate language (§ 1.2.14) where the occurrences of g are replaced by w, g. The additional semantic rules relative to the modal operators $\Box$ and $\Diamond$ are as follows.

SEMANTICS

- $\mathcal{M} \models_{w,g} \Box A$ if $\mathcal{M} \models_{w',g} A$ for every w' in W such that wRw'.
- $\mathcal{M} \models_{w,g} \Diamond A$ if there exists a w' in W such that wRw' and $\mathcal{M} \models_{w',g} A$.

A formula A is said to be *true in a model* $\mathcal{M}$ if $\mathcal{M} \models_{w,g} A$ for each assignment function g and possible world w. A formula is said to be *valid* if it is true in each model.

1.2.17 First-order dynamic logic

We shall introduce a language that combines dynamic logic (§ 1.2.11) with first-order logic. In such a first-order dynamic language, the atomic programs of the propositional dynamic language are replaced by *assignment commands*, denoted $[x := \sigma]$, where x is a variable and σ is a term. Such a command has the (deterministic) meaning 'assign to x the current value of σ'. There is a strong connection between the computational process of assignment (of a value) to a variable and the syntactic process of substitution for a variable. If $A^{x:=\sigma}$ is the result of replacing the free occurrences of x in a first-order formula A by σ, then the formula

$$[x := \sigma]A \equiv A^{x:=\sigma}$$

is valid. Therefore, we are able to use a modal formula $[x := \sigma]A$ in places where the standard theory of first-order logic uses $A^{x:=\sigma}$. In this formalism, for example, the non-clausal resolution formulas ([Thayse 84] and [Thayse 88], § 1.1.15) are written:

$$\top_x(A_1, A_2) =_{def} [x := 0]A_1 \wedge [x := 1]A_2,$$
$$\bot_x(A_1, A_2) =_{def} [x := 0]A_1 \vee [x := 1]A_2.$$

This modal formalism allows us to avoid developing a theory of syntactic substitution in formulas that contain modal operators. In this computation context, the notion of *state* (of a program) can be given a concrete interpretation. The current state of a computation is determined by the *valuation* of the variables. If we define an equivalence relation

$$s \sim_x t$$

as meaning that states s and t differ only in the value they assign to x, we see that the semantic relations relative to the quantifiers $\exists$ and $\forall$ (§ 1.2.13) become

$$\vDash_s \exists x A \quad \text{if and only if} \quad \vDash_t A \text{ for some state } t \text{ such that } s \sim_x t;$$
$$\vDash_s \forall x A \quad \text{if and only if} \quad \vDash_t A \text{ for all states } t \text{ such that } s \sim_x t.$$

Clearly, this makes the expressions $\exists x$ and $\forall x$ look like modal operators, and indeed it can be verified that formally they obey the laws of an $S5$ modal logic [Goldblatt 87]. (This is to put together with the definition in Subsection 1.2.15 of the expression $T_s(x)$ that allows us to define modal operators in terms of quantified formulas.)

Let us now give a list of axioms for a first-order dynamic logic [Goldblatt 87].

1.3. EXAMPLES AND APPLICATIONS

1. $\forall x(A \supset B) \supset (\forall x A \supset \forall x B)$
2. $A \supset \forall x A$ (x does not appear in A)
3. $\forall x A \supset [x := \sigma]A$
4. $\forall y[x := y]A \supset \forall x A$ (x does not appear in A and x differs from y)
5. $\forall x A \supset [x := \sigma]\forall x A$
6. $\forall y[x := \sigma]A \supset [x := \sigma]\forall y A$ (y does not appear in $[x := \sigma]$)
7. $<x := \sigma> A \equiv [x := \sigma]A$
8. $[x := \sigma]A \equiv A^{x:=\sigma}$
9. $[x := \sigma][y := \tau]A \supset [y := \tau^{x:=\sigma}]A$
10. $[x := \sigma][y := \tau]A \supset [y := \tau^{x:=\sigma}][x := \sigma]A$ (y does not appear in $[x := \sigma]$)
11. $\sigma = \tau \supset ([x := \sigma]A \equiv [x := \tau]A)$

One of the main advantages of the modal languages based on dynamic logic lies in their expressive power. These formal languages provide us with a coherent frame in which most of the formal properties of a programming language can be expressed. The reader will find more about this interpretation in [Goldblatt 87], [Harel 79] and [Turner 84].

1.3 Examples and applications

1.3.1 Introduction

Reasoning about the knowledge and beliefs of computer and human agents is assuming increasing importance in artificial intelligence systems for natural language understanding. We propose in this section three examples of application of logic in the domains of knowledge representation and of reasoning about knowledge. Subsection 1.3.2 introduces the 'wise man problem'; the reasoning involved in this problem hinges on the ability of an agent to reason about other agents beliefs. In this example we show how modal operators, axioms and inference rules can capture our intuitive ideas about knowledge and belief and about reasoning on knowledge and belief. Subsection 1.3.3 contains an example of proving self-utterances, in a formal theorem proving context, by use of first-order logic; this example is the widely used 'knights and knaves' problem. Finally, Subsection 1.3.4 introduces a collection of clauses for set theory, thus developing a foundation for the expression of most theorems of mathematics in a form acceptable to a resolution based automated theorem prover.

1.3.2 Modal logic and reasoning about knowledge

As an example to illustrate how the axioms of modal logic can be used to reason about knowledge agents, let us consider the wise man puzzle which McCarthy has used extensively as a test of models of knowledge [McCarthy et al. 78]. The wise man puzzle is as follows.

A king, wishing to know which of his three advisers is the wisest, paints a white spot on each of their foreheads, tells them the spots are black or white and there is at least one white spot, and asks them to tell him the colour of their own spots. After a time the first wise man says, 'I do not know whether I have a white spot'; the second hearing this, also says he does not know. The third wise man then responds, 'My spot must be white'.

The 'puzzle' part is to generate the reasoning employed by the third wise man. The reasoning involved is really quite complex and hinges on the ability of the wise men to reason about one another's beliefs. The solution of the wise man puzzle is as follows.

The third man reasons: 'Suppose my spot were black. Then the second of us would know that his own spot was white, since he would know that, if it were black, the first of us would have seen two black spots and would have known his own spot's colour. Since both answered that they did not know their own spot's colour, my spot must be white'.

The difficulty behind this puzzle seems to lie in the nature of the third wise man's knowledge reasoning about the beliefs of the first two. Not only must he pose a hypothetical situation ('suppose my spot were black'), but he must then reason within that situation about what conclusions the second wise man would come to after hearing the first wise man's answer. This in turn means that he must reason about the second wise man's reasoning about the first wise man's knowledge as revealed by his reply to the king. Reasoning about such 'nested knowledges' can quickly become confusing, especially when there are conditions present concerning what an agent does not know.

As quoted above, the wise man puzzle has been used extensively as a test of models of knowledge. Konolige has proposed an interesting version of the puzzle where the third wise man responds 'I also don't know the colour of my spot; but if the second of us were wiser, I would know it' [Konolige 84].

We shall formalize (a simplified version of) the classical three wise men puzzle by means of a multimodal logic of knowledge. In this logic language the operator $[i]$ will mean 'the adviser i knows that'; the logic must be provided with the following axiom schemata and inference rules:

1.3. EXAMPLES AND APPLICATIONS

$[i](X \supset Y) \supset ([i]X \supset [i]Y)$ (distribution axiom)
$[i]X \supset X$ (knowledge axiom)
$X \vdash [i]X$ (necessitation rule)
$X \supset Y, X \vdash Y$ (Modus Ponens rule)

The distribution axiom and the inference rules of Modus Ponens and of necessitation allow us to infer that an agent knows all the logical consequences of his knowledge. This property can be expressed as a new inference rule:

$$(X \vdash Y), [i]X \vdash [i]Y \text{ (logic omniscience)}.$$

In order to reduce the number of the inferences, we shall consider the two-man version of the wise man puzzle. In this version it is assumed that there are only two advisers, represented by A and B respectively. The information that we need for the statement of the puzzle can be described as follows.

(1) A and B know that each can see the other's forehead, and hence:

(1a) If A does not have a white spot, B will know that A does not have a white spot. If $White(x)$ is a predicate that takes the value 'true' if x has a white spot and the value 'false' otherwise, this is formalized by

$\neg White(A) \supset [B](\neg White(A))$.

(1b) A knows (1a):
$[A](\neg White(A) \supset [B](\neg White(A)))$.

(2) A and B each know that at least one of them has a white spot, and they each know that the other knows that; in particular:

(2a) A knows that B knows that if A has no white spot, then B has a white spot, i.e.,
$[A]([B](\neg White(A) \supset White(B)))$.

(3) B says that he does not know whether he has a white spot, and A thereby knows that B does not know, i.e.,
$[A](\neg[B](White(B)))$.

Statements (1b), (2a) and (3) are written as the first three lines of the following proof of the theorem $[A](White(A))$. In the list below, which constitutes a proof of this theorem, each line is the resolvent of some preceding lines and of some axioms; the numbers of these lines and the axioms are indicated on the right of the resolvent.

1. $[A](\neg White(A) \supset [B](\neg White(A)))$
2. $[A]([B](\neg White(A) \supset White(B)))$
3. $[A](\neg[B](White(B)))$
4. $\neg White(A) \supset [B](\neg White(A))$ 1, axiom of knowledge
5. $[B](\neg White(A) \supset White(B))$ 2, axiom of knowledge
6. $[B](\neg White(A)) \supset [B](White(B))$ 5, axiom of distribution
7. $\neg White(A) \supset [B](White(B))$ 4, 6, transitivity
8. $\neg[B](White(B)) \supset White(A)$ reformulation of 7
9. $[A](\neg[B](White(B)) \supset White(A))$ 1, 2, 4, 5, 8, logic omniscience
10. $[A](\neg[B](White(B))) \supset [A](White(A))$ 9, axiom of distribution
11. $[A](White(A))$ 3, 10, Modus Ponens

Line 9 of the proof can be understood as follows. Hypotheses 1 and 2 mean that A knows formulas 4 and 5. From formulas 4 and 5 we infer formula 8. The rule of logic omniscience allows us therefore to infer that A knows formula 8; we can write this as follows:

$$(1, 2 \vdash [A](4, 5)), (4, 5 \vdash 8) \vdash [A](8).$$

The formalization of hypotheses 1–3 is borrowed from [Genesereth and Nilsson 87]. The strategy of proof for the three-level version of the wise man puzzle is basically the same as for the two-level version.

1.3.3 Predicate logic and formalization of reasoning

The knights and knaves problem can be stated as follows [Smullyan 78].

An island exists whose only inhabitants are knights, knaves and a princess. The knights on the island always tell the truth, while the knaves always lie. Some of the knights are poor and the rest of them are rich. The same holds for the knaves. The princess is looking for a husband who must be a rich knave. In uttering one statement, how can a rich knave convince the princess that he is indeed a prospective husband for her ?

The solution method given hereafter is borrowed from [Miller and Perlis 87]; it is reprinted by permission of D. Reidel publishing company; copyright 1987. The solution of the problem will be obtained in the form of a proof of a theorem in first-order logic. In fact, many applications of predicate calculus in artificial intelligence can be regarded as methods for finding a proof of a theorem. The theorem proving procedures usually follow a *refutation pattern*. The hypotheses H_j of the theorem to be proved are added temporarily as new axioms to those, A_i, of the domain of discourse we are working in. A proof by contradiction is then sought, by adding the negation of the conclusion C of the

1.3. EXAMPLES AND APPLICATIONS

theorem as a further temporary axiom and trying to prove that the formula thus obtained is false. If this succeeds, then we have proved

$$(\bigwedge_j H_j \land \bigwedge_i A_i \land \neg C) \supset \mathbf{F}.$$

Procedures based on this pattern are called *refutation systems* (see [Thayse 88], § 1.1.6 and § 1.1.9). We have thus first to express the 'knights and knaves problem' as a first-order logic formula containing existentially quantified variables. This formulation will constitute the conclusion (or theorem) that we want to prove logically from the hypotheses. The instantiation of the existentially quantified variables will provide us with the solution of the problem. Let us state the problem as follows.

'What statement, when made by anyone, will convince the princess that the person making the statement is a rich knave ?'

Miller and Perlis propose to formalize this statement by means of a first-order theory containing the following elements.

- I : individual constant representing the word 'I'.

- *knave* : function constant; *knave(p)* stands for the statement 'p is a knave'.

- *knight* : function constant; *knight(p)* stands for the statement 'p is a knight'.

- *rich* : function constant; *rich(p)* stands for the statement 'p is rich'.

- *not* : function constant; *not(p)* stands for the statement 'p is not'.

- *and* : function constant; *and(p, q)* stands for the statement 'p and q'.

- *CanSay* : predicate constant; *CanSay(p, q)* means: 'p can say q'.

- *TU* : predicate constant; *TU(p, q)* means: term q would be true if the occurrences of I in q were replaced by p.

- *T* : predicate constant; *T(t)* means: term t is true.

The formula

$$C \; : \; \exists u \forall p \, (\, CanSay(p, u) \equiv V(and(rich(p), knave(p))))$$

represents the statement: 'there exists a sentence u that can only be stated by an individual p that is a rich knave'. This formula is interpreted as being the conclusion (or theorem) that we have to deduce by a theorem proving method from the following hypotheses.

(1): $\forall p \forall u\, (T(knave(p)) \equiv (CanSay(p,u) \equiv \neg TU(p,u)))$,
(i.e., p is a knave if and only if things u that p can say are precisely those which would be false if they were stated by p);

(2): $\forall y \forall z\, (T(and(y,z)) \equiv T(y) \wedge T(z))$,
(this captures the meaning of the function 'and');

(3): $\forall s\, (T(s) \equiv \neg T(not(s)))$,
(this captures the meaning of the function 'not');

(4): $\forall p\, (TU(p, f(I)) \equiv T(f(p)))$.

Hypothesis (4) is a formula schema, an instance of which is for example:

$$TU(Paul, rich(I)) \equiv T(rich(Paul)).$$

We suppose all variables range over knights, knaves, the princess and utterances. The following four hypotheses recursively establish all possible instances of schema (4) in terms of the leftmost function symbol occurring in the second argument of predicate name TU.

(4_{and}) : $\forall u \forall v \forall p\, (TU(p, and(u,v)) \equiv (TU(p,u) \wedge TU(p,v)))$.
(4_{not}) : $\forall u \forall p\, (TU(p, not(u)) \equiv \neg TU(p,u))$
(4_{rich}) : $\forall p\, (TU(p, rich(I)) \equiv T(rich(p)))$
(4_{knave}) : $\forall p\, (TU(p, knave(I)) \equiv T(knave(p)))$

Thus the hypotheses we use for the knights and knaves problem will be (1), (2), (3), (4_{and}), (4_{not}), (4_{rich}) and (4_{knave}); the conclusion (or theorem) to be stated from these hypotheses is the formula C. Hence we have to prove

$$((1) \wedge (2) \wedge (3) \wedge (4_{and}) \wedge (4_{not}) \wedge (4_{rich}) \wedge (4_{knave})) \supset C.$$

We have first to transform the hypotheses and the negation of the conclusion in terms of clauses; this will allow us to use the resolution as inference rule ([Thayse 88], § 1.2.14). Below we present the hypotheses in clause form, leaving out those clauses that are not necessary to the resolution proof.

Clauses derived from hypothesis (1):

C11 : $\neg T(knave(p)) \vee \neg CanSay(p,u) \vee \neg TU(p,u)$
C12 : $\neg T(knave(p)) \vee CanSay(p,u) \vee TU(p,u)$
C13 : $T(knave(p)) \vee \neg CanSay(p,u) \vee TU(p,u)$

Clauses derived from hypothesis (2):

C21 : $\neg T(and(y,z)) \vee T(y)$
C22 : $\neg T(and(y,z)) \vee T(z)$
C23 : $T(and(y,z)) \vee \neg T(y) \vee \neg T(z)$

Clauses derived from hypothesis (3):

1.3. EXAMPLES AND APPLICATIONS

C31 : $T(s) \vee T(not(s))$
C32 : $\neg T(s) \vee \neg T(not(s))$

Clauses derived from the four instances of schema (4):

C41 : $TU(p, and(u, v)) \vee \neg TU(p, u) \vee \neg TU(p, v)$
C42 : $\neg TU(p, and(u, v)) \vee TU(p, u)$
C43 : $\neg TU(p, and(u, v)) \vee TU(p, v)$
C44 : $TU(p, not(u)) \vee TU(p, u)$
C45 : $\neg TU(p, not(u)) \vee \neg TU(p, u)$
C46 : $TU(p, rich(I)) \vee \neg T(rich(p))$
C47 : $\neg TU(p, rich(I)) \vee T(rich(p))$
C48 : $TU(p, knave(I)) \vee \neg T(knave(p))$
C49 : $\neg TU(p, knave(I)) \vee T(knave(p))$

Let g be a Skolem function resulting from the elimination of the existential quantifier in the negation of the conclusion C (see [Thayse 88], § 1.2.8).

Clauses derived from the negation of the conclusion C:

C51 : $CanSay(g(u), u) \vee T(and(rich(g(u)), knave(g(u))))$
C52 : $\neg CanSay(g(u), u) \vee \neg T(and(rich(g(u)), knave(g(u))))$

Given these clauses we must be able to use resolution to find the desired solution. For the sake of compact presentation we abbreviate 'and' as 'a', 'rich' as 'r', 'knave' as 'k', 'not' as 'n' and 'CanSay' as 'CS'. Each line of the proof contains the resolvent clause and the labels of the resolved clauses; moreover, some lines contain the key substitutions that occur in the resolution process.

Resolvent clause	Resolved clauses
R1 : $CS(g(u), u) \vee T(r(g(u)))$	C51, C21
R2 : $CS(g(u), u) \vee \neg T(n(r(g(u))))$	C32, R1
R3 : $CS(g(u), u) \vee \neg T(a(n(r(g(u))), z))$	C21, R2
R4 : $CS(g(u), u) \vee \neg T(n(r(g(u)))) \vee \neg T(z)$	C23, R3
R5 : $CS(g(u), u) \vee \neg T(n(r(g(u)))) \vee \neg TU(p, k(I))$	C49, R4
R6 : $CS(g(u), u) \vee T(r(g(u))) \vee \neg TU(p, k(I))$	C31, R5
R7 : $CS(g(u), u) \vee TU(g(u), r(I)) \vee \neg TU(p, k(I))$	C46, R6
R8 : $CS(g(u), u) \vee \neg TU(g(u), n(r(I))) \vee \neg TU(p, k(I))$	C45, R7
R9 : $CS(g(u), u) \vee \neg TU(g(u), n(r(I))) \vee \neg TU(p, a(u', k(I)))$	C43, R8
R10 : $CS(g(u), u) \vee \neg TU(g(u), a(n(r(I), v))) \vee$ $\vee \neg TU(p, a(u', k(I)))$	C42, R9
F11 : $CS(g(u), u) \vee \neg TU(g(u), anr)$ $\{(u', n(r(I))), (v, k(I)), (p, g(u))\}$	R10, Unification

(We use anr as a shorthand notation for $and(not(rich(I)), knave(I))$.)

R12 : $CS(g(u),u) \lor CS(g(u),anr) \lor \neg T(k(g(u)))$ \hfill C12, F11

F13 : $CS(g(anr),anr) \lor \neg T(k(g(anr))), \{(u,anr)\}$ \hfill R12, Unification

R14 : $CS(g(u),u) \lor T(k(g(u)))$ \hfill C51, C22

R15 : $CS(g(anr),anr)$ \hfill F13, R14

R16 : $\neg T(a(r(g(anr)),k(g(anr))))$ \hfill C52, R15

R17 : $\neg T(r(g(anr))) \lor \neg T(k(g(anr)))$ \hfill C23, R16

R18 : $T(n(r(g(anr)))) \lor \neg T(k(g(anr)))$ \hfill C31, R17

R19 : $\neg CS(g(anr),t) \lor TU(g(anr),t) \lor T(n(r(g(anr))))$ \hfill C13, R18

R20 : $TU(g(anr),anr) \lor T(n(r(g(anr)))) \lor \neg T(k(g(anr)))$ \hfill F13, R19

R21 : $TU(g(anr),n(r(I))) \lor T(n(r(g(anr)))) \lor \neg T(k(g(anr)))$ \hfill C42, R20

R22 : $\neg TU(g(anr),r(I)) \lor T(n(r(g(anr)))) \lor \neg T(k(g(anr)))$ \hfill C45, R21

R23 : $\neg T(r(g(anr))) \lor T(n(r(g(anr)))) \lor \neg T(k(g(anr)))$ \hfill C46, R22

R24 : $T(n(r(g(anr)))) \lor \neg T(k(g(anr)))$ \hfill C31, R23

R25 : $\neg T(k(g(anr))) \lor \neg TU(g(anr),anr)$ \hfill R15, C11

R26 : $\neg T(k(g(anr))) \lor \neg TU(g(anr),k(I)) \lor$
$\lor \neg TU(g(anr),n(r(I)))$ \hfill C41, R25

R27 : $\neg T(k(g(anr))) \lor \neg TU(g(anr),k(I)) \lor$
$\lor TU(g(anr),r(I))$ \hfill C44, R26

R28 : $\neg T(k(g(anr))) \lor \neg TU(g(anr),k(I)) \lor$
$\lor T(r(g(anr)))$ \hfill C47, R27

R29 : $\neg T(k(g(anr))) \lor \neg TU(g(anr),k(I)) \lor$
$\lor \neg T(n(r(g(anr))))$ \hfill C32, R28

R30 : $\neg T(k(g(anr))) \lor \neg TU(g(anr),k(I))$ \hfill R24, R29

R31 : $\neg T(k(g(anr)))$ \hfill C48, R30

R32 : $T(k(g(anr))) \lor TU(g(anr),anr)$ \hfill R15, C13

R33 : $TU(g(anr),anr)$ \hfill R31, R32

R34 : $TU(g(anr),k(I))$ \hfill C43, R33

R35 : $T(k(g(anr)))$ \hfill C49, R34

R36 : **F** \hfill R31, R35

The theorem is thus proved; the answer to be given is: *I am not rich and I am a knave.*

1.3.4 Mathematics and predicate logic

Mathematics offers a virtually unlimited supply of difficult, or even open, problems to be studied with automated reasoning logic systems. Some conjectures have, for example, been verified by means of strategies based on resolution. Boyer and his co-authors have presented an axiomatization of set theory, expressible in a finite number of clauses, that follows from von Neumann's approach [Boyer et al. 86]. The interest of presenting a first-order logic version of set theory lies in the fact that set theory is a sufficient vehicle within which to express the entire logic of functions normally used by mathematicians. Boyer and his co-authors present a collection of clauses for set theory, thus developing a foundation for the expression of most theorems of mathematics in a form acceptable to a resolution based automated theorem prover. The expressive power of this formulation is illustrated by providing statements of some well-known open questions in number theory and by giving some intuition about how the axioms are used by including sample proofs. We give here some samples of the theory proposed in [Boyer et al. 86].

The axioms of set theory are usually divided into four classes. Here we consider only the axioms of the first class, denoted A. The initial logic formulation is given in terms of first-order quantified formulas. These formulas are then transformed so as to allow their treatment by means of the resolution rule. In these formulas we will make use of a unary predicate constant denoted M; the expression $M(e)$ means 'e is a little set'.

AXIOMS A AND DEFINITION OF A SUBSET

- Axiom A-1 : Little sets are sets.

- Axiom A-2 : Elements of sets are little sets;
  $\forall x \forall y \, (x \in y \supset M(x))$.

- Axiom A-3 : Principle of extensionality;
  $\forall x \forall y \, (\forall u \, (M(u) \supset (u \in x \equiv u \in y)) \supset x = y)$.

- Axiom A-4 : Existence of non-ordered pairs;
  $\forall u \forall x \forall y \, (u \in \{x,y\} \equiv M(u) \wedge (u = x \vee u = y))$;
  $\forall x \forall y \, (M(\{x,y\}))$.

- Definition of a subset;
  $\forall x \forall y \, (x \subset y \equiv \forall u (M(u) \supset (u \in x \supset \in y)))$.

These first-order formulas are then transformed into products of clauses, by means of a program that first Skolemizes and then converts to conjunctive normal form. The symbols fj that appear in the clauses below are Skolem function constants.

- Axiom A-2 :

  1. $x \notin y \vee M(x)$.

- Axiom A-3 :

  2. $M(f1(x,y)) \vee x = y$,
  3. $f1(x,y) \in x \vee f1(x,y) \in y \vee x = y$,
  4. $f1(x,y) \notin x \vee f1(x,y) \notin y \vee x = y$.

- Axiom A-4 :

  5. $u \notin \{x,y\} \vee u = x \vee u = y$,
  6. $u \in \{x,y\} \vee \neg M(u) \vee u \neq x$,
  7. $u \in \{x,y\} \vee \neg M(u) \vee u \neq y$,
  8. $M(\{x,y\})$.

- Definition of a subset :

  9. $x \not\subset y \vee u \notin x \vee u \in y$,
  10. $x \subset y \vee f2(x,y) \in x$,
  11. $x \subset y \vee f2(x,y) \notin y$.

Some of the clauses 1–11 allow us to prove the following lemma.
Lemma : $\forall x \forall y (x \subset y \wedge y \subset x \supset x = y)$.

We shall prove that the conjunction of some of the clauses 1–11 with the negated lemma produces the empty clause. The negated lemma is represented by the expression

$$\exists x \exists y ((x \subset y \wedge y \subset x) \wedge (x \neq y)).$$

The Skolemized form of this expression is equivalent to the product of clauses 12–14 below.

- Negated lemma :

  12. $f3 \subset f4$,
  13. $f4 \subset f3$,

1.3. EXAMPLES AND APPLICATIONS

14. $f3 \neq f4$.

Proof of the Lemma.

Resolvent clauses	Resolved clauses
15. $u \notin f3 \vee u \in f4$	9, 12
16. $u \notin f4 \vee u \in f3$	9, 13
17. $f1(f3, f4) \in f3 \vee f1(f3, f4) \in f4$	3, 14
18. $f1(f3, f4) \notin f3 \vee f1(f3, f4) \notin f4$	4, 14
19. $f1(f3, f4) \notin f3$	15, 18
20. $f1(f3, f4) \in f3$	16, 17
21. **F**	19, 20

Thus, we have represented the axioms A of set theory and the definition of a subset by means of a collection of eleven clauses. Boyer and his co-authors have proposed a clausal representation of all the axioms of set theory and of the main definitions relative to this theory (including some definitions of abstract algebra and of number theory) by means of a collection of about a thousand clauses. As mentioned above, these clauses were derived from a first-order representation of the set theory axioms and definitions by a program that first Skolemizes, then converts to conjunctive normal form. Let us cite *verbatim* the conclusion of [Boyer et al. 86] about set theory and automated theorem proving.

Anyone attempting to submit the set of clauses given here to an automated theorem prover will quickly confront many fundamental issues in theorem proving. This set of clauses is difficult to work with for several reasons.

- There are many clauses, many of them non-Horn, and few useful units. Because so few basic concepts are used, there are many pairs of complementary literals. Unrestricted binary resolution or paramodulation will rapidly result in the dreaded combinatorial explosion.

- A strategy for restricting inferences must avoid 'opening up' every definition so that mathematical arguments can be made at the appropriate level. Yet sometimes definitions must be expanded. Thus a key technique will be to layer the deductions and to identify (automatically) the proper occasions for crossing layer boundaries. This problem has been the topic of much research, and many contributions to its solution have been proposed. It has never really been solved.

- A human learns relatively quickly how to construct proofs in this system, once he knows the general outline of the proof. Such expertise needs to be represented as a strategy for a theorem prover.

- In constructing any proof in this system, one finds a need for a set of lemmas. If a large body of lemmas is built up, however, and uniformly made available to the theorem prover, the problem of a large axiom set is compounded. It is important to develop a strategy for selecting from a database the lemmas relevant to a particular problem.

Chapter 2

Intensional logic and natural language

2.1 Logic processing of natural languages

2.1.1 Introduction

The preceding chapter might suggest that contemporary logic belongs exclusively to the scope of mathematics. At first glance, its methods and results do not seem to have any bearing on the study of natural language. The sophisticated formal apparatus used by logicians today makes us forget that logic grew out of philosophy and grammar. Yet the logical formalism developed in the first two chapters of [Thayse 88] and in Chapter 1 of this book can also be used for describing and processing natural languages such as English or French. Indeed, Tarski's elaboration of a semantics for formal languages, Kripke's invention of a possible-world semantics in connection with modal logic and Montague's semantics for intensional logic provided powerful tools by means of which natural languages could be analysed rigorously.

This first section introduces the reader to the logic processing of natural languages and to Montague's recursive semantics for a fragment of natural language. (The content of this section is borrowed from [Gochet 82a].)

In Section 2.2 the semantics of a modal logic suited to the description of a natural language will be introduced formally. This modal logic will be enriched in Section 2.3 by means of the concept of intension. This will lead us to the definition of intensional logic. The translation of a fragment of natural language into intensional logic will be studied in Chapter 3.

2.1.2 Formal semantics and natural language

The method of constructing the syntax of a formal language has been well known since the appearance of the work of Carnap [Carnap 36]. We saw in Section 1.2 the requirements which a logician must meet when he defines the syntax of a formal language. These requirements are:

- to enumerate the basic symbols;

- to define the notion of a formula, that is, to state the formation rules;

- to give the axioms;

- to give the inference rules.

The method of constructing the semantics of a formal language is also well known. Tarski showed that semantic notions of truth, of logical consequence, and of validity can be defined rigorously for a formal language such as propositional logic or predicate logic [Tarski 72]. All of these definitions are stated in Section 1.2; we also introduce therein the possible-world semantics for modal propositional logic. The notion of 'truth with respect to a model' was introduced by way of this semantics (§ 1.2.3). This notion is more general than 'absolute truth'; we will see that it plays an essential role in Montague's semantics.

The inventors of formal syntax and semantics were the first to ask whether the methods which they had devised for the study of the syntax and semantics of formal languages could be applied to natural languages. Their answer was at first negative. Between 1968 and 1970, Montague, an American logician from the Tarski school, wrote three papers the goal of which was to show that no important theoretical differences exist between natural languages and the formal languages of logic [Montague 74a], [Montague 74b], [Montague 74c]. Montague's research started where Davidson's work left off [Davidson 69].

To understand the impact of the theses of Davidson, it is best to start with the problems which he sought to resolve and which may be stated as follows. Every individual who is able to speak knows the meaning of a finite number of words and knows how to apply a finite number of formation rules. How can he, so equipped, understand a virtually infinite number of sentences ? This problem is the semantic equivalent of the question that grammarians ask themselves about syntax. Every individual recognizes as members of his language's vocabulary a finite number of words and knows how to apply a finite number of formation rules. How can he, under these conditions, recognize the 'well-formed', 'grammatical' or 'sensible' character of an infinity of sentences ?

In the case of formal languages, the problem of *syntactic competence* is easily resolved. All that is needed is to introduce among the formation rules a certain number of recursive rules. On the other hand, it is difficult to formulate the formation rules of a natural language, because, while a formal language has all the properties that its creator wanted to give it, a natural language is an infinite set of well-formed sentences which existed *before* the formation rules proposed by the linguist. Let us suppose, nevertheless, that the problem of syntactic competence is resolved. The question is then to know whether, to resolve the

2.1. LOGIC PROCESSING OF NATURAL LANGUAGES

problem of *semantic competence*, it is enough to *add* an interpretation of the atomic expressions to the syntax. Davidson's answer to this crucial question was *no*. If the semantics must contain a theory of meaning as we understand it, the knowledge of the structural characteristics which convey the grammatical correctness of a sentence along with the knowledge of the meaning of terminal parts of that sentence (i.e., the words of the vocabulary) do not add up to give us the knowledge of what the sentence in question means.

2.1.3 Davidson's recursive semantics

Katz and Fodor, to whom we owe the first recursive semantics, introduced into semantics a mechanism that aims at explaining how the meanings of syntactically combined words combine semantically: this mechanism is formalized by the *projection rules* [Katz and Fodor 64].

When the matter is examined more closely, however, it is seen that *projection rules* have a purely *negative* role, that is the role of selecting some specific meanings whenever the words which, in isolation, have several meanings, are combined within a syntactic construction. In the English sentence *the man hits the colourful ball*, for instance, the projection rule that combines the sense of *colourful* and the sense of *ball* does nothing more than cancel out the sense of 'picturesqueness' when *ball* means a physical object of globular shape rather than a social assembly for dancing. The elimination of the latter sense is itself due to the combination of *ball* with *hit*. Katz and Fodor, however, did not spell out rules for the composition of senses which are really satisfactory.

What must be added to recursive syntax to obtain a recursive semantics which describes *positively* and not negatively the phenomenon of composition of meanings ? Davidson proposed to add a recursive definition of truth. A theory of meaning (for a language $\mathcal{L}$) which contains a recursive definition of truth in $\mathcal{L}$ shows how the meaning of sentences depends of the meaning of the words [Davidson 69].

Since 'meaning' is not synonymous with 'truth', a definition of truth is not necessarily a definition of the meaning. Therefore, Davidson's thesis is not quite so evident. We can however prove it easily enough. We begin by identifying the meaning of declarative sentences with their truth conditions as, for example, Cresswell did when he wrote that to understand the meaning of the sentence 'the snow is white' is, among other things, to know under what conditions (or in what circumstances) this sentence is true and under what conditions it is false [Cresswell 78].

Once we have admitted the identity 'meaning of a declarative sentence = truth conditions', we must admit that a definition that says how the *truth conditions* of a complex sentence depends upon the *truth conditions* of the elementary sentences of which it is composed also tells us how the *meaning* of a complex sentence depends upon the *meanings* of the elementary sentences of which it is composed.

That is precisely what a recursive definition of truth does, since, for each propositional connective, it provides a clause which specifies the truth conditions of the complex sentence in terms of the truth conditions of elementary sentences. We require something more, however, from a recursive definition of meaning. We require that it tells us also how the meaning of an elementary sentence, that is, its truth conditions, depends on the meaning of the *words* of which it is composed. Here, a major difficulty appears: the meaning of a syntactic constituent which is not a sentence cannot be identified to its truth conditions, because only *sentences* may be said to be true or false. We can talk about the truth conditions of the sentence 'everybody loves somebody', but not of those of the predicate 'loves' or of the open formula 'x_1 loves x_2', because a predicate (just as a noun) cannot be viewed as true or false. (In this section, the term 'predicate' has its linguistic acceptation. Within a statement, a predicate denotes what is said of a referent denoted by another expression, the subject. For example, *are mortal* is a predicate within the statement *all men are mortal*.)

Tarski discovered how to overcome this difficulty. His method consists of two steps. First, he recursively defines a notion more general than that of truth, the notion of *satisfaction*, which applies as well to open formulas (predicates) as to closed formulas (sentences). Next, he defines the notion of truth by way of the notion of satisfaction. By stating the conditions for satisfaction of each open formula, composed of an atomic predicate and of a list of variables, and by adding to these conditions some recursive rules that specify how the satisfaction of an open formula depends on the satisfaction of its constituents (predicates and variables), we can show how the conditions for satisfaction of the base elements (or atoms) contribute to the truth conditions of the sentences [Partee 75]. For example, we can show how the meanings of the predicates 'love' and 'hate' contribute to the meaning of the sentence 'everybody loves or hates somebody'. (The notion of 'satisfaction' will be defined formally in Subsection 2.2.4; see the final remark in this subsection.)

Davidson disputed that a *recursive syntax* combined with a dictionary constitute a *recursive semantics*. Tarski provided what Katz and Fodor were lacking: an *autonomous* recursive semantics. This recursive semantics is autonomous, but nevertheless *parallel* to the recursive syntax: to every syntactic rule of formation of a complex expression (of a certain type), there corresponds a semantic rule which specifies the conditions for satisfaction of that expression.

2.1.4 Montague's innovations

Davidson meant to confine recursive truth theory to first-order predicate calculus. Montague, too, believed that formal semantic methods could be applied to natural language, but he abandoned first-order predicate calculus in favor of a *categorial grammar* incorporating the categories used traditionally by grammarians in the description of natural language, such as, for example, the categories of adverb and of preposition. Contrary to Davidson, he gave an autonomous semantic value (also called *denotation*) to expressions in each category and made systematic use of set theory to build the denotations of the expressions of the language under study. Montague also differed from Davidson by the fact that he replaced the notion of absolute truth by the notion of *truth relative to a model*, which allowed him to define the notions of 'logical truth' and 'consequence' for the fragment of natural language which he formalized. Furthermore, he recognized the need to complement extensional logic with intensional logic.

Extensional logic allows evaluation of the truth value of a sentence. This truth value depends on the state of the world to which this sentence refers (formally: it depends on the *model*). Generally, we do not know the extension (reference, denotation) of the components of a sentence, but only their intension (meaning, understanding, sense). That is why we are able to understand a sentence without knowing whether it is true. Extensional logic has the potential of determining whether a sentence is true, while intensional logic only goes as far as to try to understand the sentence.

Thus, in his semantic analysis of meaning, Montague distinguishes two elements: *intension* (or sense) and *extension* (or reference). He applies these two notions to predicates, nouns, and sentences. The intension of a predicate, i.e., its sense, is identified with the property it expresses; its extension is the class of objects which possess the property or which stand in the relation expressed by the predicate. The extension of a noun, of an individual expression (such as 'the natural satellite of the Earth') is the individual which it denotes; its intension is the individual concept it expresses. The extension of a sentence is its truth value; its intension is the proposition it expresses.

Montague introduced *intensional logic* as a way of giving a formal account of the concepts of intension and extension. An extensional logic can only assign truth values to sentences while an intensional logic can, in addition, assign a meaning to these sentences. Montague's intensional logic makes use of necessity/possibility modes which in turn use past/future time. Language formalization must therefore include the corresponding modal operators.

On the other hand, Montague kept one of Davidson's fundamental requirements: that of a perfect correspondence between syntax and semantics. Indeed, he required that each syntactic combination rule be matched with a semantic combination rule for denotations. Thus, just as Davidson, he did not believe that a recursive syntax along with an interpretation of atomic expressions could constitute a recursive semantics. He even developed a conception of the correspondence which was much more technical than Davidson's.

The correspondence between syntactic and semantic rules which *work in parallel* constitutes a formalization of Frege's *compositionality principle* that can be stated as follows [Frege 52]: *The meaning of a sentence is a function of the meaning of its parts and of their mode of combination.* We shall see in Sections 2.2 and 2.3 how to formalize this principle in the framework of extensional and intensional logics. The subject can be introduced as follows.

Consider a syntactic rule of the form 'if a is an element of syntactic category A, if b is an element of syntactic category B, ..., and if m is an element of syntactic category M, then $f(a, b, \ldots, m)$ is an element of syntactic category N'. The function f specifies how the input arguments $a, b, \ldots, m$ are to be combined so as to produce an output which is an element of syntactic category N. For any expression α, we use the notation $[\![\alpha]\!]$ to represent the semantic value of α. Corresponding to the syntactic rule above there will be a semantic rule of the form 'if a is an element of syntactic category $A, \ldots$, and if m is an element of syntactic category M, then $[\![f(a, b, \ldots, m)]\!]$ is expressible as $g([\![a]\!], [\![b]\!], \ldots, [\![m]\!])$. The function g specifies the 'semantic mode of combination' of the semantic values which are its arguments. Thus, with the syntactic formation rule

$$n = f(a, b, \ldots, m)$$

is associated a semantic formation rule

$$\begin{aligned}[n]\!] &= [\![f(a, b, \ldots, m)]\!], \\ &= g([\![a]\!], [\![b]\!], \ldots, [\![m]\!]).\end{aligned}$$

Using the adequate mathematical term for this situation, we can say that there is a *homomorphism* from syntax to semantics. It should be noted that this mapping is not an isomorphism, since different syntactic structures can receive the same semantic value ([Dowty et al. 81], p. 43).

2.1.5 Montague's approach

Tarski's truth theory applies to formal languages such as propositional calculus and predicate calculus. Davidson proposed to extend it to natural language but did not go much further than spelling out a programme. Montague took it over. In this subsection we shall consider the main steps of Montague's policy which was designed to formally represent the semantics of a natural language.

2.1. LOGIC PROCESSING OF NATURAL LANGUAGES

A detailed study of these steps will be made in Sections 2.2 and 2.3 and in Chapter 3.

Firstly Montague works out a syntax made up of a set of rules S_i, generating a fragment of natural language (English, in the present case). Next, for each syntactic rule S_i he defines a translation rule T_i which allows us to recursively translate every statement of the natural language into an appropriate logic language. From the knowledge of the state of the world, the semantics associated with that logic language enables us to assign a truth value to the sentences of the formal language, and thus to those of the natural language, by means of the translation rules.

In his approach, Montague acknowledges that it is necessary to maintain a close parallelism between the grammatical structure and the semantic interpretation of sentences. This parallelism is obtained by endowing the natural language with a categorial syntax and by endowing the logic language (into which the natural language sentences are to be translated) with a semantics which is based on the theory of types. The grammar of the fragment of the English language studied by Montague is a *categorial grammar*; such a grammar supplies the means of representing the syntax of natural language with recursive rules similar to those used in the syntax of a logic language. (Montague's categorial grammar will be defined in Section 3.2.) The correspondence between the syntactic rules of natural language and the semantic rules of logic language is obtained by endowing the latter with a *type-theoretic semantics* onto which the categorial grammar can be mapped. Some logic languages endowed with a semantics of that kind are described in Sections 2.2 and 2.3.

Section 2.2 defines a simplified version of a type-theoretic logic language. In the particular case in which only the reference or extension of a sentence (i.e., its truth value) is taken into account, the appropriate formal language is an *extensional type-theoretic logic*. If, besides the extension, the sense or intension is taken into account, extensional logic has to be supplemented by an intensional component. This gives rise to an *intensional type-theoretic logic*. The intensional logic obtained from the extensional logic of Section 2.2 will be described in Section 2.3. By giving a precise definition of intension and by elaborating a semantics of extension and intension which is both recursive and compositional, Montague has carried Frege's semantic programme much further.

Thus, in order to succeed in setting up a systematic correspondence between the syntactic and semantic categories of natural language and those of formal language, Montague imposes the structure of a categorial grammar on the natural language he studies and imposes the structure of a theory of types on his intensional logic. He also brings Church's *lambda calculus* to bear on the issue. This technique plays a role in the process by which natural language is translated into formal language. It will be seen (Sect. 2.2, 2.3) that lambda calculus helps making it possible to treat natural and formal languages within one single theory, by translating the sentences of natural language into formulas of intensional logic: in this theory the sentences and the corresponding formulas are analysed in isomorphic patterns.

2.2 An extensional type-theoretic logic

2.2.1 Introduction

Montague's semantics has achieved the status of a paradigm within truth-conditional approaches to natural language semantics. Montague's contention that natural languages are susceptible to the same kind of semantic analysis as formal languages has proved to be enormously influential among linguists and logicians. This success has been marked by the appearance of an excellent book by Dowty, Wall and Peters [Dowty et al. 81]. We follow rather closely the elegant formalization of Montague's semantics that can be found in this book.

This section and the next one aim at acquainting the reader with the fundamentals of the formal language developed by Montague for representing natural languages. The formal language under consideration is the 'intensional logic' which Montague employs as an intermediate translation language connecting English to its model-theoretic interpretation. This formal language will allow us to apply the techniques developed within mathematical logic to the semantics of natural languages.

Intensional logic is a rather intricate language; we shall introduce it in a progressive way, following the approach of [Dowty et al. 81]. In this section we introduce the 'extensional logic language' that has been developed by Montague to represent the 'referential' (or 'extensional') part of English language. The representation (or translation) of an English sentence into extensional logic will allow us to determine whether this sentence is true. In Section 2.3, we shall extend the extensional logic language into an 'intensional logic language'. The representation of an English sentence into intensional logic will allow us not only to determine whether this sentence is true, but also to understand this sentence, i.e., to determine the 'meaning' or 'sense' of this sentence.

2.2.2 Syntax and semantics of a propositional language

We begin by considering a simple propositional language where the propositions are written in a 'predicate form'. This means that the propositions of the language are written as predicates whose arguments take their value in a

2.2. AN EXTENSIONAL TYPE-THEORETIC LOGIC

set of *individual constants*; this language will be denoted $\mathcal{L}_0$. The interest of this form of propositional language comes from the fact that propositions are no longer treated as the most basic elements of the language as they are in a classical propositional language (§ 1.2.2). Each proposition is made up of a predicate constant and of individual constants that become the basic elements of the language. This new formulation of a propositional language will allow us to introduce the semantic concept of *satisfaction* (of a formula) that generalizes the concept of truth (§ 2.1.3, remark in § 2.2.4).

We shall introduce the formalism of the language $\mathcal{L}_0$ by means of an example where the individual constants represent some writers and some writers' works; the propositions describe relations between authors and works. For any expression α of $\mathcal{L}_0$, we use $[\![\alpha]\!]$ to indicate the semantic value of α. As an illustration, we shall consider a rather elementary version of the language where only six particular individual constants receive a semantic value. These six semantic values constitute the 'domain of discourse' or 'domain of interpretation'. To each of the selected individual constants in $\mathcal{L}_0$ we can then assign one of these individuals as its semantic value, as in the following table.

Constant	Semantic value	
c_B	$[\![c_B]\!]$	= Baudelaire
c_F	$[\![c_F]\!]$	= Flaubert
c_P	$[\![c_P]\!]$	= Poe
c_{FM}	$[\![c_{FM}]\!]$	= Les Fleurs du mal
c_S	$[\![c_S]\!]$	= Salambô
c_{GB}	$[\![c_{GB}]\!]$	= The Gold Bug

The semantic value, noted $[\![P_k]\!]$, of an n-place predicate constant P_k is a set consisting of some n-tuples of individuals belonging to the domain of discourse. As an illustration, we shall consider the example given in the following table.

Constant	Semantic value	
P_A	$[\![P_A]\!] =$	Set of authors.
P_W	$[\![P_W]\!] =$	Set of works.
P_{AW}	$[\![P_{AW}]\!] =$	Set of pairs; the author represented by the first element wrote the book represented by the second element.
P_{AA}	$[\![P_{AA}]\!] =$	Set of pairs of authors; the author represented by the first element died before the author represented by the second element.
P_{AAW}	$[\![P_{AAW}]\!] =$	Set of triples; the author represented by the first element translated the book represented by the third element that was written by the author represented by the second element.

The individual constants c_j, the predicate constants P_k, and the syntactic rules of Subsection 1.2.2 define the syntax of the language $\mathcal{L}_0$. We can now see how the truth value of a proposition formed from a predicate constant and a collection of individual constants is to be determined. The semantic values of the individual constants and of the predicate names will allow us to compute these truth values.

SEMANTICS

- If P_1 is a one-place predicate constant and if c_j is an individual constant, then the formula $P_1(c_j)$ is true if and only if $[\![c_j]\!] \in [\![P_1]\!]$.

- If P_2 is a two-place predicate constant and if c_j, c_k are individual constants, then the formula $P_2(c_j, c_k)$ is true if and only if $<[\![c_j]\!],[\![c_k]\!]> \in [\![P_2]\!]$.

- If P_3 is a three-place predicate constant and if c_j, c_k, c_l are individual constants, then the formula $P_3(c_j, c_k, c_l)$ is true if and only if $<[\![c_j]\!],[\![c_k]\!],[\![c_l]\!]> \in [\![P_3]\!]$.

These rules, together with the semantic rules of Subsection 1.2.2 allow us to compute the truth value of any formula of a language making use (as in our example) of one-, two- and three-place predicate constants.

The knowledge that we have of the domain of discourse (i.e., the state of the world) allows us to define the sets $[\![P_A]\!]$, $[\![P_W]\!]$, $[\![P_{AW}]\!]$, $[\![P_{AA}]\!]$, $[\![P_{AAW}]\!]$, and hence to assign a truth value to any predicate or atomic formula of the language.

$P_A(c_j)$ is true if and only if $[\![c_j]\!] \in [\![P_A]\!] =_{def}$ {Baudelaire, Flaubert, Poe}.

$P_W(c_j)$ is true if and only if $[\![c_j]\!] \in [\![P_W]\!] =_{def}$ {Les Fleurs du mal, Salambô, The Gold Bug}.

$P_{AW}(c_j, c_k)$ is true if and only if $<[\![c_j]\!],[\![c_k]\!]> \in [\![P_{AW}]\!] =_{def}$ {<Baudelaire, Les Fleurs du mal>, < Flaubert, Salambô >, < Poe, The Gold Bug >}.

$P_{AA}(c_j, c_k)$ is true if and only if $<[\![c_j]\!],[\![c_k]\!]> \in [\![P_{AA}]\!] =_{def}$ {<Poe, Baudelaire>, <Poe, Flaubert>, <Baudelaire, Flaubert>}.

$P_{AAW}(c_j, c_k, c_l)$ is true if and only if $<[\![c_j]\!],[\![c_k]\!],[\![c_l]\!]> \in [\![P_{AAW}]\!] =_{def}$ {<Baudelaire, Poe, The Gold Bug >}.

As explained in the beginning of this subsection, we shall denote $\mathcal{L}_0$ any propositional language whose propositions are written in terms of n-place predicate constants followed by n individual constants. The predicate constants and the individual constants constitute the set of *non-logical constants* of $\mathcal{L}_0$. The semantic rule for the predicates is defined as follows.

2.2. AN EXTENSIONAL TYPE-THEORETIC LOGIC

- If P is an n-place predicate constant and if $c_1, c_2, \ldots, c_n$ are individual constants, then $P(c_1, c_2, \ldots, c_n)$ is true if and only if the n-tuple $<[\![c_1]\!], [\![c_2]\!], \ldots, [\![c_n]\!]>$ belongs to the set $[\![P]\!]$.

In summary, the elements of the semantic component of a language $\mathcal{L}_0$ are as follows.

1. A set of semantic values that can be:

    - objects from the domain of discourse,
    - truth values,
    - various functions constructed out of these objects and truth values by means of set theory.

2. A specification for each syntactic category of the type of semantic value that is to be assigned to expressions of that category; for example:

    - individual constants are to be assigned objects of the domain of discourse,
    - n-place predicate constants are to be assigned n-tuples of these objects,
    - formulas are to be assigned truth values.

3. A set of semantic rules specifying how the semantic value of any complex expression is determined in terms of the semantic values of its components.

4. A specific assignment of a semantic value of the appropriate type to each of the basic expressions.

Item 3 contains as implicit assumption that our semantics is to adhere to the *principle of compositionality*, which states that the semantic rules must be such that the semantic value of a syntactically complex expression is always a function of the semantic values of its syntactic components and of their mode of combination (see Subsection 1.1.4).

2.2.3 Absolute truth and truth relative to a model

To interpret a formalized theory consists generally of setting it in correspondence with a mathematical reality outside this theory. For example, the formalized axiomatic theory of measure is first interpreted in terms of real numbers that are in turn set in correspondence with equivalence classes of objects whose weight is each of these numbers. In the interpretation of a logic we also consider as intermediate model a mathematical reality that represents the actual world in an idealized way.

Montague's approach to the semantic analysis of a natural language is based on the 'theory of truth'. A truth theory of semantics is one which adheres to the following statement. To know the meaning of a (declarative) sentence is to know what the world would have to be like for the sentence to be true. In other words, to give the meaning of a sentence is to specify its truth conditions, i.e., to give necessary and sufficient conditions for the truth of that sentence. More precisely, Montague's semantics is based on the notion of 'truth in a model'. A model first specifies what types of things there are in the world, and then, with respect to this assumed ontology, specifies an interpretation of the object language. This means that the formulas of the (logic) language are interpreted in a domain of discourse, noted S, that constitutes the domain of interpretation.

The concept of a formula true in a model has been introduced in Subsection 1.2.3 in the framework of modal propositional logic. We present here a variant of this model that is made suitable for the propositional language $\mathcal{L}_0$ of Subsection 2.2.2. This model will be enriched progressively, in parallel with this language $\mathcal{L}_0$. A *model* for a language represents a state of the world. Formally, a model is an ordered pair (S, V) where S is a non-empty set, the set of individuals, and V is an interpretation function which assigns semantic values of the appropriate type to the basic expressions of $\mathcal{L}_0$.

Consider again the example of Subsection 2.2.2; the set S of individuals that constitute the domain of discourse is the following:

$S = \{$Baudelaire, Flaubert, Poe, Les Fleurs du mal, Salambô, The Gold Bug$\}$.

The interpretation function V is defined as follows:

$$V(c_j) = [\![c_j]\!] \text{ and } V(P_k) = [\![P_k]\!]$$

for any j and k. In our example, this gives

$V(c_B)$ = Baudelaire,
$V(P_A)$ = {Baudelaire, Flaubert, Poe},
$V(P_{AAW})$ = $\{<$ Baudelaire, Poe, The Gold Bug $>\}$, etc.

Formally, a model for the propositional language $\mathcal{L}_0$ is a pair $\mathcal{M} = (S, V)$ where V is a mapping of the set of the non-logical constants of $\mathcal{L}_0$ to the set $S \cup 2^S \cup 2^{S \times S} \cup \ldots$ such that:

- $V(c_j) \in S$ for every individual constant c_j,

2.2. AN EXTENSIONAL TYPE-THEORETIC LOGIC

- $V(P_k) \subset S^n$ for every n-place predicate constant P_k.

Given a language $\mathcal{L}_0$, the various choices of a model are intended to represent the various ways we might effect the mapping from basic expressions of $\mathcal{L}_0$ to individuals in the world. The semantics of a formal language is consequently formed of a *variable* part which depends on the chosen model, and of a *fixed* part. The specification of the type of semantic value that is to be associated with each syntactic category (as e.g. the truth values are to be associated with formulas) and the semantic rules specifying how the semantic value of any complex expression is determined in terms of the semantic values of its components are taken as the fixed part of the semantics for a particular language. The function V that associates with each constant an appropriate extension with respect to the domain of discourse S constitutes that part of the semantics that depends on the chosen model.

For any expression α of the language $\mathcal{L}_0$, we use the notation $[\![\alpha]\!]^{\mathcal{M}}$ instead of $[\![\alpha]\!]$ to denote the semantic value of α with respect to the model $\mathcal{M}$. The notion of model allows us to replace the semantic concept of *truth* by that of *truth in a model*. The interpretation function V allows us to give well-defined truth values to the formulas of the language, with respect to the chosen model $\mathcal{M}$. All definitions of the form 'formula $P(c)$ is true if and only if $[\![c]\!] \in [\![P]\!]$' that occur in the semantic rules of Subsection 2.2.2 have to be replaced by corresponding definitions 'formula $P(c)$ is true in the model $\mathcal{M}$ if and only if $[\![c]\!]^{\mathcal{M}} \in [\![P]\!]^{\mathcal{M}}$'.

A formula can be true in one model and false in another. A formula of a language $\mathcal{L}_0$ is said to be *valid* if it is true in every possible model for $\mathcal{L}_0$, and *inconsistent* if it is false in each of these models. Two formulas are said to be *logically equivalent* if the first is true in exactly the same models in which the second is true. A formula A is said to be a *logical consequence* of a set Σ of formulas if every model in which all the formulas in Σ are true is a model in which A is true also.

Remark. It is usual to distinguish between 'formulas' and 'sentences' of a logic language. It is the traditional practice to reserve the term *sentence* for a formula containing no free occurrences of variables, whereas *formulas* may contain or may not contain free occurrences of variables. In this sense, all the formulas of a propositional language, as the language $\mathcal{L}_0$ defined in Subsection 2.2.2, are sentences. From the next subsection (where the semantics of a first-order language is defined) we shall get into the habit of explicitly distinguishing sentences from formulas.

2.2.4 Syntax and semantics of a first-order language

We shall increase the expressiveness of the propositional language $\mathcal{L}_0$ of Subsection 2.2.2 by adding individual variables and quantifiers over these variables as basic elements; the first-order language thus obtained will be denoted $\mathcal{L}_1$.

The vocabulary of $\mathcal{L}_1$ comprises the following elements.

- A set of *individual constants* noted c_j and a set of *individual variables* noted x_i; these constants and variables constitute the set of *terms*.

- A set of *predicate constants* used as *predicate names*. The set of *non-logical constants* is formed with the individual constants and the predicate constants.

The syntax of a first-order language has been defined in Subsection 1.2.13. The novelty in the interpretation of $\mathcal{L}_1$ with respect to $\mathcal{L}_0$, lies in the notion of truth of a formula by an assignment of objects to variables. We have seen in Subsection 1.2.14 that in order to provide a semantics for a first-order language with respect to a model $\mathcal{M} = (S, V)$, we had also to introduce an *assignment function* g which assigns to each individual variable an element of S.

The need for such a notion within a compositional semantics can be grasped intuitively in the following way. Let us take again the example of Subsection 2.2.2. The syntax of $\mathcal{L}_1$ generates formulas of the form $P_{AW}(x_1, x_2)$ where x_1 and x_2 are individual variables. Remember that the propositional language $\mathcal{L}_0$ was only able to generate sentences as $P_{AW}(c_1, c_2)$ whose semantic value could be computed from the semantic values of the constants P_{AW}, c_1 and c_2. We want now also to be able to assign a truth value to the formula $P_{AW}(x_1, x_2)$: this formula must be true if and only if a work x_2 has been written by an author x_1. To assign a 'true' or 'false' value to a predicate seems difficult before assigning any element of S to its variables. A first way to proceed would be to admit that the variables of the formulas are *bound*. Quantified expressions are then treated in the following way. The sentence $\forall x_2 \exists x_1 P_{AW}(x_1, x_2)$ is read: 'every work (x_2) has been written by an author (x_1)'; it is true just in case for each work x_2 there exists an author x_1 who wrote that work.

This approach is however incompatible with the semantic principle of compositionality. Indeed, since the rules introducing quantifiers make a sentence $\forall x A$ from any formula A and any variable x, we must be able to give a general semantic rule for the truth conditions for $\forall x A$ in terms of the truth or falsity of A, no matter what A is. For example, the sentence $\forall x_2 \exists x_1 P_{AW}(x_1, x_2)$ is syntactically formed by first constructing the atomic formula $P_{AW}(x_1, x_2)$, then adding an existential quantifier to give the formula $\exists x_1 P_{AW}(x_1, x_2)$, then finally adding a universal quantifier to give the sentence $\forall x_2 \exists x_1 P_{AW}(x_1, x_2)$.

2.2. AN EXTENSIONAL TYPE-THEORETIC LOGIC

The semantic rules will retrace these stages. Thus the truth definitions corresponding to the syntactic rule

'if x_1 is a variable and if $P_{AW}(x_1, x_2)$ is a formula, then $\exists x_1 P_{AW}(x_1, x_2)$ is a formula'

will have to give the truth conditions of $\exists x_1 P_{AW}(x_1, x_2)$ in terms of the truth conditions of $P_{AW}(x_1, x_2)$. Consequently, free variables are to be dealt with as such by the semantic rules if a compositional semantics is to be given ([Dowty et al. 81], p.59; [Tarski 44], p.353). To that end, we will add to the semantic environment described in Subsections 2.2.2 and 2.2.3 a function assigning to each variable of $\mathcal{L}_1$ some value from the domain of individuals S. Such a function has already been introduced in Subsection 1.2.14; it is called an *assignment of values to variables*; it will be denoted by g (as above).

The definition of truth in a model will now be given in two stages. First, we shall give a recursive definition of *formula true in a model M and for an assignment function g*. Then, on the basis of this intermediate definition, we shall be able to state the definition of *sentence true in a model*.

Let g be an assignment of values to the individual variables of the first-order language $\mathcal{L}_1$; the function g associates with each variable x of $\mathcal{L}_1$ an individual $g(x)$ in S. An *x-variant of g*, noted g', is a function identical with g except possibly for the individual assigned to x by g'. The *semantic value* of an expression α *with respect to a model M and an assignment function g* will be denoted $[\![\alpha]\!]^{M,g}$.

SEMANTICS

- Interpretation of terms:

    - If x is a variable of $\mathcal{L}_1$, then $[\![x]\!]^{M,g} = g(x)$.
    - If c is an individual or predicate constant of $\mathcal{L}_1$ then $[\![c]\!]^{M,g} = V(c)$, where V is the interpretation function of the model $M = (S, V)$.

- Interpretation of atoms:

    - If t_1 and t_2 are terms of $\mathcal{L}_1$, then $[\![t_1 = t_2]\!]^{M,g}$ is true if and only if and only if $[\![t_1]\!]^{M,g} = [\![t_2]\!]^{M,g}$.
    - If $t_1, \ldots, t_n$ are terms and if P is an n-place predicate constant of $\mathcal{L}_1$, then $[\![P(t_1, \ldots, t_n)]\!]^{M,g}$ is true if and only if $<[\![t_1]\!]^{M,g}, \ldots, [\![t_n]\!]^{M,g}> \in [\![P]\!]^{M,g}$.

- Interpretation of formulas:

- If A and B are formulas, the semantic value of the formulas $\neg A$, $(A \wedge B), (A \vee B), (A \supset B)$ and $(A \equiv B)$ is defined as usual (§ 1.2.2).
- If A is a formula and if x is a variable, then $[\![\forall x A]\!]^{\mathcal{M},g}$ is true if and only if $[\![A]\!]^{\mathcal{M},g'}$ is true for every x-variant g' of g.
- If A is a formula and if x is a variable, then $[\![\exists x A]\!]^{\mathcal{M},g'}$ is true for at least one x-variant g' of g.

Having defined the truth analysis of a formula of $\mathcal{L}_1$ by way of semantic rules, we can introduce the notions of *formula true in a model* and of *valid formula* (§ 1.2.3).

- A formula A is said to be *true in a model* $\mathcal{M}$ if $[\![A]\!]^{\mathcal{M},g}$ is true for every assignment g.

- A formula A is said to be *valid* if it is true in every model, i.e., if $[\![A]\!]^{\mathcal{M},g}$ is true in every model $\mathcal{M}$ and for every assignment g.

Remark. Tarski's notion of *satisfaction of a formula with respect to an assignment of objects to variables* [Tarski 72] corresponds exactly to the notion of *truth of a formula with respect to an assignment g* that we are considering here. As mentioned in Subsection 2.1.3, the notion of satisfaction (with respect to an assignment) is more fundamental than that of absolute truth. It is this notion of truth of a formula with respect to an assignment g that constitutes the innovation of the interpretation of language $\mathcal{L}_1$ (compared to the interpretation of language $\mathcal{L}_0$). Truth is a particular instance of satisfaction: the satisfaction with respect to all value assignments g.

2.2.5 Example

Let us use again the example of Subsection 2.2.2. First of all we consider a set S consisting only of authors. The model $\mathcal{M}$ is the pair (S, V), where S is the set {Baudelaire, Flaubert, Poe} and where V is the function defined as follows:

$V(c_B)$ = Baudelaire
$V(c_F)$ = Flaubert
$V(c_P)$ = Poe
$V(P_A)$ = {Baudelaire, Flaubert, Poe}
$V(P_{AA})$ = {<Baudelaire, Flaubert>,<Poe, Flaubert>, <Poe, Baudelaire>}

(It is unnecessary to define the values of V for the constants that do not appear in the set S under consideration.)

2.2. AN EXTENSIONAL TYPE-THEORETIC LOGIC

Suppose we pick as our initial value assignment g some function that assigns Baudelaire to the variable x_1, Flaubert to the variable x_2 and Poe to the variable x_3. We will not worry about what g assigns to the infinitely many other variables of $\mathcal{L}_1$ since presently we will only be concerned with formulas containing these three variables. Accordingly, we may represent (the initial part of) g as follows:

$$g(x_1) = \text{Baudelaire}$$
$$g(x_2) = \text{Flaubert}$$
$$g(x_3) = \text{Poe}$$

Having given the semantic values for the basic expressions of language $\mathcal{L}_1$, we can consider how the truth or falsity of a formula, say $\forall x_1 P_A(x_1)$, will be determined with respect to $\mathcal{M}$ and g (and ultimately with respect to $\mathcal{M}$ alone) by the semantic rules of $\mathcal{L}_1$. Since $g(x_1) = $ Baudelaire and $[\![P_A]\!]^{\mathcal{M},g} = \{\text{Baudelaire, Flaubert, Poe}\}$, we conclude that $P_A(x_1)$ is true with respect to $\mathcal{M}$ and g, i.e., that $[\![P_A(x_1)]\!]^{\mathcal{M},g} = true$. To determine the semantic value of $\forall x_1 P_A(x_1)$, we must determine the semantic value of $P_A(x_1)$ not just with respect to g but also with respect to all value assignments like g except for the value assigned to x_1. Since the set S contains three elements, there are two x_1-variants of g (other than g itself); they are noted g' and g'' and are defined as follows:

x_1-variant g' x_1-variant g''
$g'(x_1) = $ Flaubert $g''(x_1) = $ Poe
$g'(x_2) = $ Flaubert $g''(x_2) = $ Flaubert
$g'(x_3) = $ Poe $g''(x_2) = $ Poe

Since $[\![P_A]\!]^{\mathcal{M},g'} = [\![P_A]\!]^{\mathcal{M},g''} = \{\text{Baudelaire, Flaubert, Poe}\}$, we deduce that $[\![P_A(x_1)]\!]^{\mathcal{M},g'} = true$ and $[\![P_A(x_1)]\!]^{\mathcal{M},g''} = true$ and, therefore, that the formula $\forall x_1 P_A(x_1)$ is true with respect to the model $\mathcal{M}$ and the assignment g.

We verify that this formula is no longer true when the model $\mathcal{M}$ has a set S formed with authors and works, i.e., when $S = \{$Baudelaire, Flaubert, Poe, Salambô, Les Fleurs du mal, The Gold Bug$\}$. Indeed, $P_A(x_1)$ is false for any x_1-variant g^* of g such that $g^*(x_1) = $ 'a work'. The formula $\forall x_1 P_A(x_1)$ is therefore not true in every model and hence is not valid.

2.2.6 Function-theoretic definition of semantics

The semantics for the propositional language $\mathcal{L}_0$ has been defined in Subsection 2.2.2 in terms of sets. For example, the semantic value of the predicate constant P_A is a set of authors, namely

$$[\![P_A]\!] = \{\text{Baudelaire, Flaubert, Poe}\}.$$

This set is a subset of the set S of individuals that constitute the domain of discourse.

If S is the set of individuals and if A is any subset of S, we define the *characteristic function of A with respect to S* as the mapping $V_A : S \to \{1, 0\}$ given by

$$V_A(s) = 1 \text{ if } s \in A,$$
$$V_A(s) = 0 \text{ if } s \in S \backslash A.$$

We can define the semantic value of the predicate P_A as the characteristic function of the set $A = \{\text{Baudelaire, Flaubert, Poe}\}$ with respect to the set S of Subsection 2.2.3. The semantic values of one-place predicates will all be characteristic functions of sets of individuals in S; for example, the predicate constant P_A has as its semantic value the following characteristic function:

$$[\![P_A]\!] = \{<\text{Baudelaire} \to 1>, <\text{Flaubert} \to 1>, <\text{Poe} \to 1>,$$
$$<\text{Les Fleurs du mal} \to 0>, <\text{Salambô} \to 0>,$$
$$<\text{The Gold Bug} \to 0>\}.$$

Here and in the sequel, we represent a function f with domain S and codomain T as the set of pairs $<s \to t>$ where s varies over S and t is the element of T defined by $t = f(s)$. The usual notation for functions allows us to write

$$[\![P_A(c_B)]\!] = [\![P_A]\!]([\![c_B]\!]) = [\![P_A]\!] \text{ (Baudelaire)} = 1,$$
$$[\![P_A(c_S)]\!] = [\![P_A]\!]([\![c_S]\!]) = [\![P_A]\!] \text{ (Salambô)} \quad = 0.$$

Generally, if P_1 is a one-place predicate constant and if c is an individual constant, we write

$$[\![P_1(c)]\!] = [\![P_1]\!]([\![c]\!]).$$

We now state the semantic rule corresponding to the two-place predicate constants. We have defined in Subsection 2.2.2 the semantics of the predicate name P_{AW} as the set of pairs

$$[\![P_{AW}]\!] = \{<\text{Baudelaire, Les Fleurs du mal}>, <\text{Flaubert, Salambô}>,$$
$$<\text{Poe, The Gold Bug}>\}.$$

We can define the same semantics as a function whose domain is S and whose codomain is a set of characteristic functions V_A, relative to some subsets A of S. This representation gives here

2.2. AN EXTENSIONAL TYPE-THEORETIC LOGIC

$$\llbracket P_{AW} \rrbracket = \{\ <\text{Baudelaire} \to V_{\{\text{Les Fleurs du mal}\}}>,$$
$$<\text{Flaubert} \to V_{\{\text{Salambô}\}}>,$$
$$<\text{Poe} \to V_{\{\text{The Gold Bug}\}}>,$$
$$<\text{Les fleurs du mal} \to 0>,$$
$$<\text{Salambô} \to 0>,$$
$$<\text{The Gold Bug} \to 0>\},$$

where 0 denotes the characteristic function of the empty set, and where the non-degenerate functions V_A are of the form

$$V_{\{\text{Les Fleurs du mal}\}} = \{\ <\text{Les Fleurs du mal} \to 1\},$$
$$<\text{Salambô} \to 0>,$$
$$<\text{The Gold Bug} \to 0>,$$
$$<\text{Baudelaire} \to 0>,$$
$$<\text{Flaubert} \to 0>,$$
$$<\text{Poe} \to 0>\}.$$

The same notation that has been used for a one-place predicate allows us to write

$$\llbracket P_{AW} \rrbracket (\text{Flaubert}) = V_{\{\text{Salambô}\}}.$$

Since the characteristic function $V_{\{\text{Salambô}\}}$ represents a one-place predicate whose name is $\llbracket P_{AW} \rrbracket (\text{Flaubert})$, we can also write

$$\llbracket P_{AW}(\text{Flaubert}, \text{Salambô}) \rrbracket = [\llbracket P_{AW} \rrbracket\ (\text{Flaubert})](\text{Salambô})$$
$$= V_{\{\text{Salambô}\}} (\text{Salambô})$$
$$= 1\ .$$

Generally, if c_1 and c_2 are individual constants, the function-theoretic definition of the semantics of a two-place predicate name P_2 is given by

$$\llbracket P_2(c_1, c_2) \rrbracket = [\llbracket P_2 \rrbracket\ (\llbracket c_1 \rrbracket)]\ (\llbracket c_2 \rrbracket).$$

An iteration on the number of places leads us to the following definition for the semantics of an n-place predicate:

$$\llbracket P_n(c_1, c_2, \ldots, c_n) \rrbracket = [\ldots[\ [\llbracket P_n \rrbracket\ (\llbracket c_1 \rrbracket)]\ (\llbracket c_2 \rrbracket)]\ \ldots\ ](\llbracket c_n \rrbracket).$$

The one-place and two-place predicates play an important role in the Montague semantics: they represent intransitive and transitive verbs respectively. The argument of a one-place predicate represents the subject of the corresponding intransitive verb while the two arguments of a two-place predicate represent the subject and the complement of the corresponding transitive verb.

The preceding definitions provide us with a function-theoretic formalization of the semantics of language $\mathcal{L}_0$. The language $\mathcal{L}_1$ can be provided with a similar kind of semantics. We can reformulate the interpretation of the predicate language $\mathcal{L}_1$ in the following way (see also the semantic rules of Subsection 2.2.4).

- If P_1 is a one-place predicate constant and if α is a term, then $[\![P_1(\alpha)]\!]^{M,g} = [\![P_1]\!]^{M,g}([\![\alpha]\!]^{M,g})$.

- If P_2 is a two-place predicate constant and if α_1 and α_2 are terms, then $[\![P_2(\alpha_1, \alpha_2)]\!]^{M,g} = [[\![P_2]\!]^{M,g}([\![\alpha_1]\!]^{M,g})]([\![\alpha_2]\!]^{M,g})$.

- If P_n is an n-place predicate constant and if $\alpha_1, \ldots, \alpha_n$ are terms, then $[\![P_n(\alpha_1, \ldots, \alpha_n)]\!]^{M,g} = [\ldots [[\![P_n]\!]^{M,g}([\![\alpha_1]\!]^{M,g})] \ldots]([\![\alpha_n]\!]^{M,g})$.

The semantics of the languages $\mathcal{L}_0$ and $\mathcal{L}_1$ can be expressed completely in function-theoretic terms by defining the semantics of the logical constants (or connectives) as follows:

$$[\![\neg]\!] = \{<1 \to 0>, <0 \to 1>\},$$

$$[\![\wedge]\!] = \{<1 \to \{<1 \to 1>, <0 \to 0>\}>,$$
$$\phantom{[\![\wedge]\!] = \{}<0 \to \{<1 \to 0>, <0 \to 0>\}>\},$$

$$[\![\vee]\!] = \{<1 \to \{<1 \to 1>, <0 \to 1>\}>,$$
$$\phantom{[\![\vee]\!] = \{}<0 \to \{<1 \to 1>, <0 \to 0>\}>\},$$

$$[\![\supset]\!] = \{<1 \to \{<1 \to 1>, <0 \to 0>\}>,$$
$$\phantom{[\![\supset]\!] = \{}<0 \to \{<1 \to 1>, <0 \to 1>\}>\},$$

$$[\![\equiv]\!] = \{<1 \to \{<1 \to 1>, <0 \to 0>\}>,$$
$$\phantom{[\![\equiv]\!] = \{}<0 \to \{<1 \to 0>, <0 \to 1>\}>\}.$$

Syntactically, the negation $\neg$ combines with a formula to form another formula. Therefore, the function-theoretic approach to semantics allows us to treat it as a function mapping a truth value to a truth value, and this is in fact just what the definition of $[\![\neg A]\!]$ represents. We can write accordingly:

$$[\![\neg A]\!] = [\![\neg]\!]([\![A]\!]).$$

The syntactic and semantic rules relative to the formulas of both languages $\mathcal{L}_0$ and $\mathcal{L}_1$ can now be reformulated in the following way.

2.2. AN EXTENSIONAL TYPE-THEORETIC LOGIC

SYNTAX

- If A is a formula and if α is a unary connective, then αA is a formula.

- If A and B are formulas and if α is a binary connective then $A\alpha B$ is a formula.

SEMANTICS

- If A is a formula and if α is a unary connective, then $[\![\alpha A]\!]^{\mathcal{M},g} = [\![\alpha]\!] ([\![A]\!]^{\mathcal{M},g})$.

- If A and B are formulas and if α is a binary connective, then $[\![A\alpha B]\!]^{\mathcal{M},g} = [[\![\alpha]\!] ([\![A]\!]^{\mathcal{M},g})] ([\![B]\!]^{\mathcal{M},g})$.

(As regards the language $\mathcal{L}_0$, the notation $[\![X]\!]^{\mathcal{M},g}$ simply means $[\![X]\!]^{\mathcal{M}}$.)

2.2.7 A type-theoretic version of the language $\mathcal{L}_1$

The first-order language $\mathcal{L}_1$ described in Subsection 2.2.4 is defined from a set of basic expressions that are individual and predicate constants and individual variables. The formulas of the language are formed from the basic expressions, logical connectives and syntactic rules. The basic expressions and the formulas constitute the *syntactic categories* of the language $\mathcal{L}_1$. The purpose of this subsection is to redefine the language $\mathcal{L}_1$ in terms of other syntactic categories: *the (logical) types*. This new formulation of the syntax of a first-order language will allow a rigorous translation of a natural language into a formal language. Before defining a suitable type-theoretic language, let us make some comments about this statement.

A central working premise of Montague's theory is that the syntactic categories of a natural language (such as: transitive verb, intransitive verb, common noun, adverb, etc.) should correspond rigorously with the syntactic categories of the formal language. Montague thought that the syntax and semantics of English would turn out to require a variety of logical and semantic categories beyond those found in first-order logic. He saw in type theory a powerful system of syntactic categories which could correspond to the system of syntactic categories of English. In this respect Montague defined a categorial grammar for a fragment of English and showed how the expressions of this fragment can systematically be translated into a type-theoretic logic language. To each syntactic category of the categorial grammar corresponds a type of that language. In fact Montague's main interest in formal languages based on types lies in the systematic way that the categories are related and semantically interpreted. Remember that another crucial working premise of Montague's theory is that the syntactic rules that determine how a sentence of a natural language is built up of smaller syntactic parts should correspond one-to-one with the semantic rules that tell how the meaning of a sentence is a function of the meanings of its parts.

Summarizing, the definition of a categorial grammar for natural languages and a theory of types for formal languages allows us first to define a one-to-one correspondence between the syntactic categories of the natural languages and the types of the formal language. The semantics of the formal type-theoretic language is compositional (the meaning of a formula is a function of the meaning of its parts and of their mode of combination). The translation of the natural language into the formal type-theoretic language allows us finally to transfer the compositional semantics of the formal language to the natural language and hence to represent the truth conditions of the sentences of this latter language.

We shall first define a collection of syntactic categories for the first-order language $\mathcal{L}_1$. The kind of language we shall consider is called *type-theoretic* because its syntax is based on Russell's theory of types.

Let t and e be two basic symbols that respectively represent 'truth' and 'entity'. The *set of types* is defined recursively as follows.

- **Base**: e and t are types.

- **Induction**: if a and b are types, then $\langle a, b \rangle$ is a type.

- **Closure**: all types are obtained by applying the base and induction rules a finite number of times.

For example, since t and e are types, the base and induction rules allow us to deduce that $\langle e, e \rangle$, $\langle e, t \rangle$, $\langle t, e \rangle$ and $\langle t, t \rangle$ are types. An iterated use of the induction produces the types $\langle e, \langle e, t \rangle \rangle$, $\langle e, \langle e, \langle e, t \rangle \rangle \rangle$, $\langle \langle e, t \rangle, t \rangle$ and so on.

The syntactic categories of a formal language are 'sets of expressions'. For example, the set of variables, the set of n-place predicates and the set of formulas constitute syntactic categories. We shall assign to each of these sets a *syntactic label* that is a *type*. As will be seen, many syntactic rules can be formulated in such a way that an expression of type $\langle a, b \rangle$ combines with an expression of type a to produce an expression of type b. This *cancellation rule* can be stated formally as follows.

If α is an expression belonging to the syntactic category (or type) $\langle a, b \rangle$ and if β is an expression belonging to the syntactic category a, then the juxtaposition (or concatenation) $\alpha(\beta)$ belongs to the syntactic category b.

This can be written symbolically:

$$\alpha_{\langle a,b \rangle} + \beta_a = (\alpha(\beta))_b.$$

2.2. An Extensional Type-Theoretic Logic

We can reformulate the syntax of $\mathcal{L}_1$ as follows.

SYNTAX

- The syntactic category of the individual constants c_j and of the variables x_i is the type e.

- The syntactic category of the predicate constants
    - is the type $\langle e, t \rangle$ for one-place predicate constants,
    - is the type $\langle e, \langle e, t \rangle \rangle$ for two-place predicate constants,
    - is the type $\langle e, \langle \ldots, e, \langle e, t \rangle, \ldots \rangle \rangle$, with n occurrences of e, for n-place predicate constants.

- The syntactic category of the formulas is the type t.

- For any types a and b, if α is an element of syntactic category $\langle a, b \rangle$ and if β is an element of syntactic category a, then $\alpha(\beta)$ is an element of syntactic category b.

The logical constants (or connectives) $\neg, \wedge, \vee, \supset, \equiv$ can also be considered as basic syntactic expressions.

- The syntactic category of the unary connective $\neg$ is the type $\langle t, t \rangle$.

- The syntactic category of the binary connectives $\wedge, \vee, \supset, \equiv$ is the type $\langle t, \langle t, t \rangle \rangle$.

- If α is a logical constant of syntactic category $\langle t, t \rangle$ and if A is an element of syntactic category t, then αA is an element of syntactic category t.

- If β is a logical constant of syntactic category $\langle t, \langle t, t \rangle \rangle$ and if A is an element of syntactic category t, then $A\beta$ is an element of syntactic category $\langle t, t \rangle$.

- If A is an element of syntactic category t and if x is a variable (which is an element of syntactic category e), then $\forall x A$ and $\exists x A$ are elements of syntactic category t.

The syntax allows us to construct formulas of the language $\mathcal{L}_1$ or to verify that a sequence A of symbols constitutes a formula of $\mathcal{L}_1$, i.e., that A belongs to the syntactic category t. Consider for example the statement 'all men are mortal', represented by the formula

$$\forall x (Man(x) \supset Mor(x)).$$

The sequence of syntactic rules which are to be used for proving that this sequence of symbols constitutes a formula of the language can be deduced from the syntactic tree of Figure 2.1.

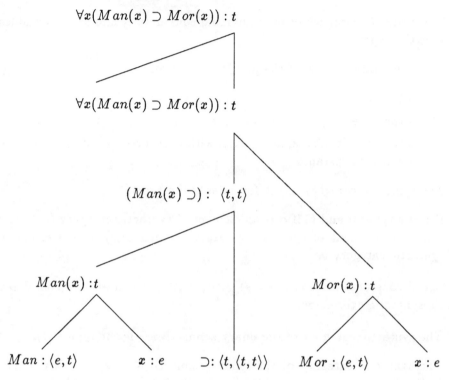

Figure 2.1 : Syntactic tree for the formula $\forall x(Man(x) \supset Mor(x))$

Remark. The syntactic rule that allows the construction of complex expressions of category $\langle t, t \rangle$ from logical constants of category $\langle t, \langle t, t \rangle \rangle$ is formally identical to the cancellation rule acting on types '$\langle a, b \rangle$ combined with a gives b' with however a little notation modification. This rule makes use of 'left simplification'; whereas the syntactic rule for logical constants of category $\langle t, \langle t, t \rangle \rangle$ uses 'right simplification'; therefore it is *not* a particular case of the former rule. However, if the logical constants were treated in *reverse Polish notation* (i.e., if a formula as $A \vee B$ was written $\vee(A, B)$) then the syntactic rules for logical constants could also be derived from the above mentioned cancellation rule for types. Note that the syntactic trees of Figures 2.1 and 2.2 make use of both kinds of rules: the general rule acting on types and the special rules acting on logical constants. Another (equivalent) formulation of the syntactic rules for logical constants is presented in the next subsection; it does not make use of the 'type-theoretic notation', and is probably easier to understand.

2.2. AN EXTENSIONAL TYPE-THEORETIC LOGIC

2.2.8 A higher-order type-theoretic language

The basic expressions of the language $\mathcal{L}_1$ are the individual constants noted c_j, the individual variables noted x_j and the predicate constants noted P_j. To each of these basic expressions has been assigned a certain type that will be introduced explicitly in the notation.

Each individual constant, which is of type e, will be a symbol of the form $c_{j,e}$, and each individual variable, which is also of type e, will be a symbol of the form $x_{j,e}$. In this notation, the first index is a natural number, which is used for enumeration, while the second index is a type.

This kind of notation will allow us to introduce, in an easy manner, a higher-order *type-theoretic logic language*, which will be denoted $\mathcal{L}_t$. The definition involves a set of types, constructed recursively as explained above. Each type (or syntactic category) will be the label of a well-defined set of expressions of $\mathcal{L}_t$, including (in principle) a denumerable set of individual constants and a denumerable set of individual variables.

SYNTAX

- The basic expressions of the language are:
  - for each type a, the individual constants noted $c_{j,a}$, with j a natural number;
  - for each type a, the individual variables noted $x_{j,a}$, with j a natural number.

- For each type a, the set E_a of expressions of type a is defined recursively as follows:
  - every constant of type a belongs to E_a;
  - every variable of type a belongs to E_a;
  - if a and b are types, and if α belongs to $E_{\langle b,a \rangle}$ and β belongs to E_b, then $\alpha(\beta)$ belongs to E_a.
  - if A and B belong to E_t, then $\neg A$, $A \vee B$, $A \wedge B$, $A \supset B$ and $A \equiv B$ belong also to E_t.
  - if A belongs to E_t and if x is a variable (of any type), then $\forall x A$ and $\exists x A$ belong also to E_t.

This syntax allows us for example to write the formula

$$\forall x_{j,\langle e,t \rangle}[x_{j,\langle e,t \rangle}(c_{1,e}) \supset x_{j,\langle e,t \rangle}(c_{2,e})]$$

which states that for each one-place predicate P, if $P(c_{1,e})$ is true, then $P(c_{2,e})$ is true. This formula belongs to a 'second-order logic', in which the predicate names can be variables (of type $\langle e,t\rangle$). In such a logic, we can quantify not only over individuals but also over sets of things. The language of a second-order logic distinguishes itself from that of first-order logic by the addition of variables for subsets, relations and functions of the domain of discourse, and by the possibility of quantification over these. An obvious example of a second-order statement is Peano's induction axiom according to which every set of natural numbers containing 0 and closed under immediate successors contains all natural numbers. Using S for successors, we may write this down as

$$\forall Y(Y(0) \wedge \forall x(Y(x) \supset Y(S(x))) \supset \forall x Y(x)),$$

where x stands for numbers, Y for sets of numbers, and $Y(x)$ says, as usual, that x is an element of Y.

The immense strength of second-order logic shows quite clearly when set theory itself is considered [Gabbay and Guenther 83]. Zermelo's *separation axiom* says that the elements of a given set sharing a given property form again a set. Knowing of problematic properties occurring in the paradoxes, Zermelo required 'definiteness' of properties to be used. In later times, Skolem replaced this by 'first-order definability', and the axiom became a first-order scheme. Nevertheless, the intruded axiom quite simply is the second-order statement

$$\forall X \forall x \exists y \forall z (z \in y \equiv z \in x \wedge X(z)).$$

Once upon the road of second-order quantification, higher-order predicates come into view. In mathematics, one wants to quantify over functions, but also over functions defined on functions, etc. But also natural languages offer such an ascending hierarchy, at least in the types of its lexical items. For example, nouns (such as 'presidents') denote properties, but then already adjectives become higher-order sentences ('former president'), taking such properties to other properties. The following formula belongs to a third-order language:

$$\forall x_{j,\langle\langle e,t\rangle,t\rangle}[x_{j,\langle\langle e,t\rangle,t\rangle}(x_{1,\langle e,t\rangle}) \supset x_{j,\langle\langle e,t\rangle,t\rangle}(x_{2,\langle e,t\rangle})].$$

This formula states that for all predicates P of predicates, if P is true of the predicate $x_{1,\langle e,t\rangle}$, then it is true of the predicate $x_{2,\langle e,t\rangle}$.

Within the Montague theory of natural language, the most important restriction on the rules and their organization concerns the way in which syntactic rules and semantic rules must be related. For every syntactic category there must be a unique semantic category, and for each syntactic rule that combines expressions of categories A and B to produce an expression of category C, there must be a unique semantic rule that operates on the corresponding

2.2. AN EXTENSIONAL TYPE-THEORETIC LOGIC

semantic interpretations to give a semantic interpretation for the resulting expression. That interpretation will be of the semantic category corresponding to the syntactic category C. (Details will be given in Section 3.2)

One could think of the grammar then as a set of ordered pairs <*syntactic rule, semantic rule*>. The syntactic rule will give the syntax-specific details of how the component sentences are to be combined, and the semantic rule will give the semantics-specific details of how the meaning of the whole is determined by the meaning of the parts. The rules in each pair must correspond in the categories of the constituent parts and in the category of the result [Partee 75]. This parallelism between syntax and semantics must also appear in the formal language in which the natural language will be translated. We shall see how this requirement is satisfied by type-theoretic logic languages.

In the context of type-theoretic languages, it is customary to make use of the term *possible denotation* to designate the *possible semantic value* of an expression whose syntactic category belongs to a certain type [Dowty et al. 81]. The symbol D_a will represent the *set of possible denotations* for the expressions of type a, with respect to a given interpretation domain S. The sets D_a are defined recursively as follows:

- $D_e =_{def} S$,

- $D_t =_{def} \{1, 0\}$,

- $D_{\langle a,b \rangle} =_{def} D_b^{D_a}$, i.e., the set of all functions from D_a to D_b, for every type a and every type b.

Consider, for example, expressions of category $\langle e, t \rangle$ (one-place predicates). The definition of denotation says that $D_{\langle e,t \rangle}$ is $D_t^{D_e} = \{1,0\}^S$, i.e., the set of all the functions that associate a truth value with each element of S. We verify also that we have

$$D_{\langle e,\langle e,t\rangle\rangle} =_{def} D_{\langle e,t\rangle}^{D_e} = (\{1,0\}^S)^S.$$

Thus it is seen that the possible denotations for a two-place predicate corresponds to the possible semantic values for this type of predicate (§ 2.2.4, 2.2.6). The unary operator ¬ combines syntactically with a formula to yield a formula (see SYNTAX in this subsection) and semantically its value is a function from truth values to truth values. Therefore, since this operator is of category $\langle t, t \rangle$, we verify that the possible denotation set,

$$D_{\langle t,t \rangle} =_{def} D_t^{\{1,0\}} = \{1,0\}^{\{1,0\}},$$

corresponds to the semantic requirement for this operator.

A semantics for the type-theoretic language $\mathcal{L}_t$ is based on the notions of model for the language (§ 2.2.3), of assignment of values to individual variables (§ 2.2.4) and of possible denotation. A *model* for $\mathcal{L}_t$ is a pair $\mathcal{M} = (S, V)$ where S is as above and V is an interpretation function that assigns to each non-logical constant $c_{j,a}$ of type a a denotation from the set D_a. An assignment of values to variables is a function g that assigns to each variable $x_{j,a}$ of type a a denotation from the set D_a. The denotation of an expression α of $\mathcal{L}_t$, relative to a model $\mathcal{M} = (S, V)$ and to an assignment function g, is noted $[\![\alpha]\!]^{\mathcal{M},g}$ and is defined recursively by the following semantic rules.

SEMANTICS

- If x is a (predicate or individual) variable, then $[\![x]\!]^{\mathcal{M},g} = g(x)$.

- If c is a non-logical (predicate or individual) constant, then $[\![c]\!]^{\mathcal{M},g} = V(c)$.

- If $\alpha \in E_{\langle a,b \rangle}$ and $\beta \in E_a$, then $[\![\alpha(\beta)]\!]^{\mathcal{M},g} = [\![\alpha]\!]^{\mathcal{M},g}([\![\beta]\!]^{\mathcal{M},g})$.

- If $A \in E_t$ and $B \in E_t$, the semantics of expressions $\neg A, A \vee B, A \wedge B$, $A \supset B$ and $A \equiv B$ is defined by the usual rules.

- If $A \in E_t$ and if x is a variable of type a, then

    - $[\![\forall x A]\!]^{\mathcal{M},g} = 1$ if and only if, for all r in $D_a, [\![A]\!]^{\mathcal{M},g^r} = 1$ where g^r is the x-variant of g obtained by assigning the value r to x (§ 2.2.4).
    - $[\![\exists x A]\!]^{\mathcal{M},g} = 1$ if and only if, for some r in $D_a, [\![A]\!]^{\mathcal{M},g^r} = 1$ where g^r is the x-variant of g obtained by assigning the value r to x.

- A formula A (i.e., an element of E_t) is said to be *true in the model* $\mathcal{M}$ if $[\![A]\!]^{\mathcal{M},g} = 1$ for every assignment function g.

2.2.9 Lambda calculus and the language $\mathcal{L}_1$

The type-theoretic language $\mathcal{L}_1$ of Subsection 2.2.7 will be expanded by adding a new operator, called the *lambda operator*, or in short λ-*operator*. This operator was briefly introduced in subsection 3.1.18 of [Thayse 88]. The reader is probably familiar with the way sets are defined by a notation such as

$$\{x \mid x \text{ is a teacher of a university}\},$$
$$\{x \mid -1 < x < 1, \; x \text{ is a real number}\}.$$

The only requirements for this way of defining sets are that we have an expression containing a variable in an unambiguous language (such as e.g. natural language, arithmetic, logic) and that we have a convention for marking which

2.2. AN EXTENSIONAL TYPE-THEORETIC LOGIC

variable x is the key for defining the set. This leads to the λ-calculus notation where a set is represented as follows:

$$\lambda x[\text{logic formula containing the variable } x].$$

In fact, we will do this by adding the λ-operator to predicate logic. If A is a logic formula, then the expression $\lambda x A$, called λ-*expression* or λ-*abstraction*, will denote a set; intuitively, this expression will characterize the set specified by A with respect to the variable x. For example, consider the model $\mathcal{M}$ where P_{AW} represents the authorship of a work and where c_F represents Flaubert; the λ-expression

$$\lambda x[P_{AW}(c_F)(x)]$$

represents the set of works that Flaubert has written. More precisely, we identify this expression with the characteristic function of this set of works with respect to the set of individuals of the model.

As λ-expressions will have the same kinds of denotations as one-place predicate constants, we will want to classify these expressions syntactically as members of the type $\langle e, t \rangle$. This means that they will combine with terms to form formulas just as any other one-place predicate constant does. The expression $\lambda x[P_{AW}(c_F)(x)]$ can be read as 'is an x such that $[P_{AW}(c_F)(x)]$'. Thus, since $\lambda x[P_{AW}(c_F)(x)]$ plays the role of a one-place predicate constant, then $\lambda x[P_{AW}(c_F)(x)](c_S)$ is a formula. This might be read as 'c_S is an x such that $[P_{AW}(c_F)(x)]$'. There are two different ways to express the same thing, namely,

$$\lambda x[P_{AW}(c_F)(x)](c_S) \quad \text{and} \quad P_{AW}(c_F)(c_S).$$

More generally, any expression of this kind can be written in two different ways, as follows:

$$\lambda x[\ldots x \ldots](\alpha) \quad \text{and} \quad [\ldots \alpha \ldots].$$

The rule that allows us to convert the first expression into the logically equivalent second expression is called the *principle of λ-conversion*. The second expression, $[\ldots \alpha \ldots]$, must be understood as the result of replacing all free occurrences of the variable x in the first formula with α; the converse transformation is called *abstraction* [Church 40], [Church 41], [Barendregt 80].

The logic formula

$$Man(x) \supset Mor(x)$$

can be read as: 'if x is a man, x is mortal'. The formula

$$\lambda x[Man(x) \supset Mor(x)](c_F)$$

can be read as: 'if Flaubert is a man, Flaubert is mortal'; this formula becomes after λ-conversion:

$$Man(c_F) \supset Mor(c_F).$$

We now turn to the formal statement of the syntactic and semantic rules for the lambda operator. We will assume that all the syntactic and semantic definitions for the type-theoretic version of language $\mathcal{L}_1$ are to be taken for the new language $\mathcal{L}_1$, to which the following rules will be added.

SYNTAX

If A is a formula (i.e., an element of the syntactic category t) and if x is a variable (of syntactic category e), then the λ-expression $\lambda x[A]$ is an element of the syntactic category $\langle e, t \rangle$.

SEMANTICS

If A is a formula and if x is a variable, then $[\![\lambda x[A]]\!]^{M,g}$ is the function $h \in \{1,0\}^S$ defined by $h(r) = [\![A]\!]^{M,g^r}$, where g^r is the x-variant of g obtained by assigning the value r to x (§ 2.2.4).

2.2.10 Lambda calculus and the language $\mathcal{L}_t$

The syntactic and semantic rules of the language $\mathcal{L}_1$ can be adapted to the more general type-theoretic language $\mathcal{L}_t$ defined in Subsection 2.2.8. The language $\mathcal{L}_t$ with λ-operator will be noted $\mathcal{L}_{\lambda t}$. The operator λ will not only act on variables of category e but also on variables of any other category.

SYNTAX

If α is an expression of syntactic category a and if x is a variable of syntactic category b, then $\lambda x[\alpha]$ is an expression of syntactic category $\langle b, a \rangle$.

SEMANTICS

If α is an expression of syntactic category a and if x is a variable of syntactic category b, then $[\![\lambda x[\alpha]]\!]^{M,g}$ is a function h from D_b into D_a such that $h(r) = [\![\alpha]\!]^{M,g^r}$ for all objects $r \in D_b$.

The objects mentioned in this definition may be of various sorts depending on what D_b is: they may be members of the set S of individuals, or they may be truth values, or functions of some kind.

2.2. AN EXTENSIONAL TYPE-THEORETIC LOGIC

Lambda calculus will be used to evaluate what semantic values should be assigned to some syntactic constituent of a larger expression whose semantic value is known. It will be seen in Chapter 3 that the λ-expressions can be a very helpful way of giving the meaning of words or sentences of natural language. In order to illustrate this use of lambda calculus we consider some examples that Montague gave in his paper '*Proper treatment of quantification in ordinary English*' [Montague 73].

English sentences	Usual translation into logic
1. Every student walks	$\forall x(S(x) \supset W(x))$
2. Every student likes Mary	$\forall x(S(x) \supset L(m)(x))$
3. Some student walks	$\exists x(S(x) \wedge W(x))$
4. Some student likes Mary	$\exists x(S(x) \wedge L(m)(x))$
5. No student walks	$\neg \exists x(S(x) \wedge W(x))$
6. No student likes Mary	$\neg \exists x(S(x) \wedge L(m)(x))$

Sentences 1 and 2 are instances of a more general sentence whose translation is a second-order logic formula:

7. Every student has the property Y $\forall x(S(x) \supset Y(x))$

where Y is a predicate variable. The λ-expression $\lambda Y[\forall x(S(x) \supset Y(x))]$ can be read 'is a Y such that each student has the property Y'; this λ-expression can be viewed as a one-place predicate name, and

1'. $\lambda Y[\forall x(S(x) \supset Y(x))](W)$
2'. $\lambda Y[\forall x(S(x) \supset Y(x))](L(m))$

are two instances of it that are equivalent to formulas 1 and 2 respectively. The expressions W and $L(m)$ being instances of one-place predicate variables, they will have $\langle e, t \rangle$ as type. Therefore, the λ-expression $\lambda Y[\forall x(S(x) \supset Y(x))]$ will have $\langle \langle e, t \rangle, t \rangle$ as type and 'each student' as equivalent English sentence.

The mechanism that transforms a logic expression into a λ-expression can be used in an iterative way. Hereunder we give a collection of English sentences that can be generated from this transformation (from an initial sentence), together with their type and their translation in a logic language.

Sentence: Each student walks; type t.
Translations: $\forall x(S(x) \supset W(x))$,
$\lambda Y[\forall x(S(x) \supset Y(x))](W)$.
Sentence: Each student; type $\langle\langle e,t\rangle,t\rangle$.
Translations: $\lambda Y[\forall x(S(x) \supset Y(x))]$,
$\lambda Z[\lambda Y[\forall x(Z(x) \supset Y(x))]](S)$.
Sentence: Each; type $\langle\langle e,t\rangle,\langle\langle e,t\rangle,t\rangle\rangle$.
Translation: $\lambda Z[\lambda Y[\forall x(Z(x) \supset Y(x))]]$.
Sentence: Each student walks; type t.
Translation: $\lambda Z[\lambda Y[\forall x(Z(x) \supset Y(x))]](S)(W)$.
Sentence: Some student walks; type t.
Translation: $\lambda Y[\exists x(S(x) \wedge Y(x))](W)$.
Sentence: Some student; type $\langle\langle e,t\rangle,t\rangle$.
Translation: $\lambda Y[\exists x(S(x) \wedge Y(x))]$.
Sentence: Some; type $\langle\langle e,t\rangle,\langle\langle e,t\rangle,t\rangle\rangle$.
Translation: $\lambda Z[\lambda Y[\exists x(Z(x) \wedge Y(x))]]$.
Sentence: No student walks; type t.
Translation: $\lambda Y[\neg\exists x(S(x) \wedge Y(x))](W)$.
Sentence: No student; type $\langle\langle e,t\rangle,t\rangle$.
Translation: $\lambda Y[\neg\exists x(S(x) \wedge Y(x))]$.
Sentence: No; type $\langle\langle e,t\rangle,\langle\langle e,t\rangle,t\rangle\rangle$.
Translation: $\lambda Z[\lambda Y[\neg\exists x(Z(x) \wedge Y(x))]]$.

The types assigned to the various λ-expressions allow us to construct the *syntactic tree* of types of Figure 2.2. This tree is associated with the following formula of the $\mathcal{L}_{\lambda t}$ language:

$$\lambda Z[\lambda Y[\forall x(Z(x) \supset Y(x))](S)](W).$$

This is logically equivalent to the following formula of the $\mathcal{L}_1$ language:

$$\forall x[S(x) \supset W(x)].$$

(They both represent the same English sentence, i.e., 'each student walks'.)

2.2. AN EXTENSIONAL TYPE-THEORETIC LOGIC

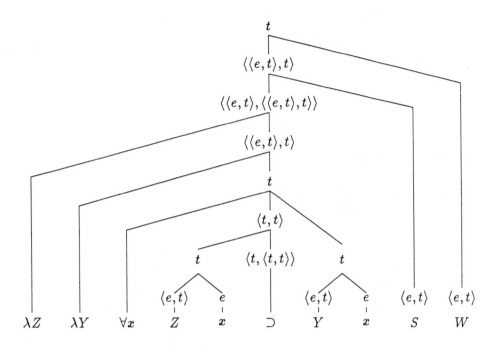

Figure 2.2 : Syntactic tree for the formula
$\lambda Z[\lambda Y[\forall x(Z(x) \supset Y(x))](S)](W)$

Figure 2.3 shows a simplified syntactic tree associated with the same sentence. In this figure, the components of the λ-expression that represent the determiner 'each' are not dissociated. On the other hand, Figure 2.3 gives the 'English translations' of the formal expressions. This allows us to illustrate one of the main requirements of Montague semantics, i.e., the necessity to maintain an isomorphism between the syntactic analysis of a sentence in natural language and its translation in formal language. (We shall see in Chapter 3 that this isomorphism holds only in the case of unambiguous sentences.)

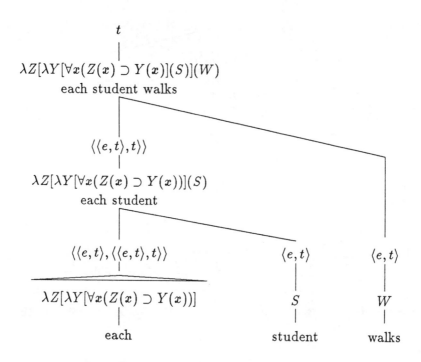

Figure 2.3 : Simplified syntactic tree

2.2.11 The multimodal language $\mathcal{L}_m$

Up to this point we have been assigning a truth value to a formula relative to a model 'once and for all', thus ignoring the fact that many sentences of natural languages may be now true, now false, or can be logically true, physically true, etc. To convey such 'modalities', we will introduce some modal operators in the formal language $\mathcal{L}_{\lambda t}$ and construct a *type-theoretic multimodal language* (§ 1.2.9) that will be noted $\mathcal{L}_m$. This language will contain three universal modal operators and their duals:

[F]	:	always true in the future,
<F>	:	sometimes true in the future,
[P]	:	always true in the past,
<P>	:	sometimes true in the past,
□	:	necessarily always true,
◇	:	possibly sometimes true.

The syntax of the language $\mathcal{L}_m$ contains all the syntactic rules of the language $\mathcal{L}_{\lambda t}$ plus the following ones.

2.2. AN EXTENSIONAL TYPE-THEORETIC LOGIC

SYNTAX

- If $[M]$ represents one of the universal modal operators $[F], [P]$ or $\square$ and if A is a formula, then $[M]A$ is a formula.

- If $<M>$ represents one of the existential modal operators $<F>$, $<P>$ or $\Diamond$ and if A is a formula, then $<M>A$ is a formula.

The semantics of the language $\mathcal{L}_m$ is based on the notion of model. The notion of model was introduced in Subsection 2.2.3 in the framework of the propositional language $\mathcal{L}_0$. Recall that a model for the language $\mathcal{L}_0$ is a pair $\mathcal{M} = (S, V)$ where S is a set of individuals and V is a mapping of the individual constants to elements of S and of the n-place predicate constants to subsets of S^n. We wish now to construct a model based semantics for the language $\mathcal{L}_m$ which contains modal operators. To that end, we will expand the definition of a model to include not only a domain of individuals S and an interpretation function V, but also a non-empty set W of *possible worlds*, and a non-empty set T of *time instants* with a linear ordering imposed upon it which corresponds to the temporal sequence of time instants in T.

The semantics of possible worlds is based on the definition of a universe W made up of (possible) worlds w. We can first define a model $\mathcal{M}$ as a triple $\mathcal{M} = (S, W, V)$ where S is the domain of individuals, W is the universe and V is a function that assigns a denotation (or semantic value) to each pair formed with an element of W and a non-logical constant of the language. The universe W is formed with the 'actual world' and with a collection of 'possible worlds' that represent each a state of the world that might have occurred had the sequence of historical events been different. We will now add a dimension to the model $\mathcal{M}$ by assuming that each world can be evaluated at different time instants.

When we consider a language that contains both tense and necessity modal operators, its interpretation will need a model with two sets besides the set of individuals: the set of worlds W and the set of times T. The model $\mathcal{M}$ is then the 4-tuple $\mathcal{M} = (S, W, T, V)$ where V is a function that assigns an appropriate denotation (§ 2.2.8) to each non-logical constant of $\mathcal{L}_m$ relative to each world w in W and to each time t in T. We can think of our semantic space as expanded in two dimensions. Any particular point in this space can be thought of as being a pair of coordinates $<w, t>$ for some w in W and some t in T; that is, as a point whose location is determined by which world it is on the one hand, and by what time it is on the other hand. We will call such a pair an *index* ([Dowty et al. 81], [Montague 74d]). Accordingly, the interpretation $V(w, t, c)$ of the non-logical constant c will be written $V(<w, t>, c)$ where w and t constitute the two coordinates of the universe $W \times T$. Any one of the indices can represent 'the actual world now'. We shall finally assume that the set T is *totally ordered*. The time is 'linear' (see Subsection 1.2.10); this means that every pair of elements s and t of T satisfies

$$(s < t) \vee (s = t) \vee (t < s),$$

where the symbol $<$ denotes the order relation. Formally, we will define a model for the multimodal language $\mathcal{L}_m$ as a 5-tuple $(S, W, T, <, V)$. The semantics of this language will be defined with respect to the model $\mathcal{M}$. The definition of denotation will now be relativized to a choice of some arbitrary index $<w, t>$. Thus the semantic rules will now give a definition of a semantic value $[\![\alpha]\!]^{\mathcal{M},w,t,g}$ for each expression α of $\mathcal{L}_m$; this semantic value is the 'denotation of α relative to a model $\mathcal{M}$, possible world w, time t and assignment of values to variables g'. Let c be a non-logical constant of $\mathcal{L}_m$; the denotation of c is defined as follows:

$$[\![c]\!]^{\mathcal{M},w,t,g} =_{def} V(<w,t>, c).$$

The semantic rules of $\mathcal{L}_m$ define recursively the denotation $[\![\alpha]\!]^{\mathcal{M},w,t,g}$; they consist of the semantic rules of $\mathcal{L}_{\lambda t}$ (where the notation $[\![\alpha]\!]^{\mathcal{M},g}$ has been replaced by $[\![\alpha]\!]^{\mathcal{M},w,t,g}$) plus the following rules.

SEMANTICS

- If A is a formula, then $[\![\Box A]\!]^{\mathcal{M},w,t,g} = 1$ if and only if $[\![A]\!]^{\mathcal{M},w',t',g} = 1$ for all w' in W and all t' in T.

- If A is a formula, then $[\![\Diamond A]\!]^{\mathcal{M},w,t,g} = 1$ if and only if $[\![A]\!]^{\mathcal{M},w',t',g} = 1$ for some w' in W and some t' in T.

- If A is a formula, then $[\![[F]A]\!]^{\mathcal{M},w,t,g} = 1$ if and only if $[\![A]\!]^{\mathcal{M},w,t',g} = 1$ for all t' in T such that $t<t'$.

- If A is a formula, then $[\![<F>A]\!]^{\mathcal{M},w,t,g} = 1$ if and only if $[\![A]\!]^{\mathcal{M},w,t',g} = 1$ for some t' in T such that $t<t'$.

- If A is a formula, then $[\![[P]A]\!]^{\mathcal{M},w,t,g} = 1$ if and only if $[\![A]\!]^{\mathcal{M},w,t',g} = 1$ for all t' in T such that $t'<t$.

- If A is a formula, then $[\![<P>A]\!]^{\mathcal{M},w,t,g'} = 1$ if and only if $[\![A]\!]^{\mathcal{M},w,t',g} = 1$ for some t' in T such that $t'<t$.

This semantics corresponds to well-determined accessibility relations R_F and R_P (inverse each of other) on the universe $W \times T$ (see Subsection 1.2.10); we have

2.2. AN EXTENSIONAL TYPE-THEORETIC LOGIC

$$<w,t> R_P <w',t'> \Longleftrightarrow (t = t') \text{ or } (t < t' \text{ and } w = w').$$

We interpret $\Box A$ so that it means 'necessarily always A'. This definition makes the operator $\Box$ 'stronger' than the tense operators $[F]$ and $[P]$ that mean 'always in the future' and 'always in the past' respectively. Therefore, the following formulas will be true in any model in which time has no beginning or end:

$$\Box A \supset [F]A,$$
$$\Box A \supset [P]A,$$
$$[F]\Box A \equiv \Box A,$$
$$[P]\Box A \equiv \Box A.$$

2.2.12 Example

As an illustration, let us consider the model

$$\mathcal{M} = (S, W = \{w_1, w_2\}, T = \{t_1, t_2, t_3\}, <, V).$$

The two possible worlds of W are w_1 and w_2 and the three times of T are t_1, t_2, t_3. We can schematize this universe $W \times T$ as in Figure 2.4.

	time t_1	time (present) t_2	time t_3
(actual) world w_1	w_1, t_1 — |	w_1, t_2 — |	w_1, t_3 |
(possible) world w_2	w_2, t_1 —	w_2, t_2 —	w_2, t_3

Figure 2.4 : Elements of the universe $W \times T$

Assume that the domain S consists just of three individuals:

$$S = \{\text{Kennedy, Johnson, Nixon}\}.$$

We will assign to the individual constants c_1, c_2 and c_3 of $\mathcal{L}_m$ the following semantic values:

$$V(<w,t>, c_1) = [\![c_1]\!]^{\mathcal{M}, w, t, g} = \text{Kennedy} \quad \text{for every } w \text{ and every } t,$$
$$V(<w,t>, c_2) = [\![c_2]\!]^{\mathcal{M}, w, t, g} = \text{Johnson} \quad \text{for every } w \text{ and every } t,$$
$$V(<w,t>, c_3) = [\![c_3]\!]^{\mathcal{M}, w, t, g} = \text{Nixon} \quad \text{for every } w \text{ and every } t.$$

Names that denote the same individual at each index are called *rigid designators*; they represent 'persons', as in the example above. We will also consider *non-rigid designators* in $\mathcal{L}_m$ as analogous in natural language to a title that different individuals hold at different times. Most of these are descriptions of functions as 'the President of the United States'. Let us represent the generic name of President of the United States by the constant c_4 of $\mathcal{L}_m$. The semantic value of c_4, i.e., the name of the individual who occupies the post of President depends on the considered world and time. In the actual world w_1, Kennedy has preceded Johnson who himself has preceded Nixon. We may also imagine an alternative sequence of presidents in a possible world w_2 in which the outcome of the 1960 presidential election would have been different from the actual outcome. The semantic value of the individual constant c_4 is given in Figure 2.5.

$V(<w,t>, c_4)$	t_1	t_2	t_3
w_1	Kennedy	Johnson	Nixon
w_2	Nixon	Nixon	Johnson

Figure 2.5 : Semantic value of the individual constant c_4

Assume now that the one-place predicate constant P_c has as semantic value the set of individuals who are in possession of a landed property in California. The semantic value of this predicate constant is defined in Figure 2.6.

$V(<w,t>, P_c)$	t_1	t_2	t_3
w_1	{Kennedy, Johnson}	{Nixon}	{Johnson, Nixon}
w_2	{Johnson, Nixon}	{Kennedy}	{Kennedy, Johnson, Nixon}

Figure 2.6 : Semantic value of the predicate constant P_c

The function V allows the evaluation of the semantic values of the formulas of the language. Let us evaluate formulas relative to the index $<w_1, t_2>$; thus we will consider w to be the actual world and t to be the present time. We deduce from the values $V(<w_1,t_2>, c_3)$ and $V(<w_1,t_2>, P_c)$ that $[\![P_c(c_3)]\!]^{\mathcal{M}, w_1, t_2, g}$

= 1, i.e., that Nixon is actually in possession of a landed property in California. We can also ask us whether 'the President is in possession of a landed property in California'; this corresponds to a search for the value of $[\![P_c(c_4)]\!]^{\mathcal{M}, w_1, t_2, g}$. Since the value $V(<w_1, t_2>, c_4) =$ Johnson does not belong to the set $V(<w_1, t_2>, P_c) = \{\text{Nixon}\}$, we deduce that $[\![P_c(c_4)]\!]^{\mathcal{M}, w_1, t_2, g} = 0$, i.e., that the actual President is not in possession of a landed property in California. The value of $[\![<F>P_c(c_4)]\!]^{\mathcal{M}, w_1, t_2, g}$ is 1 since $V(<w_1, t_3>, P_c)$ is the set $\{\text{Johnson, Nixon}\}$, which contains the value $V(<w_1, t_3>, c_4) = \{\text{Nixon}\}$ as a subset.

Consider now some formulas involving quantification, evaluated with respect to the same index $<w_1, t_2>$. The values $[\![\forall x <F> P_c(x)]\!]^{\mathcal{M}, w_1, t_2, g}$ and $[\![\forall x < P> P_c(x)]\!]^{\mathcal{M}, w_1, t_2, g}$ are 0 since neither $V(<w_1, t_3>, P_c)$, nor $V(<w_1, t_1>, P_c)$ contains the entire domain $S = \{\text{Kennedy, Johnson, Nixon}\}$. Therefore, the value $[\![\forall x<F>P_c(x) \lor \forall x<P>P_c(x)]\!]^{\mathcal{M}, w_1, t_2, g}$ is also 0. However, the value $[\![\forall x [<F>P_c(x) \lor <P>P_c(x)]]\!]^{\mathcal{M}, w_1, t_2, g}$ is 1 because every element in the domain S appears in either $V(<w_1, t_3>, P_c) = \{\text{Johnson, Nixon}\}$, or in $V(<w_1, t_1>, P_c) = \{\text{Kennedy, Johnson}\}$.

2.3 An intensional type-theoretic logic

2.3.1 Introduction

This section brings in the notion of *intension* in contrast with *extension*; the contrast is sometimes known as *sense* versus *reference*. The distinction between the sense of an expression and its reference was introduced informally in Section 2.1. The *reference* of an expression for Frege corresponds to what we have called the denotation, or semantic value, of that expression. The *sense* corresponds to what we may intuitively think of as the 'meaning' of the expression. The sense of an expression is supposed to determine what its reference is in any possible circumstances (i.e., in any 'possible worlds'). With this distinction in mind, Frege attempted to solve the problem of non-referential context (modal context of knowledge, belief, time, contingence, etc.) by saying that expressions of natural language have a sort of ambiguity in that sometimes an expression has a 'normal' denotation (i.e., the kind of denotation we have been assigning to the formal languages of Section 2.2) but in certain circumstances an expression 'denotes' what is its sense.

For example the denotation of 'President' can either refer to a particular individual (which is the reference of 'President') or to the description of a function (which is the sense of 'President'). The purpose of this section is to formalize the notion of sense (just as the notion of reference has been formalized in Section 2.2) and to incorporate the sense in a formal language, the *intensional logic*, that will be used to represent and to process the natural language.

2.3.2 Sense and reference

Let us first illustrate the difference existing between sense and reference (or denotation) by way of examples. We have seen in Section 2.2 that the reference of a sentence in a possible world is its truth value in that world. The truth value of the sentence 'Nixon has been elected President of the United States.' can vary as a function of the times with respect to which we evaluate its truth. Similarly, the reference of a one-place predicate in a particular world is the set of individuals of whom that predicate is true in that world. Thus the reference of *President* in the actual world is the set of all actual presidents. The reference of an individual constant is the individual denoted by that constant. We shall say that a sentence is *extensional* if the reference (or extension) of the whole sentence can be described as a function of the references of its parts (this is often referred to as the 'compositional semantics' principle). All of the ordinary logical connectives are extensional: to determine the truth value in a particular world of say $A \vee B$, one has only to determine the truth values of A and B in that particular world.

Let us now consider the modal operator *necessarily*, symbolized as $\Box$ in Subsection 2.2.11. *Necessarily* and *possibly* are two of the modal operators whose analysis is the central concern of modal logic, and it was this concern that provided one of the main motivations for introducing the notion of *intension*. Let us define the *intension of a sentence* as a function from possible worlds to truth values: to each possible world the function assigns the truth value of the given sentence in that world. To take a natural language example, the sentence 'The Morning Star is the Morning Star' is true since it is an instance of the axiom $(c = c)$ for any individual constant c. The intension of this sentence will then be a function that assigns the value *true* to every possible world.

The semantic interpretation rule for *necessarily* can be stated as follows (when the accessibility relation is the complete relation): '$\Box A$ is true in a given world if and only if A is true in every possible world'. Thus *necessarily* is interpreted as a function from intensions of sentences to truth values. (In the sequel, the *intension of a sentence* will also be called a *proposition*.) Accordingly, the sentence 'Necessarily the Morning Star is the Morning Star' is true.

Assume now that the phrase 'the Morning Star' denotes the same entity as the phrase 'the Evening Star' (they both denote planet Venus). Then, the sentence 'The Morning star is the Evening star' is true, while the sentence 'Necessarily the Morning Star is the Evening Star' is false, since it is a matter of contingent fact that the Morning Star is the same as the Evening Star, not

2.3. AN INTENSIONAL TYPE-THEORETIC LOGIC

a matter of logical necessity. Sentences of the form 'Necessarily A' (or $\Box A$) are a first example of a type of sentence whose reference (at a given world) cannot be described as a function of the references of its parts (here $\Box$ and A). However, if we introduce the notion of intension, we see that the reference of 'Necessarily A' can now be described as a function of the intensions of its parts. Therefore, the semantic value of such a sentence remains compositional but with regard to the intensions instead of the references.

The relevance of this question in the semantic analysis of some other types of natural language sentences will be examined in the next subsection.

2.3.3 Compositionality in non-referential context

We have seen in Subsection 2.2.6 that the denotation (or semantic value) of the expression $\neg A$ in any possible world depends only on the denotation of A in the same world. A formal way to express this is as follows: $[\![\neg A]\!] = [\![\neg]\!]([\![A]\!])$. For this reason, negation is said to be a truth-functional operator. The modal operators are said to be non-truth-functional; this means that $[\![\Box A]\!]$ cannot be written as $[\![\Box]\!]([\![A]\!])$. We have, however, seen in Subsection 2.3.2 that the modal operators become truth-functional with respect of the intensions. We could write this: $[\![\Box A]\!]_{in} = [\![\Box]\!]_{in}([\![A]\!]_{in})$ where the subscript *in* means the intension of an expression.

The semantic relation between certain verb–object combinations forces us to look beyond denotations for another class of sentences as well. A classical case discussed by Frege involves the complement of verbs like believe, think, suppose, etc. For example, the sentence 'John believes that the Morning Star is the Morning Star' might be true though the sentence 'John believes that the Morning Star is the Evening Star' is false if the individual named John is somewhat unacquainted with astronomy. The truth value of these sentences does not only depend on the truth value of the complements of verbs like believe. Such constructions are sometimes called *referentially opaque constructions* since they are 'opaque' to Leibniz law on 'substitution of co-designative names while preserving truth values' which states the following: *The result of substituting in any formula one name for another name denoting the same individual results in a formula that is true if and only if the original formula was true.*

The occurrence of certain adjectives and adverbs means that the denotations of a whole sentence cannot be explained in terms of the denotations of its parts. In many instances, it seems natural to think of an adjective as having a set as its denotation. For example, consider a treatment based on model theory of expressions such as 'the president is young'. It is natural to take this expression as true at a given index only in case where the individual denoted by 'the president' belongs to the set of individuals denoted by the adjective 'young'. It is possible to take the noun phrase 'young president' as having as its denotation (at each index) the set of individuals which are in the intersection of the set denoted by the adjective 'young' with the set denoted by the noun 'president'. However, adjectives such as 'former' cannot have as denotation the intersection of such sets: 'former president' cannot be denoted as the intersection of the set of 'presidents' with the set of 'individuals that are former' (at a given index). The definition of the denotation of 'former president' in terms of the denotation of 'president' will clearly not be adequate. What is required for the denotation of 'former president' at a given index is the denotation of 'president' at indices with earlier temporal coordinates. Again, the solution proposed by Frege to this sort of problem is to distinguish between the sense of an expression and its reference.

We shall see in this section how the various problems that have been mentioned, relative to the semantic analysis of non-truth-functional operators such as necessarily ($\Box$), always in the future ($[F]$) or in the past ($[P]$), relative to referentially opaque constructions such as sentences containing verbs like 'believe', 'know', 'think', and relative to intensional adjectives such as 'former', can be circumvented by using the concept of sense instead of reference.

2.3.4 Intension and extension

After having determined the distinction between 'sense' and 'reference', Frege attempted to solve the problem of non-referential contexts (modal operators, referentially opaque constructions, intensional adjectives) by saying that expressions of natural language have a sort of ambiguity in that sometimes an expression has a 'normal' denotation (its reference) but in some circumstances an expression 'denotes' what is its sense.

The first attempt to formalize the notion of sense was made by Carnap [Carnap 36]. (In the context of a formal language we shall generally use the terms *intension* and *extension* for *sense* and *reference* respectively.) Carnap suggested that the intension of an expression is nothing more than all the varying extensions the expression can have. In other words the intension (as formalized by Montague) is simply a function with all possible states of affairs as arguments and the appropriate extensions as values.

Let us quote Dowty and his co-authors [Dowty et al. 81]:

With the advent of Kripke's semantics for modal logic (taking possible worlds as indices), it became possible for the first time to give an unproblematic formal definition of intension *for formalized languages. When we have done this for a formal language resembling English, we will have produced a theoretical construct* (intension *of an expression) that 'does what a meaning does', insofar as*

2.3. AN INTENSIONAL TYPE-THEORETIC LOGIC

a meaning of an expression is something that determines, for any type, place and possible situation, the denotation of the expression in that time, place and situation.

For the formal modal language that make use of indices $<w,t>$, we shall take an intension as a function from indices to extensions.

We have seen in Subsection 2.2.11 that the modal language $\mathcal{L}_m$ contains non-rigid designators as analogous in natural language to a title that different individuals hold at different times (e.g. the title of *President of the United States*). If c is an individual constant of $\mathcal{L}_m$ which represents a non-rigid designator (of natural language), the constant c denotes an individual at each index. Hence, the intension of the constant c is a function from indices to individuals. For predicate constants, the intension is similarly defined as a function from indices to extensions. A one-place predicate constant denotes a set of individuals, at each index. Therefore, the intension of a one-place predicate constant will be the function that gives, for each index, the set denoted at that index. Such an intension is sometimes called a *property*. The intension of a formula can be defined as a function from indices to truth values. Such an intension is sometimes called a *proposition*.

One thing that should be immediately clear is that intensions are more general than extensions: if the intension of an expression is given, one can determine its extension with respect to a particular index but not vice versa.

We introduce a notation for representing the intension.

- If α is an expression, if $\mathcal{M}$ is a model and if g is an assignment function for variables, the notation $[\![\alpha]\!]_{in}^{\mathcal{M},g}$ represents the intension of α with respect to the model $\mathcal{M}$ and the assignment g.

2.3.5 Example

Let us illustrate the concept of intension by continuing the example of Subsection 2.2.12. From Figures 2.5 and 2.6 we deduce that the intension of the individual constant c_4 and of the predicate name P_c are the following functions:

$$[c_4]_{in}^{\mathcal{M},g} = \{<w_1,t_1> \to \text{Kennedy}, <w_1,t_2> \to \text{Johnson},$$
$$<w_1,t_3> \to \text{Nixon}, <w_2,t_1> \to \text{Nixon},$$
$$<w_2,t_2> \to \text{Nixon}, <w_2,t_3> \to \text{Johnson}\},$$

$$[P_c]_{in}^{\mathcal{M},g} = \{<w_1,t_1> \to \{\text{Kennedy}, \text{Johnson}\}>, <w_1,t_2> \to \{\text{Nixon}\},$$
$$<w_1,t_3> \to \{\text{Johnson}, \text{Nixon}\},$$
$$<w_2,t_1> \to \{\text{Johnson}, \text{Nixon}\}, <w_2,t_2> \to \{\text{Kennedy}\},$$
$$<w_2,t_3> \to \{\text{Kennedy}, \text{Johnson}, \text{Nixon}\}\}.$$

From these two intensions we deduce the intension of the formula $P_c(c_4)$, that is:

$$[P_c(c_4)]_{in}^{\mathcal{M},g} = \{<w_1,t_1> \to 1, <w_1,t_2> \to 0, <w_1,t_3> \to 1,$$
$$<w_2,t_1> \to 1, <w_2,t_2> \to 0, <w_2,t_3> \to 1\}.$$

2.3.6 Formal definition of intension

From the extensions of an expression at all indices we can determine the intension by assembling these extensions in the form of a function. The extension of an expression α, with respect to a model $\mathcal{M}$, a world w, a time t and an assignment g, is noted $[\alpha]^{\mathcal{M},w,t,g}$; its intension is defined as follows:

$$[\alpha]_{in}^{\mathcal{M},g} =_{def} \{<w,t> \to [\alpha]^{\mathcal{M},w,t,g} \mid w \in W, t \in T\}.$$

Conversely, from the intension $[\alpha]_{in}^{\mathcal{M},g}$ of any expression α we can determine the extension of α at any index we are interested in by evaluating the intension at this index:

$$[\alpha]^{\mathcal{M},w,t,g} = [\alpha]_{in}^{\mathcal{M},g}(<w,t>).$$

In particular, if c is a non-logical constant (individual constant or predicate constant), the definition of the extension of c is as follows (§ 2.2.11):

$$[c]^{\mathcal{M},w,t,g} =_{def} V(<w,t>, c).$$

Hence, the intensions of the non-logical constants are the following functions:

$$[c]_{in}^{\mathcal{M},g} = \{<w,t> \to V(<w,t>, c) \mid w \in W, t \in T\}.$$

The extension of a variable is supplied by the value assignment g only and thus does not differ from one index to the other; if x is a variable we have

$$[x]^{\mathcal{M},w,t,g} = g(x).$$

2.3. AN INTENSIONAL TYPE-THEORETIC LOGIC

Thus the intension of a variable will be a constant function on indices which corresponds to its extension. The intension of every expression of the formal language can be computed from the definitions above and from the semantic rules of intensional logic (see Subsection 2.3.9).

The Montague intensional logic is a formal language formed with expressions that can denote intensions as well as extensions. In this language, there exists a formation rule for deriving from any expression α a second expression, written as $^\wedge\alpha$, denoting the intension of α. The symbol $^\wedge$ will be referred to as the *intension operator*. The definition is the following.

- If α is an expression, then $^\wedge\alpha$ is an expression characterized by the fact that its denotation $[\![^\wedge\alpha]\!]^{\mathcal{M},w,t,g}$ is equal to $[\![\alpha]\!]^{\mathcal{M},g}_{in}$

For example, since the expression $P_c(c_4)$ is a formula of the language $\mathcal{L}_m$, the expression $^\wedge P_c(c_4)$ is a formula of the intensional language derived from $\mathcal{L}_m$. The denotation of this formula, $[\![^\wedge P_c(c_4)]\!]^{\mathcal{M},w,t,g}$, is defined to be $[\![P_c(c_4)]\!]^{\mathcal{M},g}_{in}$; it is independent of w and t (see Subsection 2.3.5).

As the notion of intension is defined for every expression of the language, we can form an expression such as $^{\wedge\wedge}\alpha$ from $^\wedge\alpha$. Since $[\![^\wedge\alpha]\!]^{\mathcal{M},w,t,g}$ is independent of the indices $<w,t>$, the denotation of $^{\wedge\wedge}\alpha$, i.e., $[\![^{\wedge\wedge}\alpha]\!]^{\mathcal{M},g}_{in}$, is a constant function the value of which is $[\![\alpha]\!]^{\mathcal{M},g}_{in}$. Therefore, there is no interest in building higher-order intensions such as $^{\wedge\wedge}\alpha$, $^{\wedge\wedge\wedge}\alpha$, etc.

Montague introduced a second syntactic tool into his intensional logic: the *extension operator* $^\vee$ that is the converse of the intension operator $^\wedge$, and that can be defined as follows.

- If α is an expression that denotes an intension, then $^\vee\alpha$ is an expression characterized by the fact that its intension equals the extension of α. Thus, for any index $<w,t>$, one has

$$[\![^\vee\alpha]\!]^{\mathcal{M},g}_{in} =_{def} [\![\alpha]\!]^{\mathcal{M},g,w,t}.$$

From this definition and from the definition of the denotation of $^\wedge\alpha$, we deduce that for all expressions α, all models $\mathcal{M}$, all indices $<w,t>$ and all assignments g, we have

$$[\![^{\vee\wedge}\alpha]\!]^{\mathcal{M},w,t,g} = [\![\alpha]\!]^{\mathcal{M},w,t,g}.$$

This identity justifies the $^{\vee\wedge}$ *cancellation rule* which will play an important role in the simplification of complex intensional logic expressions produced as translations of natural language sentences (see Chapter 3).

We have seen in Subsections 2.3.2 and 2.3.3 that the compositionality principle, as it was stated in Subsection 2.1.4, no longer holds when dealing with the semantic analysis of non-truth-functional operators (such as modal operators), with referentially opaque constructions (containing verbs as 'believe', 'know') and with intensional adjectives (such as 'former'). In a referential (or extensional) context, this compositionality principle can be stated as follows: '*the denotation of a sentence is a function of the denotations of their parts and of their mode of combination*'. In a non-referential (or intensional) context, this principle will be replaced by the more general statement: '*the denotation of a sentence is a function of the extensions of some of their parts, of the intensions of some others of their parts and of their mode of combination*'.

Let us summarize how the various concepts of intensional logic that have been introduced in this subsection will allow us to put this compositionality principle to work. In the intensional logic into which some English expressions will be translated (see Chapter 3), every expression has an *extension* with respect to each index and an *intension* that is a function from indices to extensions. The device that makes it possible to show how the extension of a complex expression may depend on the extensions of some of its parts and on the intensions of some others of its parts, is the introduction of some expressions that have as their extensions the intensions of other expressions, i.e., expressions of the kind $^\wedge\alpha$ such that $[\![^\wedge\alpha]\!]^{\mathcal{M},w,t,g} = [\![\alpha]\!]^{\mathcal{M},g}_{in}$. The compositionality principle can now remain in its original form: '*the denotation of a sentence is a function of the denotations of their parts and of their mode of combination*'.

2.3.7 Syntax of intensional logic

The intensional logic language, as it has been developed by Montague, is a formal language that combines all the syntactic and semantic devices that have been gradually introduced in Sections 2.2 and 2.3. The intensional language is a type-theoretic language based on a recursive definition of types (§ 2.2.7), which contains universal and existential quantifiers, the lambda operator (§ 2.2.9), the modal operators $\Box, \Diamond, [F], <F>, [P]$ and $<P>$ (§ 2.2.11), and the operators of intension $^\wedge$ and of extension $^\vee$ (§ 2.3.6).

As in the case of the type-theoretic version of the language $\mathcal{L}_1$ (§ 2.2.7), the definition of a syntax begins with a recursive definition of a set of types. The basic symbols are t (for truth), e (for entity) and s (for sense). The symbol s will allow us to associate with every type a a new type $\langle s, a \rangle$. The expressions of type $\langle s, a \rangle$ have intensions as extensions, namely, the intensions of expressions of type a. Thus, the expressions of type $\langle s, t \rangle$ have *propositions* (i.e., functions from indices to truth values) as extensions; the expressions of type $\langle s, e \rangle$ have individual concepts as extensions; the expressions of type $\langle s, \langle e, t \rangle \rangle$ have sets

2.3. AN INTENSIONAL TYPE-THEORETIC LOGIC 101

of individual concepts (or *properties*) as extensions, etc.

The meaning of these new types will be explained more precisely in the next subsection, devoted to the semantics of intensional logic. The formation rules for types are as follows.

- **Base**: e and t are types.

- **Induction**:
    - if a and b are types, then $\langle a, b \rangle$ is a type;
    - if a is a type, then $\langle s, a \rangle$ is a type.

- **Closure**: all types are obtained by applying the base and induction rules a finite number of times.

The syntax of the *intensional language*, noted $\mathcal{L}_I$, contains all the syntactic rules of the language $\mathcal{L}_m$ plus the syntactic rules associated with the intensional operator $\char`\^$ and with the extensional operator $\vee$. Since the intensional logic language $\mathcal{L}_I$ is the formal language that has been used by Montague to represent the English language, it is useful to give the whole collection of syntactic rules that constitute the syntax of $\mathcal{L}_I$.

- The basic expressions of the language are:
    - for every type a, the non-logical constants noted $c_{j,a}$, where j is a natural number;
    - for every type a, the variables noted $x_{j,a}$, where j is a natural number.

- For every type a, the *set E_a of expressions of type a* is defined recursively as follows:
    - every constant of type a belongs to E_a;
    - every variable of type a belongs to E_a;
    - if α is an expression in E_a and if x is a variable in E_b, then $\lambda x[\alpha]$ belongs to $E_{\langle b, a \rangle}$;
    - if α belongs to $E_{\langle a, b \rangle}$ and if β belongs to E_a, then $\alpha(\beta)$ belongs to E_b;
    - if α and β belong to E_a, then $(\alpha = \beta)$ belongs to E_t;
    - if A belongs to E_t, then $\neg A$ also belongs to E_t;
    - if A and B belong to E_t, then $A \vee B, A \wedge B, A \supset B$ and $A \equiv B$ also belong to E_t;

- if A belongs to E_t and x is a variable of any type, then $\forall x A$ and $\exists x A$ also belong to E_t;
- if A belongs to E_t, then $\Box A, \Diamond A, [F]A, <F>A, [P]A$ and $<P>A$ also belong to E_t;
- if α belongs to E_a, then $\char94\alpha$ belongs to $E_{\langle s,a\rangle}$;
- if α belongs to $E_{\langle s,a\rangle}$, then $\char86\alpha$ belongs to E_a.

In particular, E_t is the set of formulas and $E_{\langle s,t\rangle}$ is the set of propositions.

2.3.8 Model-theoretic interpretation of intensional logic

In his formal treatment of the semantics of natural language, Montague subscribes to a principle which he formulates in these terms: 'the extension of a sentence is a function of the extensions of its parts that appear in a referential context, of the intensions of its parts that appear in a non-referential context (§ 2.3.3, § 2.3.6), and of the mode of combination of intensions and extensions'. (By 'non-referential context', Montague refers to syntactic constructions with modals (*necessarily*), and the complements of verbs (*believe, seek*) and more generally to those constructions for which the Leibniz law does not hold (§ 2.3.3). The 'compositionality principle' requires that non-referential contexts be defined in a precise way for each fragment of natural language.)

The model that has been defined for the multimodal language of Subsection 2.2.11 can also serve for the language of intensional logic. This model $\mathcal{M}$ is a 5-tuple $(S, W, T, <, V)$ with S, W and T non-empty sets of individuals, possible worlds and time instants respectively, $<$ a linear ordering on the set T, and V an interpretation function mapping the non-logical constants of the language $\mathcal{L}_I$ to appropriate denotations.

For a given type a, we represent by D_a the *set of possible denotations for expressions of type a*, with respect to the triple (S, W, T). The definition, which generalizes that of Subsection 2.2.8, is given recursively as follows:

- $D_e =_{def} S$,
- $D_t =_{def} \{1, 0\}$,
- $D_{\langle a,b\rangle} =_{def} D_b^{D_a}$, i.e., the set of all functions from D_a to D_b, for every type a and every type b;
- $D_{\langle s,a\rangle} =_{def} D_a^{W\times T}$, i.e., the set of all functions from $W \times T$ to D_a, for every type a.

The relation $D_{\langle a,b\rangle} = D_b^{D_a}$ has to be read as follows: the denotation of an expression of type $\langle a, b\rangle$ is a function whose arguments lie in D_a and whose

2.3. AN INTENSIONAL TYPE-THEORETIC LOGIC

values lie in D_b. The relation $D_{\langle s,a \rangle} = D_a^{W \times T}$ has to be read as follows: the denotation of an expression of type $\langle s, a \rangle$, i.e., of an expression which denotes an intension, is a function whose arguments lie in the Cartesian product of the set of possible worlds and the set of time moments, and whose values lie in D_a.

Consider, for example, the sentence 'the Earth is round', which is an intensional expression of type $\langle s, t \rangle$. The denotation of this sentence, i.e., its *sense* (since an intensional expression is at stake) is a *proposition*, i.e., a function from $W \times T$ to $\{1, 0\}$. This function takes value 1 in all possible worlds and at all time moments in which the Earth is round, for instance in our world at the present time, and it takes value 0 in all possible worlds in which the Earth is not round, for instance, in the world of Egyptian mythology in which it was a flat disk.

By the same procedure, it can be shown that the intension of a one-place predicate, or of an intransitive verb or common noun, is a function whose arguments lie in the Cartesian product of possible worlds and time moments and whose values lie in the extension of the predicate under consideration, i.e., in its characteristic function (§ 2.2.6). Owing to this distinction between extension and intension, it can be shown how coextensive predicates such as, for instance, 'the passengers of the Mayflower' and 'the founders of Plymouth, Massachusetts' are not synonymous. They fail to have the same intension, for in other possible worlds they might correspond to different subsets of individuals.

The definition of the set of types now includes a recursive rule to define for each type a a new type $\langle s, a \rangle$ for the intensions corresponding to each type a. The set of intensions of type a is defined as $D_{\langle s,a \rangle}$. The function V will assign to each non-logical constant of $\mathcal{L}_I$ of type a a member of $D_{\langle s,a \rangle}$. For reasons of convenience we shall slightly modify the form of the function V. We have defined in Subsection 2.2.11 the interpretation function V as a function with two arguments:

$$V(<w, t>, c) = [\![c]\!]^{\mathcal{M}, w, t}.$$

We transform this function V into another function, that we also note V, whose single argument is the non-logical constant c. The value $V(c)$ will thus be an intension, i.e., a function from indices to appropriate denotations; the definition is as follows:

$$V(c)(<w, t>) = [\![c]\!]^{\mathcal{M}, w, t}.$$

As in the case of extensional logic languages, an *assignment of values to variables g* is a function having as domain the set of all variables and giving as value for each variable of type a a member of D_a. Observe that the function g assigns to each variable an *extension*, while the (one-argument) function V assigns to each constant an *intension*.

2.3.9 Semantics of intensional logic

The functions V and g allow us to define recursively the extension of every expression α of the language $\mathcal{L}_I$, with respect to a model $\mathcal{M}$, a world $w \in W$, a time $t \in T$, and an assignment g; this extension will, as usually, be noted $[\![\alpha]\!]^{\mathcal{M},w,t,g}$. The definition is as follows.

- If c is a non-logical constant, the extension of c with respect to $\mathcal{M}, w, t, g$ is obtained by evaluating the intension $V(c)$ at the index $<w,t>$, i.e.,

$$[\![c]\!]^{\mathcal{M},w,t,g} = V(c)(<w,t>).$$

- If x is a variable, then $[\![x]\!]^{\mathcal{M},w,t,g} = g(x)$.

- If α is an element of E_a and if x is a variable of type b, then $[\![\lambda x[\alpha]]\!]^{\mathcal{M},w,t,g}$ is the function $h : D_b \to D_a$ defined by $h(r) = [\![\alpha]\!]^{\mathcal{M},w,t,g^r}$, for each $r \in D_b$, where g^r is the x-variant of g obtained by assigning the value r to x.

- If α is an element of $E_{\langle a,b \rangle}$ and if β is an element of E_a, then

$$[\![\alpha(\beta)]\!]^{\mathcal{M},w,t,g} = [\![\alpha]\!]^{\mathcal{M},w,t,g}([\![\beta]\!]^{\mathcal{M},w,t,g}).$$

- If α and β are elements of E_a, then $[\![\alpha = \beta]\!]^{\mathcal{M},w,t,g}$ is 1 or 0 according as $[\![\alpha]\!]^{\mathcal{M},w,t,g}$ equals $[\![\beta]\!]^{\mathcal{M},w,t,g}$ or not.

- If A is an element of E_t, then $[\![\neg A]\!]^{\mathcal{M},w,t,g}$ is 1 or 0 according as $[\![A]\!]^{\mathcal{M},w,t,g}$ is 0 or 1.

- If A and B are elements of E_t, then $[\![A \vee B]\!]^{\mathcal{M},w,t,g} = 1$ if $[\![A]\!]^{\mathcal{M},w,t,g} = 1$ or if $[\![B]\!]^{\mathcal{M},w,t,g} = 1$; the semantics of the formulas $A \wedge B, A \supset B$ and $A \equiv B$ is defined in a similar way.

- If A is a element of E_t and if x is a variable (of any type), then $[\![\forall x A]\!]^{\mathcal{M},w,t,g} = 1$ if and only if $[\![A]\!]^{\mathcal{M},w,t,g'} = 1$ for all assignments g' that are x-variants of g. The extension $[\![\exists x A]\!]^{\mathcal{M},w,t,g}$ is defined in a dual way.

- The semantics of the modal formulas $\Box A, \Diamond A, [F]A, <F> A, [P]A$ and $<P> A$, where A is an element of E_t, has been given in Subsection 2.2.11.

- If α is an element of E_a, then $[\![{}^\wedge\alpha]\!]^{\mathcal{M},w,t,g}$ is the function $h : W \times T \to D_a$ defined by $h(<w',t'>) = [\![\alpha]\!]^{\mathcal{M},w',t',g}$ for all $w' \in W, t' \in T$.

- If α is an element of $E_{\langle s,a \rangle}$, then $[\![{}^\vee\alpha]\!]^{\mathcal{M},w,t,g} = [\![\alpha]\!]^{\mathcal{M},w,t,g}(<w,t>)$.

2.3. AN INTENSIONAL TYPE-THEORETIC LOGIC

- If α is an element of E_a, the *intension* of α with respect to a model $\mathcal{M}$ and an assignment g, noted $[\![\alpha]\!]_{in}^{\mathcal{M},g}$, is the function $h : W \times T \to D_a$ defined by $h(<w,t>) = [\![\alpha]\!]^{\mathcal{M},w,t,g}$ for all $w \in W$, $t \in T$.

This last rule and the definition of $[\![^\wedge\alpha]\!]^{\mathcal{M},w,t,g}$ allow us to write two equalities already seen in Subsection 2.3.6:

$$[\![\alpha]\!]_{in}^{\mathcal{M},g} = [\![^\wedge\alpha]\!]^{\mathcal{M},w,t,g} \quad \text{and} \quad [\![\alpha]\!]_{in}^{\mathcal{M},g}(<w,t>) = [\![\alpha]\!]^{\mathcal{M},w,t,g}.$$

If A is a formula, i.e., an element of E_t, then A is said to be *true in the model $\mathcal{M}$ and at the index $<w,t>$* if it satisfies $[\![A]\!]^{\mathcal{M},w,t,g} = 1$ for all assignments g.

2.3.10 Summary of intensional logic

This subsection contains a summary of the main definitions relative to extensional and intensional logics that have been introduced progressively in Sections 2.2 and 2.3.

If α is an expression of an extensional logic language, the *semantic value* of α is represented as

$$[\![\alpha]\!]^{\mathcal{M},w,t,g}.$$

This semantic value is evaluated with respect to a model $\mathcal{M}$, an assignment of values to variables g, a world w and a time t. The model $\mathcal{M}$ is a 5-tuple $(S, W, T, <, V)$ for which:

- S is a non-empty set of individuals or of objects; it constitutes the *interpretation domain* of the formal language (the set S is also called *domain of discourse*).

- W is a non-empty set of *possible worlds* w (the set W is called *universe*).

- T is a non-empty set of *time moments* t.

- $<$ is a *linear order relation* on T.

- V is an *interpretation function* that assigns elements of S to individual constants and sets of n-tuples of elements of S to n-place predicate constants; this function V is parametrized by an *index* $<w,t>$, with $w \in W$ and $t \in T$.

In addition to a model $\mathcal{M}$, we need an *assignment function* g, which assigns suitable denotations to the variables of each type. For example, if x is an individual variable, then its denotation $g(x)$ is an element of the set S, and if

x is a predicate variable, then its denotation $g(x)$ is an element of the set $\{1, 0\}^S$. Further details will be given below.

The semantic value of every expression α of the formal language is recursively obtained from the semantic values of the basic elements of the language (the non-logical constants and the variables) and from the rules that define the semantics of the logical connectives (or logical constants) $\neg$, $\vee$, $\wedge$, $\supset$ and $\equiv$, of the quantifiers $\forall$, $\exists$ and of the modal operators $\square$, $\Diamond$, $[F]$, $<F>$, $[P]$ and $<P>$. The semantics of the basic elements is defined as follows.

- If c is a non-logical constant:

$$[\![c]\!]^{\mathcal{M},w,t,g} =_{def} V(<w,t>,c).$$

- If x is a variable:

$$[\![x]\!]^{\mathcal{M},w,t,g} =_{def} g(x).$$

In the framework of an *extensional type-theoretic logic*, the *syntactic category* of an expression is defined as being a *type*. The semantic value of an expression is called its *denotation*. With every type (and hence with every syntactic category) is associated a set of 'possible denotations', i.e., a set of 'possible semantic values'. This set constitutes what we could call the *semantic category* associated with the syntactic category.

The types are defined recursively as follows:

- e is a type (individual constants and variables);

- t is a type (formulas);

- if a and b are types, then $\langle a, b \rangle$ is a type.

One-place predicate constants have $\langle e, t \rangle$ as type, two-place predicate constants have $\langle e, \langle e, t \rangle \rangle$ as type, etc.

The set of possible denotations for expressions of type a is written as D_a. The sets of possible denotations are defined recursively as follows:

- $D_e = S$;

- $D_t = \{1, 0\}$;

- $D_{\langle a,b \rangle} = D_b^{D_a}$.

2.3. AN INTENSIONAL TYPE-THEORETIC LOGIC

One-place predicate constants have $\{1,0\}^S$ as the set of possible denotations, two-place predicate constants have $((\{1,0\})^S)^S$ as the set of possible denotations, etc.

In the framework of an *intensional type-theoretic logic* we have to distinguish between the *sense* of an expression and its *reference*. The reference of an expression corresponds to what we have termed the denotation of this expression in an extensional type-theoretic logic. The sense corresponds to what we may intuitively think of as the 'meaning' of an expression independently of any context. The terms in each of the following two lines are synonyms when they are used for characterizing an expression of a formal language:

reference = semantic value = denotation = extension ;
meaning = sense = intension.

With every expression α of an intensional language are now associated two semantic concepts. The *extension* of α corresponds exactly to the denotation of α in an extensional logic context. The *intension* of α is written as

$$[\![\alpha]\!]_{in}^{M,g} ,$$

and is defined as a function from the set $W \times T$ to the set of extensions of α as follows:

$$[\![\alpha]\!]_{in}^{M,g}(<w,t>) =_{def} [\![\alpha]\!]_{in}^{M,w,t,g} \text{ for all } w \in W, t \in T .$$

With every expression α we associate an expression $^\wedge\alpha$ that is defined through the following semantic characterization:

$^\wedge\alpha$ is an expression (of an intensional language) whose extension is the intension of α.

This is written formally:

$$[\![^\wedge\alpha]\!]^{M,w,t,g} =_{def} [\![\alpha]\!]_{in}^{M,g} .$$

The symbol $^\wedge$ is called the *intension operator*. We also introduce the *extension operator* $^\vee$. If α denotes an intension, then we define the intension of $^\vee\alpha$ as the extension of α, i.e.,

$$[\![^\vee\alpha]\!]_{in}^{M,g} =_{def} [\![\alpha]\!]^{M,w,t,g} .$$

(If α denotes an intension, then $[\![\alpha]\!]^{M,w,t,g}$ is independent of $<w,t>$.) The result of combining the operator $^\vee$ with an expression of the form $^\wedge\alpha$ has the effect of cancelling out the two operators ˇ and ˆ; more precisely, we have the identity

$$[\![\check{}\hat{}\alpha]\!]^{\mathcal{M},w,t,g} = [\![\alpha]\!]^{\mathcal{M},w,t,g} .$$

If c is a non-logical constant, we deduce from $[\![c]\!]^{\mathcal{M},w,t,g} = V(<w,t>,c)$ and from the definition of intension, the relation

$$[\![c]\!]_{in}^{\mathcal{M},g}(<w,t>) = V(<w,t>,c) \text{ for all } w \in W, t \in T .$$

It is interesting to rewrite the extension $V(<w,t>,c)$ as the value taken by a function $V(c)$ at the point $<w,t>$; the definition is

$$V(c)(<w,t>) =_{def} V(<w,t>,c) .$$

This allows us to write the intension of a non-logical constant c as follows:

$$[\![c]\!]_{in}^{\mathcal{M},g} = V(c) .$$

The extension of an expression of the intensional logic language is obtained recursively from the extensions of the basic elements of the language (which are the same as those of the extensional logic language) and from the semantic rules for the logical connectives, quantifiers, modal operators (as in the extensional logic), and the intension and extension operators.

In the framework of an *intensional type-theoretic logic*, the *syntactic category* of an expression is still defined as being a *type*. Types are defined recursively from three basic elements t, e and s. The first two of them are elementary types and have already been defined in the framework of extensional languages. The third element is a new symbol that is associated with the concept of intension. The recursive definition of types given above is completed by means of the following rule:

- if a is a type, then $\langle s, a \rangle$ is a type.

If α is an expression of type a, then $\hat{\alpha}$ is an expression of type $\langle s, a \rangle$; if α is an expression of type $\langle s, a \rangle$, then $\check{\alpha}$ is an expression of type a. The sets of possible denotations D_a are defined recursively by means of the rules seen above and of the following additional rule:

- $D_{\langle s,a \rangle} = D_a^{W \times T}$.

If α is an expression of type a, its extension $[\![\alpha]\!]^{\mathcal{M},w,t,g}$ is an element of D_a.

With every expression α of type a is associated an expression $\hat{\alpha}$ of type $\langle s, a \rangle$ whose extension $[\![\hat{\alpha}]\!]^{\mathcal{M},w,t,g}$ is an element of the set $D_{\langle s,a \rangle}$; this extension

2.3. AN INTENSIONAL TYPE-THEORETIC LOGIC

is identical to the intension $[\![\alpha]\!]_{in}^{M,g}$ of α. With every expression α that denotes an intension and whose type is $\langle s, a \rangle$ is associated an expression $^{\vee}\alpha$ of type a; the intension $[\![^{\vee}\alpha]\!]_{in}^{M,g}$ of $^{\vee}\alpha$ is the extension $[\![\alpha]\!]^{M,w,t,g}$ of α (it is independent of the index $<w,t>$).

As mentioned in the beginning of this section, by giving a formal definition to the concept of intension, by developing an intensional language and its model-theoretic interpretation based on the ideas of Kripke's semantics for modal logic, and by applying this formal language to the analysis of intensional phenomena in English, Montague completed the semantic programme of Frege. The sense or intension of an expression is now defined as the set of all extensions this expression can have, organized as a function with all possible indices as arguments and the corresponding extensions as values. Conversely, from the intension of any expression we can determine the extension of the expression at any possible index by just evaluating the intension at the index we are interested in. The distinction between intension and extension allowed Montague to preserve Frege's compositionality principle, even in non-referential (or intensional) contexts, by saying that expressions of natural languages have sometimes an extension as semantic value and in other circumstances have an intension as semantic value. Frege's principle in a general context can be stated as follows: *the extension of a sentence is a function of the extensions of its parts that appear in a referential context, of the intensions of its parts that appear in non-referential contexts and of the mode of combination of intensions and extensions.*

A key point in Montague's approach is the introduction of the intensional logic formalism where some expressions, written as $^{\wedge}\alpha$, have as their extensions the intensions of the corresponding expressions α. As a result, in the framework of the formal language of intensional logic, Frege's compositionality principle can again be stated in terms of extensions only: *the extension (or denotation) of a formula is a function of the extensions of its parts, of their mode of combination by means of logical connectives and of the semantic rules associated with these connectives.*

Examples that illustrate the use of the intension operator are given in the next two subsections, which contain an introduction to the formal treatment of the adverb 'necessarily' and of intensional adjectives (that were presented in Subsection 2.3.2 to illustrate the limitation of extensional logic).

In the sequel, we shall often use the notation $(D \to C)$ to represent the set of functions having D as domain and C as co-domain; thus we have

$$(D \to C) =_{def} C^D$$

2.3.11 The adverb 'necessarily'

Sentences of the form 'necessarily p' can be considered as constructions where the extension of the whole is a function of the extensions of the parts if the word 'necessarily' is treated as a function that combines with $\wedge p$ rather than with p. This means that the expression in $\mathcal{L}_I$ in which the adverb 'necessarily' is translated must be an expression of the type $\langle\langle s,t\rangle,t\rangle$ instead of an expression of the type $\langle t,t\rangle$ as could be inferred from the syntax proposed in Subsection 2.2.11. Let us represent by **Nec** the expression which is the translation of 'necessarily' in the logic language $\mathcal{L}_I$; since it is of the type $\langle\langle s,t\rangle,t\rangle$, its denotation is a function f that belongs to the set

$$((W \times T \to \{1,0\}) \to \{1,0\}),$$

that is, $f : \{h : W \times T \to \{1,0\}\} \to \{1,0\}$. We can now give a semantic value for **Nec** by defining $[\![\mathbf{Nec}]\!]^{\mathcal{M},w,t,g}$ as that function from propositions to truth values that gives the value 1 when applied to any proposition h just in case h itself maps every index into 1 (see [Dowty et al. 81], p. 163).

In his PTQ grammar, Montague makes use of the modal operator $\square$ for translating the adverb 'necessarily' (which is said to be treated syncategorematically). We can then consider the expression $\square p$ as an abbreviation for **Nec**($\wedge p$). This treatment will be explained in Subsection 3.2.15.

2.3.12 Extensional and intensional adjectives

Owing to the concept of intension, Montague's semantics is equipped to account for the difference between the inferential potential which is associated with the grammatical construction: *attributive adjective + common noun* ('good president') and that which is associated with the construction *common noun + predicative adjective* ('American president'). Obviously, 'American president' means 'American and president' whereas 'good president' does not mean 'good and president'. Syntactically speaking, adjectives (whether attributive such as 'good' or predicative such as 'vegetarian' and 'american') are expressions which, when combined with common nouns, generate phrases which again belong to the category of common nouns (CN in the PTQ grammar of Subsection 3.2.2).

Semantically, predicative adjectives such as 'vegetarian' are functions whose arguments are classes, or, equivalently characteristic functions belonging to the set $(S \to \{1,0\})$, and whose values are also classes. In other words, 'vegetarian' is a function in the set $((S \to \{1,0\}) \to (S \to \{1,0\}))$. Combined with a common noun, 'president' for example, which denotes a class (or more precisely the characteristic function thereof), this function makes up a complex common noun 'vegetarian president' whose extension is again a class (the class of those

2.3. AN INTENSIONAL TYPE-THEORETIC LOGIC

individuals who are both vegetarian and presidents, or more precisely, the corresponding characteristic function).

We cannot extend this analysis to attributive adjectives and say that their extension is made up of functions whose arguments are classes. If this were the case, it would indeed be possible to replace a common noun which is the argument of an attributive adjective by a coextensional common noun without any alteration of the semantics. This, however, is not the case. Even if 'president' had the same extension as 'car driver', it would not follow that 'good president' would have the same extension as 'good car driver', whereas it would be valid in this case to conclude that 'vegetarian president' has the same extension as 'vegetarian car driver'. The kind of semantics which fits attributive adjectives runs as follows. The extensions of an adjective will not be functions belonging to the set $((S \to \{1,0\}) \to (S \to \{1,0\}))$ but will be functions belonging to the set $((W \times T \to (S \to \{1,0\})) \to (S \to \{0,1\}))$, i.e., functions whose *arguments* are *properties* and whose *values* are *classes of individuals*. (By definition, a property is a function whose arguments are indices, composed of a possible world and a time instant, and whose values are classes of individuals.)

The attributive adjective 'former' for instance is defined as follows by Dowty and his co-authors. The extension of 'former' can be viewed as a function whose argument is a property (e.g. the property of being president) and whose value is the class of individuals who are no longer, but were once, members of the extension which corresponds to that property. We can then define the extension of sentences such as 'former president' on the basis of the extensions of 'president' and 'former': we have to combine 'former', not with an expression which represents a predicate (expression of type $\langle e, t \rangle$) but with an expression which represents a property (expression of type $\langle s, \langle e, t \rangle \rangle$). To do this, we introduce an expression, noted **For** (for 'formerly', [Dowty et al. 81]), which is of type $\langle \langle s, \langle e, t \rangle \rangle, \langle e, t \rangle \rangle$ and whose extension thus belongs to the set

$$((W \times T \to (S \to \{1,0\})) \to (S \to \{1,0\})).$$

Hence $[\![\mathbf{For}]\!]^{\mathcal{M},w,t,g}$ is a function (analogous with $[\![\mathbf{Nec}]\!]^{\mathcal{M},w,t,g}$) which has access not only to the standard extension of the one-place predicate to which it is combined, but also to all the other extensions of this predicate. More precisely, $[\![\mathbf{For}]\!]^{\mathcal{M},w,t,g}$ is a function whose argument is a property (e.g. the property of being president) and whose value is the set of individuals which belonged to the extension of that property in the past. We can then define the intension $[\![\mathbf{For}]\!]_{in}^{\mathcal{M},g}$ as the function h such that, for every property k, the set $h(<w,t>)(k)$ coincides with the set $k(<w,t'>)$ for some $t' < t$. In other words, for every individual $r \in S$, we have $[[h(<w,t>)](k)](r) = 1$ if and only if $k(<w,t'>)(r) = 1$ for some $t' < t$.

Let us return to the examples of Subsections 2.2.12 and 2.3.5. The semantic value of the one-place predicate constant P_c is the set of individuals owning a landed property in California. In Subsection 2.3.5 we computed the function $[\![P_c]\!]_{in}^{\mathcal{M},g}$, and we found

$$[\![P_c]\!]_{in}^{\mathcal{M},g} = \{\ <w_1,t_1> \to \{\text{Kennedy}, \text{Johnson}\}, <w_1,t_2> \to \{\text{Nixon}\},$$
$$<w_1,t_3> \to \{\text{Johnson}, \text{Nixon}\},$$
$$<w_2,t_1> \to \{\text{Johnson}, \text{Nixon}\}, <w_2,t_2> \to \{\text{Kennedy}\},$$
$$<w_2,t_3> \to \{\text{Kennedy}, \text{Johnson}, \text{Nixon}\}\}.$$

From this we deduce

$$[\![\text{For }^\wedge P_c]\!]_{in}^{\mathcal{M},g} = \{\ <w_1,t_1> \to \emptyset,\ <w_2,t_1> \to \emptyset,$$
$$<w_1,t_2> \to \{\text{Kennedy}, \text{Johnson}\},$$
$$<w_2,t_2> \to \{\text{Johnson}, \text{Nixon}\},$$
$$<w_1,t_3> \to \{\text{Kennedy}, \text{Johnson}, \text{Nixon}\},$$
$$<w_2,t_3> \to \{\text{Kennedy}, \text{Johnson}, \text{Nixon}\}\}.$$

2.3.13 Uniformization of the semantic treatment

We said above that Montague set up a rigorous correspondence, a homomorphism, between syntax and semantics, i.e., a function which correlates *types of entities* with *syntactic categories*. Distinct syntactic categories can correspond to the same type of entities but not conversely, since if this were possible the correspondence would no longer be a function. It looks at though we had ignored this warning as far as the syntactic category of adjective is concerned, in so far as we have associated two types of entities with the same category, namely, entities taken from the set of functions $((S \to \{1,0\}) \to (S \to \{1,0\}))$ and entities taken from the set of functions $((W \times T \to (S \to \{1,0\})) \to (S \to \{1,0\}))$. We thus need to restore uniformity to the semantic treatment of attributive adjectives and predicative adjectives. Shall we match up the former on the latter or the latter on the former? The first alternative has to prevail. As the notion of property is richer than that of class, one will never be able to derive the former from the latter. The converse operation, however, can be performed; one can always derive what is simple from what is complex.

Montague's *meaning postulates* come into the picture at this very point. Meaning postulates are axioms in which (extra-logical) terms of natural language occur as primitive terms. Let us give two examples of meaning postulates:

$\forall x \forall y\ [x \text{ is the father of } y \supset x \text{ is older than } y]$,
$\forall y\ [x \text{ is a bachelor} \supset x \text{ is unmarried}]$.

Meaning postulates serve to erase the excess of wealth which we had to put up with in order to have a uniform treatment of a class of linguistic facts.

2.3. AN INTENSIONAL TYPE-THEORETIC LOGIC

For instance, all transitive verbs are given a semantics which ascribes them intensional entities as objects. This is due to the existence of verbs like 'seek' whose object may be a purely fictional entity. But other transitive verbs such as 'find' demand real objects. The required uniformization is obtained through a meaning postulate; its role will be to spell out the fact that for all the verbs of a given category, the intensional object can be reduced to the status of extensional object. It is a solution of this kind that seems to recommend itself for predicative adjectives.

Chapter 3

Montague's semantics

3.1 Formal representation of natural languages

3.1.1 Introduction

We begin this chapter with a section that extends the contents of Section 2.1 by describing the philosophical and logical environment in which Montague's semantics has been developed, and by situating Montague's work in a historical perspective. The reader interested only in the continuation of the statement of logic formalism of Sections 2.2 and 2.3 can skip this section.

Russell maintained that the persistent dependence of our thoughts upon ordinary language is one of the principal obstacles to progress in philosophy. He accused this language of being full of ambiguities and of having an abominable syntax [Schilpp 44], [Russell 05]. Philosophers who share this opinion are confined to the construction of formalized languages, admitting along with Kemeny that ordinary language is not appropriate to logical arguments [Kemeny 57]. The most tolerant among these have introduced a canonical notation to restrain natural language to an existing formal language such as predicate calculus. That is what Quine was doing [Quine 72a], [Quine 72b].

Russell's two complaints are not of equal importance. One can certainly admit that, in so far as it is ambiguous, natural language must be amended, but it is not obvious that all of the things that distinguish the syntax of natural languages from that of formal languages are abominable. Restraining natural language to a canonical notation, however justifiable it may be for metaphysicians or epistemologists, nevertheless impoverishes unduly the set of valid inferences of natural language.

Montague put forward a procedure which opposes to such restraint, that is the procedure of modelling logical calculations according to the syntax of natural language, with the aim of formalizing valid reasoning processes which resist being so restrained. Quine already had the same idea, in the definition that he gave for logical truth, a definition which clears the way for an extension of the formalization of valid reasoning [Quine 72b]. Montague deserves the credit for showing that such an objective was achievable. He did it in preparing the way with three seminal papers, the most widely known being 'The proper treatment of quantification in ordinary English' [Montague 73]. This section is meant to serve as an introduction to the reading of that paper, in presenting its main points, and to prepare the formal definition of a *categorial grammar* for a fragment of English language.

The reader interested in the formal representation of the semantics of natural language will find more complete statements on this matter in [Gochet 82b], [Gochet 86], [Guenthner and Guenthner-Reutter 78], [Jayez 88], [Kamp 84], [Keenan and Faltz 85] and [Lepore 87].

3.1.2 Montague's programme

Montague means to formulate a syntax and a semantics of natural language which has the same rigour and precision as the syntax and semantics of formal languages. Montague's programme may be summarized in five points.

1. The differences between natural language and formal language can be reduced in such a way that semantic notions of the latter may be applied to the former.

2. For both natural and formal languages, semantics must be compositional and recursive (Tarski's requirements).

3. We can extend reference (or extension) theory to form a sense (or intension) theory which is as rigorous as reference theory.

4. We can assign to meaning postulates the role of providing general reasons of semantic facts.

5. When we translate natural inferences into formal logic language, we must stay as close as possible to the surface grammatical form.

These five points are studied in Chapters 2 and 3. Here are some details about the contents.

1. We have built in Chapter 2 a formal language, namely intensional logic, whose expressive power is near to that of natural language. In this chapter, we shall adapt natural language in order to bring it into a state in which its translation into intensional logic can be considered as an algebraic operation.

2. The process of translating natural language into logic language is described in Section 3.2. This process allows the semantics of natural language to enjoy the recursivity and compositionality properties inherent in the semantics of intensional logic language.

3.1. FORMAL REPRESENTATION OF NATURAL LANGUAGES

3. The theory of intension has been developed in Section 2.3.

4. The role of meaning postulates, which has been alluded to in Subsection 2.3.14, and their formalization into a logic language will be studied in Subsections 3.2.13 and 3.2.14.

5. A categorial grammar for a fragment of English language is introduced in Subsections 3.2.2 and 3.2.3. This type of grammar allows us to represent sentences of natural language by means of logic formulas that are as close as possible to the surface structure of these natural sentences. Otherwise stated, the categorial grammar allows us to render the syntactic analysis of a sentence of natural language isomorphic to the syntactic analysis of the formula that represents this sentence in a logic language.

3.1.3 Natural and formal languages

Montague does not deny the evidence of the differences between natural and formal languages. He knows perfectly well that the two evolved differently and that they fill different functions, but he denies that these practical differences indicate a theoretical difference. He begins his article 'English as a formal language' with the statement: 'I reject the notion that there is a significant theoretical difference between natural and formal languages' [Montague 74d].

However unimportant they may be, the differences between natural and formal languages exist none the less. Natural language, unlike formal languages, contain nonsensical sentences such as 'the green colourless ideas sleep furiously' and ambiguous expressions such as the word 'crane'.

These peculiarities of natural language stand in our way when we attempt to spell out a definition of truth for natural language which satisfies Tarski's requirements. These obstacles, however, can be removed. One solution, which Montague himself adopted in 'The proper treatment of quantification in ordinary English', consists in limiting one's treatment to a *fragment* of natural language from which all the lexically ambiguous expressions such as 'crane' have been excluded [Montague 73]. Alternatively, one might also settle, from the very beginning, for a semi-formal language, as Quine does [Quine 60]. This language could be predicate calculus enriched with some terms borrowed from natural language.

Some would object that to regularize and normalize natural language beforehand as Montague proposes is to start down a dangerous path. Where does normalizing natural language end and where does restraining it to a canonical semi-formal notation begin ? From the moment that we introduce variables into our vocabulary, we have stepped decidedly out of the domain of natural language. Montague enriched his fragment of English through the use of numbered pronouns which are, in fact, variables (see Subsection 3.2.2). Doesn't he risk, in so doing, rendering insubstantial his assertion that there is no important theoretical difference between natural and formal language, and transforming a statement which he holds as a synthetic statement into a mere tautology completely devoid of interest ?

The answer to that objection depends upon criteria of identity and individuation of languages. If one holds that the syntax is more important than the vocabulary in the definition of a language, then the objection that we just raised can be dismissed. Indeed, we observe that the English fragment studied in 'The proper treatment of quantification in ordinary English' contains the syntactic categories that one uses classically to describe natural language: common nouns, intransitive verbs, transitive verbs, adverbs, prepositions, etc., rather than the syntactic categories of predicate calculus (one-place predicates, two-place predicates, etc.).

Montague's criterion for assessing a semantics is as follows: the theory must be able to systematically and correctly match declarative sentences with their truth conditions and their entailment conditions. Logicians have always sought to account for the inferential potentials of declarative sentences, but those who follow Russell's tradition, like Quine, think that they must, to succeed at it, distinguish the *grammatical form* from the *logical form* and disregard the former in order to concentrate on the latter, since the logical form is the grammatical form of an ideal language.

Montague does not in any way dispute Russell's analysis. On the contrary, he integrates it into his semantics. On the other hand, Montague believes that classical predicate logic is too restricted to account for the contrast between the following inferences which are formally valid:

'all soldiers are motorists, therefore all *English* soldiers are *English* motorists',

'all soldiers are motorists, therefore all *vegetarian* soldiers are *vegetarian* motorists',

and the following inferences which are not valid:

'all soldiers are motorists, therefore all *former* soldiers are *former* motorists',

'all soldiers are motorists, therefore all *experienced* soldiers are *experienced* motorists'.

Just as in passing from propositional calculus to predicate calculus we increase the number of reasonings that we can formalize, also in establishing a formal semantics for natural language, we obtain a tool capable of dealing with inferences which elude classical logic.

3.1.4 Conditions for the suitability of a semantics

One needs to establish the conditions whereby a semantics may be deemed suitable. Since Chomsky's work, one usually requires of a syntax it to generate all well-formed sentences and it to be able to describe them structurally. Similarly, one requires a semantics to map the infinite set of well-formed sentences and expressions to the interpretation that would be given them by a linguistically competent person. To meet this requirement, the semantics must satisfy the following two conditions.

- It must be *compositional*, that is, conform, to a certain extent, to Frege's compositionality principle according to which the meaning of the whole is a function of the meaning of the parts.

- It must be *recursive*, that is, contain rules which may be applied in an iterative manner.

The reason is that it must have the potential to account for the fact that every speaking person is able to understand new sentences; such an ability is described by Russell as follows: 'Given the meaning of separate words, and the rules of the syntax, the meaning of a sentence is determined'. This is why we can understand a sentence that we have never heard before. You have probably not yet heard the proposition 'the inhabitants of the Adaman Islands eat hippopotamus stew for lunch', but you have no problem understanding this proposition. Schlick, the founder of the Vienna Circle, has further detailed the mechanism of composition: 'The same set of signs that was used to describe a certain state of things may, by rearrangement (of the order of the signs), be used to describe a completely different state of things in such a way that we understand the meaning of the new combination without its having been explained to us'.

Syntactic arrangement, that is, the order of words, constitutes a mechanism which undeniably plays a semantic role. It even happens that the syntactic contribution is a simple question of order and arrangement. So, for example, the difference in the *meaning* between the following two true sentences: 'the Earth attracts the Moon', and 'the Moon attracts the Earth', rests entirely upon the difference in the order of the noun phrases 'the Earth' and 'the Moon'.

But the syntactic contribution may also have other aspects. Let us look at the sentence 'the shooting of the hunters was a disgrace'. This sentence can convey two different propositions depending on whether the verb 'to shoot' is assigned to the syntactic category of transitive verbs or to that of intransitive verbs, even though the word order is the same in both cases. Syntax is therefore not only a matter of *order* and *arrangement* as Schlick thought it to be, but also a matter of membership in *syntactic categories*.

We just finished saying why semantics must be *compositional*, i.e., why the composition mechanism must be able to produce the whole from the parts, or to decompose the whole into its parts; but this requirement is accompanied by another, that of *recursivity*. This requirement is dictated not by the ability of human beings to understand new sentences, but by their ability, as Chomsky frequently mentions, 'to understand an infinite number of sentences' [Chomsky 65]. Let us admit, along with Chomsky, that the fundamental goal of the analysis of a language $\mathcal{L}$ is to separate the grammatical sequences which are sentences of $\mathcal{L}$, from the non-grammatical sequences which are not sentences of $\mathcal{L}$, and to study the structure of grammatical sequences. How can we, using a finite equipment, generate the infinite set of well-formed sentences ? The answer is clear: we need a 'generative grammar', that is, to quote Chomsky, 'a set of rules which may be iterated to produce an indefinitely large number of structures' [Chomsky 57], [Chomsky 65].

The objective of the semantics can now be defined precisely. It is concerned with assigning a *meaning* to an infinite set of sentences and phrases. Like the syntax, the semantics must therefore be recursive. However, one might imagine that semantics can borrow from syntax its recursive construction mechanisms and limit its own contribution to the production of meaning for isolated words. Davidson evoked this expeditious fashion of constructing a recursive semantics as follows: Suppose that we have a satisfactory theory of the syntax of our language, a theory consisting of an effective, that is, recursive, way of telling us for any given expression whether it is an expression that has an autonomous meaning (it is a sentence), and suppose that this implies that each sentence is considered to be composed from a finite set of syntactic elements; then we hope that the syntax, so conceived, will produce the semantics when a dictionary giving the meaning of each syntactic atom is added to it [Davidson 69].

It is undeniable that the recursive rules that generate well-formed expressions partially explain the way in which the meanings of words are combined to form the overall meaning of a sentence. We note that the syntactic rules for sentence formation sometimes make a selection between homonyms. So, for example, in the sentence 'I like tennis', the syntactic constraints select the transitive verb 'like' and reject the homonymous adjective. On the other hand, syntactic constraints are unable to explain why one of the possible meanings of the word 'crane' is selected in the sentence 'the crane and the stork differ in their wingspans'. Indeed, in replacing 'crane' with 'machine for lifting weights' we have a sentence which is semantically absurd, but still syntactically well-formed. We must therefore admit along with Davidson that knowledge of the structural characteristics of a sentence plus knowledge of the meaning of the parts do not add up to give the meaning of the sentence, and conclude with him that recursive syntax together with a dictionary is not necessarily a recursive semantics [Davidson 69].

As a conclusion, linguists are forced to provide the semantics with a recursive mechanism to generate composite meanings that is distinct from the recursive sentence-generation mechanism.

3.1.5 A formal semantics for natural language

Two ways of providing natural language with a rigorous semantics, along the lines of the standard semantics of formal languages, have been considered. The first provides a direct interpretation of a fragment of natural language in a model [Montague 74c], [Keenan and Faltz 85]. The second, called indirect interpretation, proceeds in two steps. Firstly, a fragment of natural language is being translated into a formal language; then, the latter is interpreted. In the second method, the rigorous semantics of predicate calculus and that of intensional logic can be seen to *indirectly* provide an interpretation of the expressions of natural language which are translational correlates of the interpreted expressions of formal language. The natural language semantics will inherit the rigorous character of the formal language in so far as the translation itself is rigorous and systematic. Translating natural language into formal language however is generally held to be unsystematic. According to Quine, such a translation work cannot be viewed as a computation work [Quine 72a]. The natural language semantics will entirely inherit the rigour of formal language semantics only if the translation itself is a calculation as e.g. the operation of switching from decimal to binary number system. Can we construe the translation of natural language, or of a substantial part of it, as an operation of this kind ? Montague attempts to answer in the affirmative. To ensure that the translation of natural language into formal language is as regular and systematic as it should be, Montague imposes an extremely strong constraint on it: once the fragment of natural language which one aims at translating into formal language has been purged of all lexical ambiguity, one demands that each simple expression of natural language be translated into a single expression of formal language (the latter expression may be simple or complex), and one demands that each syntactic operation of natural language be associated with a finite combination of syntactic operations of formal language.

We have just seen that, through a suitable translation, the semantics of formal language could serve as a semantics for a fragment of natural language. The kind of formal semantics Montague has in mind is obviously Tarski's semantics. Here, however, some further details should be given. Tarski presented two definitions of truth which are, in fact, close to one another; he defined the concept of 'absolute truth' and the concept of 'truth relative to a model'. Montague takes up the second one: '*Like Donald Davidson I regard the construction of a theory of truth — or rather, of the more general notion of truth under an arbitrary interpretation — as the basic goal of serious syntax and semantics [...]*' [Montague 73]. The difference lies in the following fact: the clauses of the recursive definition of truth do something more than just provide an account (within natural language) of the way primitive predicates, connectives and quantifiers contribute to the meaning of the sentences of formal language which contain them; they assign to the expressions of formal language, extra-linguistic entities built from the entities which are in the model (see Subsections 3.2.5–3.2.7). Montague's main contribution lies in his imposing the very strong constraint of *homomorphism* not only on the translation relation holding between (the fragment of) natural language and formal language but also on the interpretation function that assigns entities belonging to the model and combinations of such entities to isolated expressions and combined expressions.

The homomorphism requirement is the keystone of the whole construction. It determines the way Montague conceives the syntax of natural language, the syntax of formal language, the semantics and even the extra-linguistic entities which semantics assigns to linguistic items. The set of expressions of natural language, the set of expressions of formal language correlated to them by translation, and the set of extra-linguistic entities assigned to these formal expressions are provided with *operations*. Each of these three sets, equipped with the corresponding operators, is regarded as an *algebra*. In order to have a homomorphism between these algebras, not only the syntax but also the entities assigned to expressions as their interpretation, have to be conceived in an appropriate way. The homomorphism requirement can be satisfied by adopting a syntax constructed as a *categorial grammar* and by adopting a *theory of types* whose members are functions. This does not mean, however, that there is no other way of meeting the requirement (§ 2.2.7, § 3.2.2, [Cooper and Parsons 76]).

3.1.6 A categorial grammar

When Montague began the research which resulted in the kind of grammar described in 'The proper treatment of quantification in ordinary English', several kinds of compositional and recursive syntaxes of natural language were already in existence [Montague 73]. *Categorial grammar*, for example, has the following two features. It is *compositional* in so far as the grammaticality of the whole depends on the grammaticality of the parts and on the mode of composition. In a categorial grammar, the phrases of the language are

3.1. Formal representation of natural languages

indeed gathered together in *syntactic categories* which are either *basic* or *derived* categories. Significantly, the latter have been called 'functor categories' [Ajdukiewicz 35]. Functor categories gather expressions together on the basis of their combination potential. The operation of syntactic construction which generates composite expressions from their components is defined as an operation of *functional application* in the mathematical sense (see Subsection 3.2.2).

Let $\mathcal{A}$ and $\mathcal{B}$ be syntactic categories of the language. The combination rule for syntactic categories reads as follows: $\mathcal{A}/\mathcal{B}$ combined with $\mathcal{B}$ returns $\mathcal{A}$. If an expression of category $t/(t/e)$, e.g. 'Aldebaran', is combined with an expression of category t/e, e.g. 'is shining', an expression belonging to category t is obtained, namely the sentence 'Aldebaran is shining'.

This syntax is not only *compositional*, it is also *recursive*. It contains, among others, certain rules governing operations for which the result of the operation belongs to the same category as one of its arguments. One can repeat such an operation indefinitely. For instance, a sentence adverb, i.e., an element of category t/t, combined with a sentence, i.e., with an element of category t, yields a sentence. Thus, 'truly', of category t/t, combined with 'Aldebaran is shining', of category t, yields 'truly Aldebaran is shining', which belongs to category t. The last sentence can again be combined with 'truly'.

3.1.7 A recursive semantics

Categorial grammar fulfils the conditions we set: it is compositional and recursive. These two requirements are constraints for semantics also and direct the choice of entities which will stand as meanings.

Under what conditions will it be legitimate to consider the meaning of the sentence 'Aldebaran is shining' as a function (in the mathematical sense) whose arguments are the meaning of 'Aldebaran', the meaning of 'is shining', and the meaning of the syntactic operation in which a noun is combined with an intransitive verb ? The answer is clear; the conjunction of the following three conditions is both necessary and sufficient:

1. The meaning assigned to the sentence must correspond to the *value* of a function.

2. The meanings assigned to the proper pronoun 'Aldebaran' and to the verb phrase 'is shining' must correspond to *arguments* of a function.

3. It must be possible for the *dependence* between the meaning of 'Aldebaran is shining' and the meanings of 'Aldebaran' and 'is shining' to be described with an *operation* or *function* which has the meanings of 'Aldebaran' and

'is shining' as *arguments* and the meaning of 'Aldebaran is shining' as *result* or *value*.

Whoever claims that the meaning of the whole is a function of the meaning of the parts, is forced to take meanings out of a set of entities that have the appropriate mathematical properties. In order to merit the status of arguments or values of mathematical functions, meanings have to satisfy a precise criterion of identification which enables us to differentiate a case in which the *same* meaning occurs twice from a case in which two different but close meanings are involved.

If one confines oneself to the part of meaning which comes under what Quine describes as the 'theory of reference', i.e., to *extension*, one observes that standard extensions fulfil the conditions required of entities in order for them to be regarded as arguments or values of a function. Usually, an individual term such as 'Aldebaran' is assigned an *individual object* which belongs to the reference domain (or domain of discourse) denoted by S (Chap. 2). An intransitive verb such as 'is shining' — which would be treated as a one-place predicate in predicate calculus — is normally assigned a *class* which is a subset of S, namely the class of shining objects. Equivalently, the same intransitive verb can be assigned the *characteristic function* which takes the value *true* when it is given a member of the class of shining objects as argument, and the value *false* otherwise (§ 2.2.6). Let us finally mentions the third case for syntactic categories: a sentence is assigned a *truth value* as extension. Individual objects, characteristic functions and truth values give rise to no problem as far as identification criteria are concerned. These entities fit perfectly as arguments and values of a function.

As long as we remain at the extension level, we immediately obtain a compositional semantics if we assign an extension of the adequate logical type (individuals, characteristic functions, truth values) to the expressions of a syntactic category (proper nouns, intransitive verbs (e.g.), sentences). This however does not suffice, since, as Davidson emphasized, a recursive syntax combined with a dictionary does not automatically result in a recursive semantics. We have to introduce something analogous to Katz and Fodor's projection rules, i.e., semantic rules governing the composition of meanings [Katz and Fodor 64] (see Subsection 2.1.3).

As we saw, Montague insists that semantics should be homomorphic to syntax. This goal will be achieved if the conditions given below are fulfilled.

1. There must be a mapping from the *syntactic categories* of expressions (simple or composite) to the *logical types* of entities assigned to these expressions as meanings.

2. There must be a one-to-one mapping between the *syntactic rules* that

govern the combination of expressions and the *semantic rules* that govern the composition of meanings (reduced to extensions at this stage).

Montague puts 'Aldebaran' into the category $t/(t/e)$, and 'is shining' into the category t/e. Thus, the intransitive verb 'is shining' denotes a characteristic function that maps shining individuals into *true* and non-shining individuals into *false*. On the other hand, the proper noun 'Aldebaran' denotes a function that maps the sets of individuals to which the star belongs into *true* and those to which it does not belong into *false*. It therefore follows that if the set of shining individuals is given as an argument to the function denoted by 'is shining' and if Aldebaran belongs to this set, then the value of the function *Aldebaran* applied to the argument *is shining* will be *true* as it should be.

The semantics sketched above is not only compositional but also recursive in so far as recursive rules of semantic composition are associated with recursive rules of syntactic composition. Consider for example the syntactic composition rule defined as: $(\text{Truly})_{t/t} + (\text{Aldebaran is shining})_t = (\text{Truly Aldebaran is shining})_t$. The corresponding semantic composition rule is obtained as follows. The extension of 'truly' is a function whose arguments and values lie in the set $\{true, false\}$; if this function is given as an argument the extension of 'Aldebaran is shining', i.e., a truth value, then it yields as value the extension of 'Truly Aldebaran is shining', i.e., a truth value. The operation of semantic composition therefore has the status of a functional application, just as the operation of syntactic composition.

3.2 Montague's grammar

3.2.1 Introduction

Montague's celebrated paper 'The proper treatment of quantification in ordinary English' (PTQ, in short) describes a categorial grammar for a well-defined fragment of English language [Montague 73]. This grammar is referred to in the sequel as the PTQ grammar. It consists of a set of recursive syntactic rules specifying how complex expressions are to be formed out of simpler ones. Montague's PTQ grammar uses a basic dictionary made up of nine categories of words : intransitive verbs, terms (proper nouns, numeral adjectives, pronouns), transitive verbs, verb-phrase adverbs, common nouns, sentence adverbs, prepositions, sentence-complement verbs (believe, assert, etc.) and infinitive-complement verbs (try, wish, etc.). The elements of a tenth class, containing the determiners 'every', 'the', 'a', 'an', are introduced syncategorematically in the same way as the logical connectives and the quantifiers (see chapter 1 of [Thayse 88]). Montague's categorial grammar will be described in

Subsections 3.2.2–3.2.4.

As mentioned in Subsection 3.2.5, Montague does perform a direct interpretation of the fragment of English he studies. Instead, he proceeds indirectly by translating each natural language basic expression into an expression of the intensional logic presented in Section 2.3. The interpretation of the intensional logic language provides us (indirectly) with an interpretation of the PTQ fragment of English. The translation of this fragment of English into intensional logic language is described in Subsections 3.2.5–3.2.15.

3.2.2 Montague's categorial grammar

Montague's PTQ grammar belongs to the family of categorial grammars that have been developed by Ajdukiewicz [Ajdukiewicz 35]. In a categorial grammar, the concept of *syntactic category* is defined recursively as follows:

- t is a syntactic category, that of the expressions to which a truth value can be assigned, i.e., the category of sentences;

- e is a syntactic category, that of the expressions to which entities can be assigned;

- if $\mathcal{A}$ and $\mathcal{B}$ are syntactic categories, then $\mathcal{A}/\mathcal{B}$ and $\mathcal{A}//\mathcal{B}$ are syntactic categories.

In a categorial grammar, an expression of category $\mathcal{A}/\mathcal{B}$ (or of category $\mathcal{A}//\mathcal{B}$; we shall see the difference below) combines with an expression of category $\mathcal{B}$ to produce an expression of category $\mathcal{A}$. This rule is referred to as the *categorial cancellation rule*.

In the fragment he studies, Montague uses a basic dictionary made up of words distributed into ten categories. The first category is the *intransitive verb category* whose elements are run, walk, talk, etc. This category is represented by t/e. The second category is that of *terms*; it is made up of proper nouns such as John, Mary, etc., of numeral adjectives such as twenty, ninety, etc., and of 'subscripted pronouns' such as he_0, he_1, etc. This category is noted $t/(t/e)$. As an example, an expression of category $t/(t/e)$ such as 'John' combines with an expression of category t/e such as 'walk' to yield an expression of category t, i.e., a sentence, such as 'John walks'. The categorial cancellation rule is sometimes written in terms of the symbols $+$ and $=$ as illustrated by the following example:

$$\text{John}_{t/(t/e)} + \text{walk}_{t/e} = (\text{John walks})_t.$$

We see that the categorial cancellation rule allows us to characterize in a rigorous way the concept of *sentence*. A sentence is any recursive combination

3.2. Montague's Grammar

of basic expressions that produces, after a finite number of applications of the cancellation rule, an expression of category t. A sentence does not belong to a basic category: we do not find sentences as entries in a dictionary, but only words.

The syntax of a categorial grammar is not only compositional but also recursive. It contains indeed some cases where the categorial cancellation rule produces complex expressions whose syntactic category is identical to the syntactic category of one of its components. Consider for example the sentence adverb 'necessarily': it is treated as a member of category t/t and so it combines with a sentence to give a new sentence. For example:

$$\text{Necessarily}_{t/t} + (\text{John walks})_t = (\text{Necessarily John walks})_t.$$

We see that t appears as a syntactic category of both the resultant expression and of one of its components. It is possible to use this rule iteratively:

$$\text{Necessarily}_{t/t} + (\text{Necessarily John walks})_t =$$
$$(\text{Necessarily necessarily John walks})_t.$$

A more interesting example is given by sentence-complement verbs that are used in knowledge and belief context. The verb 'believe' is treated as a member of the syntactic category $(t/e)/t$. This allows us to derive constructions such as

$$\text{John}_{t/(t/e)} + (\text{believes that})_{(t/e)/t} + (\text{Mary walks})_t =$$
$$(\text{John believes that Mary walks})_t.$$

Again this construction scheme contains a component whose syntactic category (t) is the same as that of the resultant expression. It allows an indefinite iteration of the form: 'John believes that Mary knows that ...'.

Let us note that the categorial cancellation rule that allows the construction of complex expressions from simpler ones is formally identical to the cancellation rule for types (§ 2.2.6) with, however, a little difference in notation. A logic expression of type $\langle a, b \rangle$ combines with an expression of type a to give an expression of type b ('left cancellation'), whereas an expression of syntactic category $\mathcal{A}/\mathcal{B}$ combines with an expression of syntactic category $\mathcal{B}$ to give an expression of syntactic category $\mathcal{A}$ ('right cancellation').

It is the parallelism between the syntactic categories of a natural language and the types of a formal language that will allow a systematic translation of the former into the latter. There is however a factor that can complicate this translation. Montague observed that there exist certain cases where syntactically distinct categories of English will correspond to the same logical type, and to make this distinction he introduced the notation $\mathcal{A}//\mathcal{B}$; the symbols $\mathcal{A}/\mathcal{B}$ and $\mathcal{A}//\mathcal{B}$ designate two different syntactic categories that correspond to the same logical type.

The syntactic categories of the *PTQ* grammar are represented in Figure 3.1; the mnemonic abbreviations (for the complex categories) that are used in this figure are due to Montague. For example, the category t/e is represented by IV (which is an abbreviation for 'intransitive verb'), the category $t/IV = t/(t/e)$ is represented by T (which is an abbreviation for 'term phrase') and the category $IV/T = ((t/e)/t/(t/e))$ is represented by TV (which is an abbreviation for 'transitive verb'). The basic expressions, or words of the dictionary, are represented in bold-face notation. This notation will be used throughout the remaining of this section.

Syntactic categories			Basic expressions
Name	Categorial definition	Usual definition	
e			-
t		Sentence	-
IV	t/e	Verb phrase, intransitive verb	run, walk, talk, rise, change
T	t/IV	Noun phrase, proper name (Term)	John, Mary, Bill, he$_0$, he$_1$,...; him$_0$, him$_1$, ...
TV	IV/T	Transitive verb	find, lose, eat, love, date, seek, conceive
IAV	IV/IV	Verb-phrase adverb (modifying an element of category IV)	rapidly, slowly, voluntarily, allegedly
CN	$t//e$	Common noun	man, woman, park, fish, unicorn, price, temperature
t/t		Sentence adverb (modifying the sense of a sentence)	necessarily
IAV/T		Preposition	in, about
IV/t		Sentence-complement verb	believe, assert
$IV//IV$		Infinitive-complement verb	try, wish
DET	T/CN	Determiner	every, the, a, an

Figure 3.1 : Syntactic categories and basic expressions.

(Reprinted by permission of D. Reidel publishing company; copyright 1981.)

3.2.3 Syntax of the PTQ categorial grammar

Let $\mathcal{A}$ be a syntactic category of the fragment of English considered in the preceding subsection. We shall designate by $B_\mathcal{A}$ the set of 'words in the dictionary' or 'basic expressions' of category $\mathcal{A}$. In particular, if there exists no basic expression of category $\mathcal{A}$, then $B_\mathcal{A}$ is the empty set. Such is the case for the syntactic category t; since a dictionary does not contain sentences as entries, B_t is the empty set. The set of groups of words (called expressions or phrases) of category $\mathcal{A}$ is denoted $P_\mathcal{A}$; it includes the set $B_\mathcal{A}$ of basic expressions of category $\mathcal{A}$ together with the set of complex expressions of category $\mathcal{A}$ that are formed with the syntactic rules given below.

The first of these rules, denoted $S1$, initializes the recursion: it constitutes the base of the formal (recursive) definition of the grammar.

- $S1$: For each syntactic category $\mathcal{A}$, the set $B_\mathcal{A}$ is included in the set $P_\mathcal{A}$.

The remaining syntactic rules deal with complex expressions. Each of these rules contains three sorts of information:

- The specification of the syntactic categories of the components.
- The definition of the syntactic operation that is performed.
- The specification of the syntactic category of the resulting expression.

Hence, all the syntactic rules forming complex expressions have the following general form:

- If $\alpha \in P_\mathcal{A}$ and if $\beta \in P_\mathcal{B}$, then $F(\alpha, \beta) \in P_\mathcal{C}$. After that comes a definition for the function F.

Some of the syntactic rules proposed by Montague are given hereafter. They are the basic rules ($S2$–$S3$), the functional application rules ($S4$–$S10$) and the quantification rule ($S14$).

The functional application rules must obey the categorial cancellation rule: $\mathcal{A}/\mathcal{B}$ (or $\mathcal{A}//\mathcal{B}$) combined with $\mathcal{B}$ gives $\mathcal{A}$. We use the same numbering as in [Montague 73].

- $S2$: Complex expressions of category T (terms).
  If $\alpha \in P_{T/CN}$ and if $\beta \in P_{CN}$, then $F_2(\alpha, \beta) \in P_T$.
  The function $F_2(\alpha, \beta)$ is $\alpha^*\beta$, where α^* is α except in the case where α^* is a and the first word in β begins with a vowel; in this case, α^* is an.

- $S3$: Complex expressions of category CN (common nouns).
  If $\alpha \in P_{CN}$ and if $A \in P_t$, then $F_{3,n}(\alpha, A) \in P_{CN}$.

The function $F_{3,n}(\alpha, A)$ is α **such that** A^*, where A^* comes from A by replacing each occurrence of **he**$_n$ or **him**$_n$ by **he, she** or **it**, or by **him, her** or **it**, respectively, according as the first common noun in α is of masculine, feminine or neuter gender.

- $S4$: Complex expressions of category t (sentences).
  If $\alpha \in P_T$ and if $\beta \in P_{IV}$, then $F_4(\alpha, \beta) \in P_t$.
  The function $F_4(\alpha, \beta)$ is $\alpha\beta^*$ where β^* is the result of replacing the first verb (or member of one of the categories $B_{IV}, B_{TV}, B_{IV/T}$ or $B_{IV//IV}$) in β by its third person singular present form.

- $S5$: Complex expressions of category IV (intransitive verbs).
  If $\alpha \in P_{TV}$ and if $\beta \in P_T$, then $F_5(\alpha, \beta) \in P_{IV}$.
  The function $F_5(\alpha, \beta)$ is $\alpha\beta$ if β does not have the form **he**$_n$; $F_5(\alpha, \text{he}_n)$ is α **him**$_n$

- $S6$: Complex expressions of category IAV (verb-phrase adverbs).
  If $\alpha \in P_{IAV/T}$ and if $\beta \in P_T$, then $F_6(\alpha, \beta) \in P_{IAV}$; F_6 is defined in the same way as F_5.

- $S7$: Complex expressions of category IV (intransitive verbs).
  If $\alpha \in P_{IV/t}$ and if $A \in P_t$, then $F_7(\alpha, A) \in P_{IV}$.
  The function $F_7(\alpha, \beta)$ is α **that** β.

- $S8$: Complex expressions of category IV (intransitive verbs).
  If $\alpha \in P_{IV//IV}$ and if $\beta \in P_{IV}$, then $F_8(\alpha, \beta) \in P_{IV}$.
  The function $F_8(\alpha, \beta)$ is α **to** β.

- $S9$: Complex expressions of category t (sentences).
  If $\alpha \in P_{t/t}$ and if $A \in P_t$, then $F_9(\alpha, A) \in P_t$.
  The function $F_9(\alpha, A)$ is αA.

- $S10$: Complex expressions of category IV (intransitive verbs).
  If $\alpha \in P_{IV/IV}$ and if $\beta \in P_{IV}$, then $F_{10}(\alpha, \beta) \in P_{IV}$.
  The function $F_{10}(\alpha, \beta)$ is $\beta\alpha$.

- $S14$: Quantification rule.
  If $\alpha \in P_T$ and if $A \in P_t$, then $F_{14,n}(\alpha, A) \in P_t$.
  The value $F_{14,n}(\alpha, A)$ has two different definitions according to the value of α:

  - If α does not have the form **he**$_k$, then $F_{14,n}(\alpha, A)$ comes from A by replacing the first occurrence of **he**$_n$ or **him**$_n$ by α and all other occurrences of **he**$_n$ or **him**$_n$ by **he, she** or **it**, or by **him, her** or **it**, respectively, according as the gender of the first element of category B_{CN} or B_t in α is masculine, feminine or neuter.

- If α is he_k, then $F_{14,n}(\alpha, A)$ comes from A by replacing all occurrences of he_n or him_n by he_k or him_k, respectively.

3.2.4 Examples

As a first illustration, let us verify that the sequence 'every woman such that she walks talks' is an English sentence, i.e., a member of P_t. (Note that sentences formed with such that, which are characteristic of a kind of mathematical jargon, can hardly be considered as a part of colloquial English. It is for reasons of expository simplicity that Montague did not choose to include the pronouns who, whom, whose in the PTQ grammar; sentences such as 'every woman who walks talks' are therefore not elements of the language.)

The way in which a sentence is constructed syntactically can be displayed in the form of an *analysis tree* or *parse tree*. The analysis tree for 'every woman such that she walks talks' is depicted in Figure 3.2; it can be constructed by using the syntactic rules of Subsection 3.2.3 in a ' bottom up' way.

By $S1$: walk $\in P_{IV}$ and $he_5 \in P_T$
By $S4$: $F_4(he_5, walk) = he_5$ walks $\in P_t$
By $S1$: woman $\in P_{CN}$
By $S3$: $F_{3,5}(woman, he_5$ walks$) =$ woman such that she walks $\in P_{CN}$
By $S1$: every $\in P_{T/CN}$
By $S2$: $F_2(every,$ woman such that she walks$) =$ every woman such that she walks $\in P_T$
By $S1$: talk $\in P_{IV}$
By $S4$: $F_4(every$ woman such that she walks, talk$) =$ every woman such that she walks talks $\in P_t$

A second example will allow us to put into evidence that some English sentences such as 'John seeks a unicorn' can be understood in two ways. In the sequel, we shall see how different formulas of intensional logic can be associated with the different ways this sentence can be understood. There will be as many possible translations of a sentence in intensional logic as there are possible meanings of this sentence. Otherwise stated, to choose a particular translation in intensional logic is equivalent to choosing a particular meaning of an ambiguous sentence. The translation operation will thus remove the ambiguity of the natural language.

In the first reading, called *de dicto* reading, of the sentence 'John seeks a unicorn', it will be assumed that John seeks an animal that is named unicorn, but he does not necessarily know which individual the name unicorn actually denotes. This reading is sometimes described as the one in which a unicorn is 'John's description' of the animal in question. The sentence '**John seeks a unicorn**' can thus be true even in a world (the actual world) where unicorns do not exist. The *de dicto* reading of this sentence could for example be represented by the first-order predicate formula *Seek(John, unicorn)*.

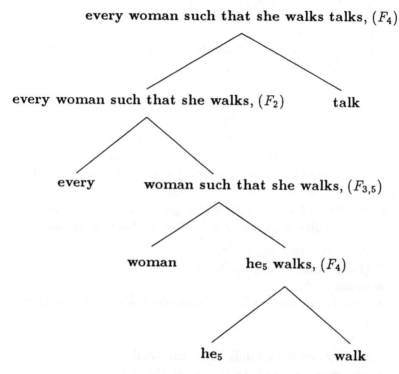

Figure 3.2 : Analysis tree for the sentence
'every woman such that she walks talks'

There is a second reading of '**John seeks a unicorn**': the *specific* or *de re* reading which means that there exists some particular but unnamed (or unreferenced) unicorn that John is looking for. On this reading, the example could describe a given animal that John has already seen and that John called **unicorn**. Thus, the *de re* reading implies the existence of a unicorn. This reading could be represented by the first-order predicate formula $\exists x (Unicorn(x) \land Seek(John, x))$, which means that there exists an individual x, of the species 'unicorn', and that John seeks this individual.

Let us now see how meanings of our sentence can be constructed by means of the syntactic rules given in the preceding subsection. First, we consider the following analysis.

3.2. MONTAGUE'S GRAMMAR

By $S1$: $\mathbf{a} \in P_{T/CN}$ and **unicorn** $\in P_{CN}$
By $S2$: $F_2(\mathbf{a}, \mathbf{unicorn}) = \mathbf{a\ unicorn} \in P_T$
By $S1$:   **seek** $\in P_{IV/T}$
By $S5$: $F_5(\mathbf{seek}, \mathbf{a\ unicorn}) = \mathbf{seek\ a\ unicorn} \in P_{IV}$
By $S1$:   **John** $\in P_T$
By $S4$: $F_4(\mathbf{John}, \mathbf{seek\ a\ unicorn}) = \mathbf{John\ seeks\ a\ unicorn} \in P_t$

This first analysis of the sentence is represented by the parse tree of Figure 3.3. We shall see in Subsection 3.2.12 that with this parse tree is associated a translation of the *de dicto* reading of the sentence.

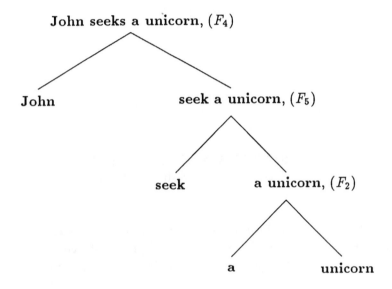

Figure 3.3 : Analysis tree for the *de dicto* reading of the sentence **'John seeks a unicorn'**

The same sentence can be analysed in another way, which corresponds to its *de re* reading, as follows.

By $S1$:   **seek** $\in P_{IV/T}$ and $\mathbf{he}_1 \in P_T$
By $S5$: $F_5(\mathbf{seek}, \mathbf{he}_1) = \mathbf{seek\ him}_1 \in P_{IV}$
By $S4$: $F_4(\mathbf{John}, \mathbf{seek\ him}_1) = \mathbf{John\ seeks\ him}_1 \in P_t$
By $S2$: $F_2(\mathbf{a}, \mathbf{unicorn}) = \mathbf{a\ unicorn} \in P_T$
By $S14$: $F_{14,1}(\mathbf{a\ unicorn}, \mathbf{John\ seeks\ him}_1) =$
           John seeks a unicorn $\in P_t$

This second analysis is represented by the parse tree of Figure 3.4.

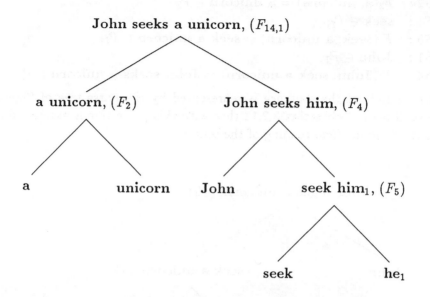

Figure 3.4 : Analysis tree fo the *de re* reading of the sentence **'John seeks a unicorn'**

Finally, we analyse a third sentence according to both *de dicto* and *de re* readings. The sentence to be analysed is **'John believes that a woman talks'**. In the *de dicto* reading, the meaning of the sentence is that 'John believes that there exists at least one woman who talks' but this belief does not refer to a particular woman, or, stated otherwise, John does not know any name for that woman. This reading can be analysed as follows.

By $S1$:   **a** $\in P_{T/CN}$ and **woman** $\in P_{CN}$
By $S2$: $F_2(\mathbf{a}, \mathbf{woman}) = \mathbf{a\ woman} \in P_T$
By $S1$:   **talk** $\in P_{IV}$
By $S4$: $F_4(\mathbf{a\ woman}, \mathbf{talk}) = \mathbf{a\ woman\ talks} \in P_t$
By $S1$:   **believe** $\in P_{IV/t}$
By $S7$: $F_7(\mathbf{believe}, \mathbf{a\ woman\ talks}) =$
          believe that a woman talks $\in P_{IV}$
By $S1$:   **John** $\in P_T$
By $S4$: $F_4(\mathbf{John}, \mathbf{believe\ that\ a\ woman\ talk}) =$
          John believes that a woman talks $\in P_t$

This first analysis is represented by the parse tree of Figure 3.5.

3.2. MONTAGUE'S GRAMMAR

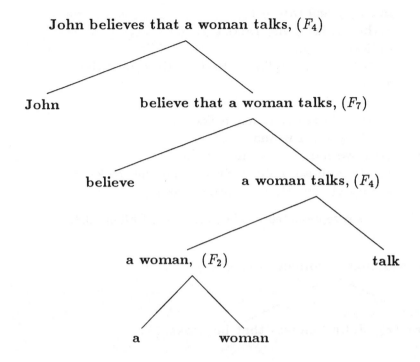

Figure 3.5 : Analysis tree for the *de dicto*
reading of the sentence '**John believes that a woman talks**'

The *de dicto* reading of our sentence can be translated into the following formula of predicate logic:

$$Believe(John, \exists x(Woman(x) \wedge Talk(x))).$$

In the *de re* reading, the meaning of the sentence is that 'John believes that there exists a well-determined woman who is talking'. The corresponding analysis is performed as follows.

By $S1$: $he_5 \in P_T$ and $\textbf{talk} \in P_{IV}$
By $S4$: $F_4(he_5, talk) = he_5$ talks $\in P_t$
By $S1$: $\textbf{believe} \in P_{IV/t}$
By $S7$: $F_6(believe, he_5\ talks) = $ believe that he_5 talks $\in P_{IV}$
By $S1$: $\textbf{John} \in P_T$
By $S4$: $F_4(John, believe\ that\ he_5\ talks) = $
         John believes that he_5 talks $\in P_t$
By $S1$: $\textbf{a} \in P_{T/CN}$ and $\textbf{woman} \in P_{CN}$
By $S2$: $F_2(a, woman) = $ a woman $\in P_T$
By $S14$: $F_{14,5}(a\ woman, John\ believes\ that\ he_5\ talks) = $
         John believes that a woman talks

This second analysis is represented by the parse tree of Figure 3.6.

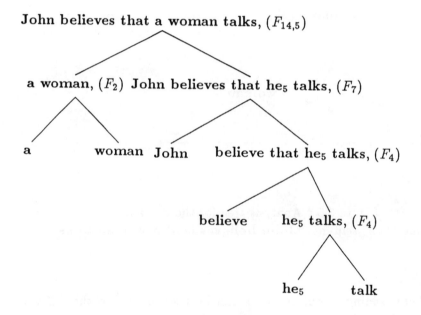

Figure 3.6 : Analysis tree fo the *de re*
reading of the sentence 'John believes that a woman talks'

The *de re* reading of our sentence can be translated into the following formula of predicate logic:

$$\exists x [Woman(x) \land Believe(John, Talk(x))].$$

3.2.5 Introduction to the translation formalism

Traditionally, translating natural language into a formal language was regarded as pertaining to art, not to a rigorous science. On the other hand, there do exist cases of rigorous translation from a formal language to another formal language, such as, for example, the shift from the decimal number system to the binary number system. Montague aims at giving the same rigour to the translation of natural language to formal language. He cannot, however, construe translation as a function in the mathematical sense, as a given expression in natural language might well correspond to several logic expressions. (The *de dicto* and *de re* readings mentioned in the preceding subsection exemplify this point.) This will be the case if the natural language expression is ambiguous. It was precisely this risk of ambiguity which led Montague to translate natural language (or the fragment he studied) into formal language (intensional logic). He wanted to define the notions of logical consequence and logical validity for expressions in natural language. This goal remains unattained as long as such expressions can be ambiguous. For example, the following inference cannot be declared valid without reservation: 'John seeks a fish, therefore John seeks a fish'. Nothing in natural language prevents us from interpreting the first occurrence of 'John seeks a fish' in the *de dicto* sense (which does not imply the existence of a fish), while interpreting the second occurrence in the *de re* sense (which does imply the existence of a fish).

The translation of Montague's fragment of English, noted $\mathcal{L}_E$, into the language of intensional logic, $\mathcal{L}_I$, is performed within several steps. We first establish a correspondence between syntactic categories of $\mathcal{L}_E$ and types of $\mathcal{L}_I$ in the form of a recursive definition of a function f which maps syntactic categories into types. The definition of f is as follows:

1. $f(t) = t$,

2. $f(e) = e$,

3. $f(\mathcal{A}/\mathcal{B}) = f(\mathcal{A}//\mathcal{B}) = \langle\langle s, f(\mathcal{B})\rangle, f(\mathcal{A})\rangle$ for all categories $\mathcal{A}$ and $\mathcal{B}$.

For example, an expression of $\mathcal{L}_E$ of syntactic category t/e as '**walk**' will be translated into an expression of $\mathcal{L}_I$ having $\langle\langle s, f(e)\rangle, f(t)\rangle = \langle\langle s, e\rangle, t\rangle$ as type; a proper noun such as '**John**', of category $t/(t/e)$, will be translated into a logic expression of type $\langle\langle s, \langle\langle s, e\rangle, t\rangle\rangle, t\rangle$.

The definition of function f introduces 'intensional types' $\langle s, a \rangle$ in cases where they are not motivated by the syntax of the natural language. Intransitive verbs and common nouns, for example, will translate into expressions denoting sets of *individual concepts* rather than the expected sets of *individuals*. In order to avoid this unnecessary complication, Bennett has proposed a way to make intransitive verbs and common nouns simply denote sets of individuals rather than sets of individual concepts without disturbing the rest of the recursively defined correspondence between syntactic categories and types ([Bennett 76], see also [Dowty et al. 81], p.188). The basic syntactic categories in Bennett's definition are t, IV and CN; any combination $\mathcal{A}/\mathcal{B}$ formed recursively out of these is also a syntactic category. The correspondence between syntactic categories and logical types is now stated from the modified function f defined as follows.

1. $f(t) = t$,

2. $f(CN) = f(IV) = \langle e, t \rangle$,

3. $f(\mathcal{A}/\mathcal{B}) = f(\mathcal{A}//\mathcal{B}) = \langle \langle s, f(\mathcal{B}) \rangle, f(\mathcal{A}) \rangle$ for all categories $\mathcal{A}$ and $\mathcal{B}$.

The correspondence between syntactic categories and types, defined by means of the Bennett function f, is indicated in the table of Figure 3.7, borrowed from [Dowty et al. 81]. We observe that this definition — that will be adopted henceforth — will allow us to avoid the introduction of intensional types where they are unnecessary.

3.2.6 Translation of basic expressions

After having correlated classes of expressions of natural language (Montague's fragment of English) with classes of expressions of formal language (intensional logic) we have to explain how the expressions themselves are translated from the former language into the latter.

We begin by defining a translation function, noted ', that maps each basic expression of the natural language $\mathcal{L}_E$ to a non-logical constant (of the appropriate logical type) of the intensional logic language $\mathcal{L}_I$. For example, the basic expressions **man** and **walk** are translated into constants of $\mathcal{L}_I$, of type $\langle e, t \rangle$, as follows:

    man   is translated into   **man**',
   walk is translated into   **walk**'.

More generally, if α is a basic expression of syntactic category $\mathcal{A}$ in the language $\mathcal{L}_E$, then the translation of α will be noted α'; by definition, α' is a non-logical constant of type $f(\mathcal{A})$ in the language $\mathcal{L}_I$.

The translation rule that we have just described is defined for all the basic expressions except for the proper nouns **John** and **Mary**. These proper nouns belong to the syntactic category $T = t/IV$. They must thus translate into some expressions of type $f(t/IV) = \langle \langle s, f(IV) \rangle, f(t) \rangle = \langle \langle s, \langle e, t \rangle \rangle, t \rangle$.

3.2. Montague's grammar

Syntactic category	Type	Denotation of this type
t	t	Truth value
CN	$\langle e,t \rangle$	Set of individuals
IV	$\langle e,t \rangle$	Set of individuals
T	$\langle\langle s,\langle e,t\rangle\rangle,t\rangle$	Set of properties of individuals
IAV	$\langle\langle s,\langle e,t\rangle\rangle,\langle e,t\rangle\rangle$	Function from properties of individuals to sets of individuals
TV	$\langle\langle s,\langle\langle s,\langle e,t\rangle\rangle, t\rangle\rangle, \langle e,t\rangle\rangle$	Function from properties of properties of individuals to sets of individuals
T/CN	$\langle\langle s,\langle e,t\rangle\rangle,\langle\langle s, \langle e,t\rangle\rangle,t\rangle\rangle$	Function from properties of individuals to sets of properties of individuals
t/t	$\langle\langle s,t\rangle,t\rangle$	Set of propositions
IV/t	$\langle\langle s,t\rangle,\langle e,t\rangle\rangle$	Function from propositions to sets of individuals
$IV//IV$	$\langle\langle s,\langle e,t\rangle\rangle,\langle e,t\rangle\rangle$	Function from properties of individuals to sets of individuals
IAV/T	$\langle\langle s,\langle\langle s,\langle e,t\rangle\rangle, t\rangle\rangle,\langle\langle s,\langle e,t\rangle\rangle, \langle e,t\rangle\rangle\rangle$	Function from properties of properties of individuals to functions from properties of individuals to sets of individuals

Figure 3.7 : Correspondence between syntactic categories and logical types

(Reprinted by permission of D. Reidel publishing company; copyright 1981.)

Expressions of this type denote sets of properties of individuals (see Figure 3.7). Thus **John** will translate into some expression denoting a set of properties. Let P be the representation in the $\mathcal{L}_I$ language of a predicate variable of type $\langle s, \langle e,t\rangle\rangle$ which ranges over properties of individuals, let j and m be the representations of non-logical constants of type e which denote the persons **John** and **Mary** respectively, and let x_0, x_1, etc., be individual variables. The translations of the proper nouns and of the subscripted pronouns he$_n$ are defined as follows:

John   is translated into $\lambda P[{}^\vee P(j)]$,
Mary   is translated into $\lambda P[{}^\vee P(m)]$,
he$_n$ is translated into $\lambda P[{}^\vee P(x_n)]$.

Observe that we must have ${}^\vee P(.)$, and not simply $P(.)$, in these expressions. Indeed, P denotes a *property*, i.e., a function from indices to functions from individuals to truth values. In order that this function could take individuals as arguments, we have to transform it into a function from individuals to truth values, i.e., into a function that is represented by ${}^\vee P$. Therefore, it is the expression ${}^\vee P$ we have to combine with j, m or x_n, and not simply the expression P.

In summary, **John** is an element of the syntactic category T and must have an expression of type $\langle\langle s, \langle e, t\rangle\rangle, t\rangle$ as translation. This expression is built up by means of symbols $P, {}^\vee, j$ and λ according to the following scheme.

Formal expression	Associated type
P	$\langle s, \langle e, t\rangle\rangle$
${}^\vee P$	$\langle e, t\rangle$
j	e
${}^\vee P(j)$	t
$\lambda P[{}^\vee P(j)]$	$\langle\langle s, \langle e, t\rangle\rangle, t\rangle$

3.2.7 Translation of the rules of functional application

We have listed in Subsection 3.2.3 a collection of syntactic *rules of functional application*. These are the rules $S4$–$S10$; they are characterized by the following general form:

Sj : If $\alpha \in P_{A/B}$ and if $\beta \in P_B$, then $F_j(\alpha, \beta) \in P_A$.

Each syntactic functional application rule Sj has associated with it a translation rule Tj defined as follows:

Tj : If $\alpha \in P_{A/B}$ and if $\beta \in P_B$, and if α and β translate into α' and β', respectively, then $F_j(\alpha, \beta)$ translates into $\alpha'({}^\wedge\beta')$.

The translation rule for a syntactic rule of functional application is always of the same form: it states that we have to take the translation α' of the expression of syntactic category A/B and combine it with the translation β' of the expression of syntactic category B, prefixed with the intensional operator $^\wedge$, to give $\alpha'({}^\wedge\beta')$. Note that the basic syntactic rule $S2$ also has the form of a functional application rule; hence, it has associated with it a translation rule $T2$ of the form above. Observe again that the translation rules give an expression of the type $\alpha'({}^\wedge\beta')$ rather than $\alpha'(\beta')$. Indeed, the translation in

3.2. Montague's grammar

the intensional logic language of an expression of syntactic category $\mathcal{A}/\mathcal{B}$ has $\langle\langle s, f(\mathcal{B})\rangle, f(\mathcal{A})\rangle$ as type. For this reason, we have to combine an expression of category $\mathcal{A}/\mathcal{B}$ with an expression whose translation has $\langle s, f(\mathcal{B})\rangle$ as type, i.e., with an expression that denotes the intension of an expression of syntactic category $\mathcal{B}$.

Let us consider an elementary example.

- We have **Mary** $\in B_T$ and thus, by $S1$, **Mary** $\in P_T$; **talk** $\in B_{IV}$ and thus, by $S1$, **talk** $\in P_{IV}$; hence, by $S4$, **Mary talks** $\in P_t$.

- The translation of **Mary** is $\lambda P[{}^\vee P(m)]$; the translation of **talk** is $\mathbf{talk}'$; the translation of **Mary talks** is $\lambda P[{}^\vee P(m)](^\wedge\mathbf{talk}')$.

- Since $^\wedge\mathbf{talk}'$ is an expression of type $\langle s, \langle e, t\rangle\rangle$ and $\lambda P[{}^\vee P(m)]$ is an expression of type $\langle\langle s, \langle e, t\rangle\rangle, t\rangle$, it is seen that $\lambda P[{}^\vee P(m)](^\wedge\mathbf{talk}')$ is an expression of type t, i.e., a formula of intensional logic. This formula can be simplified, in two steps, as follows:

$\lambda P[{}^\vee P(m)](^\wedge\mathbf{talk}') \iff {}^{\vee\wedge}\mathbf{talk}'(m)$.

${}^{\vee\wedge}\mathbf{talk}'(m) \iff \mathbf{talk}'(m)$.

The first simplification step uses the λ-conversion principle (§ 2.2.9), and the second simplification step uses the ${}^{\vee\wedge}$ cancellation rule (§ 2.3.6). The original translation, $\lambda P[{}^\vee P(m)](^\wedge\mathbf{talk}')$, is a formula of intensional logic. The final translation, $\mathbf{talk}'(m)$, is an equivalent formula of first-order predicate logic.

The proper nouns, such as **Mary** and **John** and the intransitive verbs, such as **walk** and **talk**, belong to the *extensional part of the fragment of English* we consider[1]. For this reason, one could normally expect that the sentence **Mary talks** could be translated into a formula of the predicate logic language. This is indeed the case: the process of performing simplifications by appeal to such principles as lambda conversion, simplification of ${}^{\vee\wedge}\alpha$ to α, and other such rules allow us to transfer intensional logic translations of extensional sentences into equivalent predicate logic formulas. We shall see that the same is true even of more complicated sentences we will encounter later.

3.2.8 Adequacy criterion for a logic of natural language

There is such a thing as a logic immanent to natural language, i.e., a set of relations of logical consequence inherent in natural language. The task

---

[1]By '*extensional part of the fragment of English*' we mean the collection of sentences that do not appear in non-referential constructions of English and whose translation can therefore be an extensional logic formula. These sentences are called *extensional sentences*; the other sentences of the fragment of English are called *intensional sentences*.

facing the semanticist or the theoretician dealing with what has sometimes been called 'natural logic', consists in constructing a theory that accounts for the truth conditions of declarative sentences and for the relations of logical consequence existing between them. The minimal requirement imposed upon a semantics and a logic of natural language can be spelled out in the following terms: one has to formulate a finite number of rules which make it possible to assign truth conditions to all the sentences of as large as possible a fragment of natural language, and to define the relations of logical consequence on this fragment.

This is the minimal requirement of descriptive adequacy that a semantics and a logic of natural language must fulfil. Keenan and Faltz added a new criterion which enables us to understand the motivation behind Montague's undertaking and to assess what he has achieved. This criterion reads as follows: *'Given two logics for English of the same or comparable descriptive adequacy, prefer that one which yields a better correspondence between the elements (constituents) of the logical structures and the elements of the surface forms they adequately represent the logical properties of'*. More simply, one should prefer a logic which — the descriptive adequacy requirements being satisfied — departs as little as possible from the surface form of the sentences belonging to the fragment of English which the logic is meant to represent [Keenan and Faltz 85]. Ideally, the parse trees of natural language sentences and of their translations in logic should be isomorphic.

Montague's treatment of quantified expressions such as 'a horse', 'all unicorns', etc. displays his concern to comply with the two precepts we have just formulated. As is well known, classical formal logic, such as predicate calculus, bestows a status on proper nouns which is different from that bestowed on common nouns prefixed by an indefinite or definite determiner. The former are constructed as individual constants, the latter as quantified expressions. The proper noun 'John' is represented in logic by an individual constant whereas 'all men [are]' is represented by a formula such as $\forall x Man(x)$, in which the individual variable x is bound by a universal quantifier $\forall$. One is faced here with a discrepancy between the grammatical form and the logical form.

Logicians base this distinction on the fact that valid inferences vary considerably depending on whether we are dealing with a proper noun or with a common noun attached to an indefinite determiner. Let us consider the following example. From the fact that the two sentences 'John knows Hillary' and 'Hillary is the conqueror of Everest' are true, it follows that the sentence 'John knows the conqueror of Everest' is true. However, from 'John knows some man' and 'some man is the conqueror of Everest' it does not follow 'John knows the conqueror of Everest'.

Montague should be given the credit for showing that logic can be reconciled

3.2. MONTAGUE'S GRAMMAR

with grammar, i.e., that the gap between logical form and grammatical form can be bridged without renouncing the results of formal logic and without jeopardizing the descriptive adequacy of logic which is required to account for the difference between valid and invalid inferences. In this respect, let us especially emphasize the fact that, in Montague's grammar, common nouns prefixed by determiners belong to the same syntactic category as proper nouns.

3.2.9 Translation rules for determiners

We have seen in Subsection 3.2.6 how Montague translates some of the basic expressions of natural language into non-logical constants of formal language. Let us now see how the innovation mentioned in the preceding subsection enters the scene at this very point. It is concerned in particular with the assignment of a logic translation to the indefinite determiner **a** that allows us to classify, along with grammarians, '**a man**' as an expression of the same syntactic category as '**John**', and to recognize, along with logicians, that the quantifiers have inference capabilities different from those of the proper nouns. In this respect, the determiners will not be translated into constants of the intensional logic language but will be treated as kinds of logical words to which complex logical expressions are assigned as translation. Let Y and Z be two variables of type $\langle s, \langle e, t \rangle \rangle$. The determiners **every**, **a** and **the** are translated as follows:

the	is translated into	$\lambda Z[\lambda Y \exists y[\forall x[\check{}Z(x) \equiv x = y] \wedge \check{}Y(y)]]$,
every	is translated into	$\lambda Z[\lambda Y \forall x[\check{}Z(x) \supset \check{}Y(x)]]$,
a	is translated into	$\lambda Z[\lambda Y \exists x[\check{}Z(x) \wedge \check{}Y(x)]]$.

(These translations were already suggested at the end of Subsection 2.2.10 but differ slightly here in that they take into account intensions in the appropriate places.)

Though these expressions are complex, a careful reading can confirm that they actually are intensional logic formulas of the appropriate type to serve as translations of English expressions of category T/CN, which is the syntactic category of determiners. We have indeed:

$$\begin{aligned}
f(T/CN) &= f((t/IV)/CN), \\
&= \langle \langle s, f(CN) \rangle, f(t/IV) \rangle, \\
&= \langle \langle s, \langle e, t \rangle \rangle, \langle \langle s, f(IV) \rangle, f(t) \rangle \rangle, \\
&= \langle \langle s, \langle e, t \rangle \rangle, \langle \langle s, \langle e, t \rangle \rangle, t \rangle \rangle.
\end{aligned}$$

Let us now verify that the proposed translations are of this type. The expression $\forall x[\check{}Z(x) \supset \check{}Y(x)]$ is obviously a formula, i.e., an expression of type t. Since Y is a variable of type $\langle s, \langle e, t \rangle \rangle$, the expression $\lambda Y \forall x[\check{}Z(x) \supset \check{}Y(x)]$ is of type $\langle \langle s, \langle e, t \rangle \rangle, t \rangle$. Since Z is a variable of the same type as Y, the expression $\lambda Z[\lambda Y \forall x[\check{}Z(x) \supset \check{}Y(x)]]$ is of type $\langle \langle s, \langle e, t \rangle \rangle, \langle \langle s, \langle e, t \rangle \rangle, t \rangle \rangle$, i.e., of type

$f(T/CN)$. A similar reasoning can be made for the other two translations.

For computing the translation of complex sentences that make use of the translations of the determiners **the**, **every**, and **a**, we have first to complete the statement of the translation rules associated with the syntactic rules that were defined in Subsection 3.2.3. The translations of the functional application rules $S4$–$S10$ and of the basic rule $S2$ have already been given in Subsection 3.2.7. In the next subsection, we define the translation rules that are associated with the remaining syntactic rules, $S3$ and $S14$.

3.2.10 Additional syntactic and translation rules

We first define the translation rules $T3$ and $T14$ associated with the syntactic rules $S3$ and $S14$ respectively.

$T3$: If $\alpha \in P_{CN}$, if $A \in P_t$, and if α and A translate into α' and A' respectively, then $F_{3,n}(\alpha, A)$ translates into $\lambda x_n[\alpha'(x_n) \wedge A']$.

$T14$: If $\alpha \in P_T$, if $A \in P_t$, and if α and A translate into α' and A' respectively, then $F_{14,n}(\alpha, A)$ translates into $\alpha'(^\wedge \lambda x_n A')$.

Next, we examine the translation of the sentence adverb **necessarily**. We have seen in Subsection 3.2.2 that this adverb belongs to the syntactic category t/t (Figure 3.1); it combines with a sentence to form a new sentence. We have also seen in Subsection 2.3.11 that, in order to comply with Frege's compositionality principle in the interpretation of the sentence 'necessarily A' (where A represents a sentence of natural language), we have to apply the translation of **necessarily** to an expression that denotes the intension of the translation A' of A rather than to the translation A' itself. In order to conform to the logic tradition, the adverb **necessarily** does not translate into a non-logical constant (as alluded to in Subsection 2.3.12) but is given in terms of the modal operator $\Box$. Let p be a variable that represents a formula prefixed by the intensional operator $^\wedge$; thus p is a variable of type $\langle s, t \rangle$ and represents a proposition. Accordingly, the translation of **necessarily** is as follows:

necessarily is translated into $\lambda p[\Box ^{\vee} p]$.

Summarizing, **necessarily**, which is an element of category t/t, must have as translation an expression of type $f(t/t) = \langle \langle s, t \rangle, t \rangle$; this expression is constructed in terms of the symbols $p, ^{\vee}, \Box$ and λ according to the following scheme.

Formal expression	Associated type
p	$\langle s, t \rangle$
$^{\vee}p$	t
$\Box ^{\vee}p$	t
$\lambda p[\Box ^{\vee}p]$	$\langle \langle s, t \rangle, t \rangle$

3.2. Montague's grammar

We finally give the remaining syntactic rules of the PTQ grammar, together with their associated translation rules. Contrary the rules of Subsection 3.2.3, these rules will not be illustrated by way of examples.

- $S11a$: If $A \in P_t$ and if $B \in P_t$, then $F_{11a}(A, B) \in P_t$.
  The function $F_{11a}(A, B)$ is A **and** B.

- $T11a$: If A and B translate into A' and B' respectively, then $F_{11a}(A, B)$ translates into $(A' \wedge B')$.

- $S11b$: If $A \in P_t$ and if $B \in P_t$, then $F_{11b}(A, B) \in P_t$.
  The function $F_{11b}(A, B)$ translates into A **or** B.

- $T11b$: If A and B translate into A' and B' respectively, then $F_{11b}(A, B)$ translates into $(A' \vee B')$.

- $S12a$: If $\alpha \in P_{IV}$ and if $\beta \in P_{IV}$, then $F_{12a}(\alpha, \beta) \in P_{IV}$.
  The function F_{12} is defined in the same way as F_{11a}.

- $T12a$: If α and β translate into α' and β' respectively, then $F_{12a}(\alpha, \beta)$ translates into $\lambda x[\alpha'(x) \wedge \beta'(x)]$.

- $S12b$: If $\alpha \in P_{IV}$ and if $\beta \in P_{IV}$, then $F_{12b}(\alpha, \beta) \in P_{IV}$.
  The function F_{12b} is defined in the same way as F_{11b}.

- $T12b$: If α and β translate into α' and β' respectively, then $F_{12b}(\alpha, \beta)$, i.e., α **or** β, translates into $\lambda x[\alpha'(x) \vee \beta'(x)]$.

- $S13$: If $\alpha \in P_T$ and if $\beta \in P_T$, then $F_{13b}(\alpha, \beta) \in P_T$.
  The function F_{13} is defined in the same way as F_{11b}.

- $T13$: If α' and β' translate into α' and β' respectively, then $F_{13}(\alpha, \beta)$ translates into $\lambda P[\alpha'(P) \vee \beta'(P)]$, with P a variable of type $\langle s, \langle e, t \rangle \rangle$.

- $S15$: If $\alpha \in P_T$ and if $\beta \in P_{CN}$, then $F_{15,n}(\alpha, \beta) \in P_{CN}$.
  The function $F_{15,n}$ is defined in the same way as $F_{14,n}$.

- $T15$: If α and β translate into α' and β' respectively, then $F_{15,n}(\alpha, \beta)$ translates into $\lambda y \alpha'(\hat{}\lambda x_n[\beta'(y)])$.

- $S16$: If $\alpha \in P_T$ and if $\beta \in P_{IV}$, then $F_{16,n}(\alpha, \beta) \in P_{IV}$.
  The function $F_{16,n}$ is defined in the same way as $F_{14,n}$.

- $T16$: If α and β translate into α' and β' respectively, then $F_{16,n}(\alpha, \beta)$ translates into $\lambda y \alpha'(\hat{}\lambda x_n[\beta'(y)])$.

- $S17$: If $\alpha \in P_T$ and if $\beta \in P_{IV}$, then $F_{17a}(\alpha,\beta), F_{17b}(\alpha,\beta), F_{17c}(\alpha,\beta)$, $F_{17d}(\alpha,\beta)$ and $F_{17e}(\alpha,\beta) \in P_t$;
  $F_{17a}(\alpha,\beta)$ is $\alpha\beta^a$ where β^a is the result of replacing the first verb in β by its negative third person singular present;
  $F_{17b}(\alpha,\beta)$ is $\alpha\beta^b$ where β^b is the result of replacing the first verb in β by its third person singular future;
  $F_{17c}(\alpha,\beta)$ is $\alpha\beta^c$, where β^c is the result of replacing the first verb in β by its negative third person singular future;
  $F_{17d}(\alpha,\beta)$ is $\alpha\beta^d$, where β^d is the result of replacing the first verb in β by its third person singular perfect;
  $F_{17e}(\alpha,\beta)$ is $\alpha\beta^e$ where β^e is the result of replacing the first verb in β by its negative third person singular present perfect.

- $T17$: If α and β translate into α' and β' respectively, then
  $F_{17a}(\alpha,\beta)$ translates into $\neg \alpha'(^\wedge\beta')$;
  $F_{17b}(\alpha,\beta)$ translates into $<F> \alpha'(^\wedge\beta')$;
  $F_{17c}(\alpha,\beta)$ translates into $\neg <F> \alpha'(^\wedge\beta')$;
  $F_{17d}(\alpha,\beta)$ translates into $<P> \alpha'(^\wedge\beta')$;
  $F_{17e}(\alpha,\beta)$ translates into $\neg <P> \alpha'(^\wedge\beta')$.

3.2.11 Examples of extensional sentence translations

We first consider a simple example that illustrates an advantage of intensional logic over predicate logic. By applying the translation rules that were defined in the preceding subsections, we are able to represent the sentence 'a woman talks' of the language $\mathcal{L}_E$ by the intensional logic formula

$$\lambda Z[\lambda Y \exists x[^\vee Z(x) \wedge {^\vee Y}(x)]](^\wedge\text{woman}')(^\wedge\text{talk}').$$

This formula of $\mathcal{L}_I$ has the same syntactic tree as the sentence of $\mathcal{L}_E$ that it represents (see Figure 3.8). On the other hand, it could easily be verified that the predicate logic formula $\exists x(Woman(x) \wedge Talk(x))$ that represents the same sentence has a very different syntactic tree. It has been shown that this rather complex formula of intensional logic can be converted to the simpler predicate logic formula by applying principles such as lambda conversion and $^\vee{^\wedge}$ cancellation (see Subsection 3.2.7). It turns out that all translations of English sentences into intensional logic yield rather complex formulas that can be transformed to simpler logically equivalent formulas by appeal to these principles. Let us emphasize that the intensional logic formula has exactly the same model-theoretic interpretation as the corresponding predicate logic formula and thus represents the meaning of, for example, the English sentence 'a woman talks' in exactly the same way.

3.2. MONTAGUE'S GRAMMAR

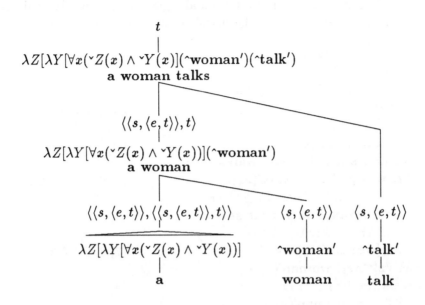

Figure 3.8 : Syntactic tree

Summarizing, we see that the intensional logic translation does not present any drawback with respect to the predicate logic translation and has a significant advantage over it: instead of describing natural language by means of an ill suited formalism, we translate it into a formalism that satisfies the adequation criterion mentioned in Subsection 3.2.8.

Let us now consider the translation of the more complicated sentences that were seen in Subsection 3.2.4. We first translate the sentence '**every woman such that she walks talks**'. In writing out translations and their simplifications from this point on, we will adopt two conventions from [Partee 75] which have since been used by other authors [Dowty et al. 81]. First, the steps of the translation will be written out as a 'proof', listing expressions of intensional logic by number in the left-hand column and giving in the corresponding line of the right-hand column the justification of the translation. We also often indicate the subsection where the justification has been defined. For example, (′; § 3.2.6) at the end of a line indicates that the translation in this line makes use of the translation function ′ that has been defined in Subsection 3.2.6. The principle of lambda conversion is simply indicated by λ while the ˇ^ cancellation rule is indicated by ˇ^. The second convention is the use of the symbol $\Rightarrow$ for 'translates into'. Thus anything written to the left of this symbol is an expression of $\mathcal{L}_E$, anything to the right is an expression of $\mathcal{L}_I$. Any line which does not contain the symbol ⇒ is understood to be a simplification of the line just above it, i.e., an expression of intensional logic that is logically equivalent to the preceding line according to the justification named on the right.

1. $he_5 \Rightarrow \lambda P[{}^\vee P(x_5)]$ (§ 3.2.6)
2. $walk \Rightarrow walk'$ ($'$; § 3.2.6)
3. $he_5\ walks \Rightarrow \lambda P[{}^\vee P(x_5)]({}^\wedge walk')$ (1, 2, T_4; § 3.2.7)
4. ${}^\vee{}^\wedge walk'(x_5)$ (λ; § 2.2.9)
5. $walk'(x_5)$ (${}^\vee{}^\wedge$; § 2.3.6)
6. $woman \Rightarrow woman'$ ($'$)
7. $woman\ such\ that\ she\ walks \Rightarrow$
   $\lambda x_5[woman'(x_5) \wedge walk'(x_5)]$ (5, 6, T_3; § 3.2.10)
8. $every \Rightarrow \lambda Z \lambda Y \forall x[{}^\vee Z(x) \supset {}^\vee Y(x)]$ (§ 3.2.9)
9. $every\ woman\ such\ that\ she\ walks$
   $\Rightarrow \lambda Z \lambda Y \forall x[{}^\vee Z(x) \supset {}^\vee Y(x)]$
   $({}^\wedge \lambda x_5[woman'(x_5) \wedge walk'(x_5)])$ (7, 8, T_2; § 3.2.10)
10. $\lambda Y \forall x[{}^\vee{}^\wedge \lambda x_5[\ woman'(x_5) \wedge]]$
    $walk'(x_5)](x) \supset {}^\vee Y(x)]$ (λ)
11. $\lambda Y \forall x[\lambda x_5[woman'(x_5) \wedge$
    $walk'(x_5)](x) \supset {}^\vee Y(x)]$ (${}^\vee{}^\wedge$)
12. $\lambda Y \forall x[(woman'(x) \wedge walk'$
    $(x)) \supset {}^\vee Y(x)]$ (λ)
13. $talk \Rightarrow talk'$ ($'$)
14. $every\ woman\ such\ that\ she\ walks$
    $talks \Rightarrow \lambda Y \forall x[(woman'(x) \wedge walk'(x)) \supset$
    ${}^\vee Y(x)]({}^\wedge talk')$ (13, 14, T_4)
15. $\forall x[(woman'(x)) \supset talk'(x)]$ ($\lambda, {}^\vee{}^\wedge$)

3.2.12 Examples of intensional sentence translations

We have seen in Subsection 3.2.4 that the sentence 'John seeks a unicorn' can be understood in two different ways. We can first give this sentence its *de dicto* meaning, which does not imply the existence of a unicorn. The translation is based on the syntactic analysis that has been represented by the parse tree of Figure 3.3. Let us examine this analysis in some detail. In the translation process we use the following rewriting principle:

$$P(x)(y) \rightarrow P(y, x),$$

where P is a predicate name. This principle, which we shall call P *transformation* means that the binary predicate $P(y, x)$ is interpreted as the unary predicate $P(x)$ with the argument y. When the P transformation principle

3.2. Montague's grammar

is used in the translation process, we write simply (P) in the corresponding justification column.

1. seek $\Rightarrow$ seek' $\qquad$ (′; § 3.2.6)
2. a $\Rightarrow \lambda Z[\lambda Y \exists x[^\vee Z(x) \wedge {^\vee Y}(x)]]$ $\qquad$ (§ 3.2.9)
3. unicorn $\Rightarrow$ unicorn' $\qquad$ (′)
4. a unicorn $\Rightarrow$
   $\lambda Z[\lambda Y \exists x[^\vee Z(x) \wedge {^\vee Y}(x)]](^\wedge\text{unicorn}')$ $\qquad$ $(2, 3, T_2)$
5. $\lambda Y \exists x\,[\text{unicorn}'(x) \wedge {^\vee Y}(x)]$ $\qquad$ $(\lambda, {^{\vee\wedge}})$
6. seek a unicorn $\Rightarrow$
   seek'$(^\wedge\lambda \exists x\,[\text{unicorn}'(x) \wedge {^\vee Y}(x)])$ $\qquad$ $(1, 5, T_5)$
7. John $\Rightarrow \lambda P[^\vee P(j)]$ $\qquad$ (§ 3.2.6)
8. John seeks a unicorn $\Rightarrow$
   $\lambda P[^\vee P(j)]\,(^\wedge\text{seek}'(^\wedge\lambda Y \exists x\,$
   $[\text{unicorn}'(x) \wedge {^\vee Y}(x)]))$ $\qquad$ $(6, 7, T_4;\ §\ 3.2.7)$
9. seek'$(^\wedge\lambda Y \exists x[\text{unicorn}'(x) \wedge {^\vee Y}(x)])(j)$ $\qquad$ $(\lambda, {^{\vee\wedge}})$
10. seek'$(j, ^\wedge\lambda Y \exists x[\text{unicorn}'(x) \wedge {^\vee Y}(x)])$ $\qquad$ (P)

As an alternative, we can give the same sentence its *de re* meaning, which implies the existence of a unicorn. The translation is based on the syntactic analysis that has been represented by the parse tree of Figure 3.4.

1. he$_1 \Rightarrow \lambda P[^\vee P(x_1)]$ $\qquad$ (§ 3.2.6)
2. seek $\Rightarrow$ seek' $\qquad$ (′)
3. seek him$_1 \Rightarrow$ seek'$(^\wedge\lambda P[^\vee P(x_1)])$ $\qquad$ $(1, 2, T_5)$
4. John $\Rightarrow \lambda P[^\vee P(j)]$
5. John seeks him$_1 \Rightarrow$
   $\lambda P[^\vee P(j)](^\wedge\text{seek}'(^\wedge\lambda P[^\vee P x_1]))$ $\qquad$ $(3, 4, T_4;\ §\ 3.2.7)$
6. seek'$(^\wedge\lambda P[^\vee P(x_1)])(j)$ $\qquad$ $(\lambda, {^{\vee\wedge}})$
7. seek'$(j, ^\wedge\lambda P[^\vee P(x_1)])$ $\qquad$ (P)
8. a unicorn $\Rightarrow$
   $\lambda Y \exists x[\text{unicorn}'(x) \wedge {^\vee Y}(x)]$ $\qquad$ (preceding example)
9. John seeks a unicorn $\Rightarrow$
   $\lambda Y \exists x[\text{unicorn}'(x) \wedge {^\vee Y}(x)](^\wedge\lambda x_1$
   $[\text{seek}'(j, ^\wedge\lambda P[^\vee P(x_1)])])$ $\qquad$ $(7, 8, T14;\ §\ 3.2.10)$
10. $\exists x[\text{unicorn}'(x) \wedge \lambda x_1[\text{seek}'(j,$
    $^\wedge\lambda P[^\vee P(x_1)](x)]$ $\qquad$ $(\lambda, {^{\vee\wedge}})$
11. $\exists x[\text{unicorn}'(x) \wedge \text{seek}'(j,$
    $^\wedge\lambda P[^\vee P(x)])]$ $\qquad$ (λ)

Summarizing, we have obtained two different translations for the sentence 'John seeks a unicorn'.

Translation of the *de dicto* meaning: seek'$(j, ^\wedge\lambda Y \exists x[\text{unicorn}'(x) \wedge {^\vee Y}(x)])$.

Translation of the *de re* meaning: $\exists x[\text{unicorn}'(x) \wedge \text{seek}'(j, ^\wedge\lambda P[^\vee P(x)])]$.

In the *de dicto* translation we see that what John seeks is prefixed by the intensional operator ⌢: there exists something x that has the property to be a unicorn and that belongs to the domain of 'intensional search' of John. On the contrary, in the *de re* translation, both the existential quantifier that binds the variable x and the expression **unicorn**$'(x)$ are in an 'extensional position', which means that a unicorn actually exists; hence the *de re* translation must be equivalent to a first-order formula.

In order to convince ourselves that such is actually the case, we shall perform a series of transformations on the intensional logic formula. Two successive applications of the principle of lambda conversion produce the formulas

$$\exists x[\mathbf{unicorn}'(x) \wedge [\lambda z\, [\mathbf{seek}'(^\wedge \lambda P[^\vee P(x)])(z)](j)]],$$
$$\exists x[\mathbf{unicorn}'(x) \wedge [\lambda y[\lambda z\, \mathbf{seek}'(^\wedge \lambda P[^\vee P(y)])(z)]](x)(j)]].$$

Given a predicate constant δ that belongs to the set $E_{f(TV)}$ of expressions of type $f(TV)$, we define the λ-expression δ_* as follows:

$$\delta_* =_{def} \lambda y \lambda x [\delta(^\wedge \lambda P[^\vee P(y)])(x)].$$

Montague has shown that δ_* can be interpreted in first-order logic as a predicate constant (see also [Dowty et al. 81], p. 223). This allows us to obtain the equivalent first-order formula

$$\exists x[\mathbf{unicorn}'(x) \wedge \mathbf{seek}'_*(x)(j)]],$$

and finally, by the P transformation principle, the equivalent formula

$$\exists x[\mathbf{unicorn}'(x) \wedge [\mathbf{seek}'_*(j, x)]].$$

Thus, we have written the *de re* translation of the sentence '**John seeks a unicorn**' as a first-order logic formula.

3.2.13 Meaning postulates

The categorial grammar allows us to generate two different structural descriptions of each of the following two sentences:

(1) John seeks a unicorn,

(2) John finds a unicorn.

The *de dicto* readings of these sentences are

(1a) John seeks a (hypothetical) unicorn,

(2a) John finds a (hypothetical) unicorn.

The *de re* readings of these sentences are:

3.2. Montague's Grammar

(1b) There exists a unicorn and John searches it,

(2b) There exists a unicorn and John finds it.

We have seen in Subsection 3.2.4 that to the two different meanings of the intensional verb **seek** correspond two different syntactic analyses and hence also two different translations (§ 3.2.12). Now, for the case of extensional verbs such as **find**, the grammar will also produce two different syntactic analyses, and extensional sentences such as **John finds a unicorn** will have the same two patterns of translations as the intensional sentence **John seeks a unicorn**. But unlike the case of **seek**, there seems to be no ambiguity in the case of **find**: if John finds a unicorn it cannot be a hypothetical unicorn, but on the contrary, a unicorn that actually exists. Consequently, the two translations of the preceding subsection, but where the verb **seek** is replaced by the verb **find**, must be logically equivalent. Thus, the formula

$$\text{find}'(j, \char`\^\lambda Y \exists x[\text{unicorn}'(x) \land \check{\ } Y(x)])$$

must have the same interpretation as the formula

$$\exists x[\text{unicorn}'(x) \land \text{find}'(j, \char`\^\lambda P[\check{\ } P(x)])].$$

There is no straightforward way to restrict the PTQ grammar so that the syntactic tree which leads to the *de dicto* reading cannot be produced. Instead, Montague restricted the semantic interpretation of **find**$'$ and of the constants that translate the other extensional verbs in such a way that the formulas that translate *de dicto* readings will always be interpreted in exactly the same way as the corresponding formulas that translate *de re* readings. To capture this restriction formally, Montague used the device of so-called *meaning postulates* (see also [Dowty et al. 81], p. 224). This device, introduced by Carnap, is best thought of as a kind of constraint on possible models. If *Bachelor* represents the predicate 'is a bachelor' and if *Married* represents the predicate 'is married', then Carnap's example of meaning postulate is the formula

$$\forall x[Bachelor(x) \supset \neg Married(x)].$$

The intent of this meaning postulate is that in considering possible models for the formal language, we must restrict ourselves to models in which this formula is true.

Let us now return to the case of extensional verbs such as **find**. Unlike the case of bachelor and married, the meaning postulate we want here will not relate the meanings of two (or several) words but will restrict the meaning of an individual word in order that the different translations of a sentence in which this word occurs be equivalent. To extend such a type of requirement to all two-place predicate constants δ that translate extensional transitive verbs and to all individuals x that can be first arguments of these predicate constants, Montague introduced the following meaning postulate:

$$MP1 : \forall x \forall \mathcal{P} \square [\delta(x, \mathcal{P}) \equiv {}^\vee\mathcal{P}({}^\wedge\lambda y[\delta_*(x, y)])],$$

where δ translates **find**, **love**, **lose**, etc., δ_* is the expression defined in the preceding subsection, and $\mathcal{P}$ is a variable of type $\langle s, \langle\langle s, \langle e, t\rangle\rangle, t\rangle\rangle$. (In fact, the meaning postulate $MP1$ was stated by Montague in a somewhat different form; the $MP1$ meaning postulate in Montague's form and the way to obtain the above writing of $MP1$ can be found in [Dowty et al. 81], p. 226.)

An instance of this meaning postulate is:

$$\forall x \forall \mathcal{P} \square [\mathbf{find}'(x, \mathcal{P}) \equiv {}^\vee\mathcal{P}({}^\wedge\lambda y[\mathbf{find}'_*(x, y)])].$$

This instance will allow us to state formally the equivalence of the two translations of the sentence '**John seeks a unicorn**', and thus to show formally that this sentence is unambiguous, in the following way:

1. $\mathbf{find}'(j, {}^\wedge\lambda Y \exists x[\mathbf{unicorn}'(x) \wedge {}^\vee Y(x)])$ *de dicto*
2. ${}^\vee{}^\wedge\lambda Y \exists x\ [\mathbf{unicorn}'(x) \wedge {}^\vee Y(x)]\ ({}^\wedge\lambda y[\mathbf{find}'_*(j, y)])$ ($MP1$)
3. $\exists x[\mathbf{unicorn}'(x) \wedge (\lambda y[\mathbf{find}'_*(j, y)])(x)]$ ($\lambda, {}^\vee{}^\wedge$)
4. $\exists x[\mathbf{unicorn}'(x) \wedge \mathbf{find}'_*(j, x)]$ *de re*

3.2.14 Examples of meaning postulates

As mentioned above, Montague assigns to meaning postulates a general way of formalizing very precise and detailed relationships between word meanings of English, which cannot be as accurately stated in any other known semantic theory. If logic relations (for example, implications) exist between some word meanings, then we have to state a meaning postulate for each of these relations. For example, if our universe of discourse contains the words 'bachelor' and 'married' and if we discover that these words have opposite meanings, we must include a meaning postulate such as

$$\forall x[Bachelor(x) \supset \neg Married(x)].$$

So far, we lack a general theory of meaning and the appeal to meaning postulates triggered Chomsky's criticism: 'And if he is concerned about the question why *an oculist is an eye-doctor* is a meaning postulate, while *Washington is the capital of the United States* is not, he can be answered that there is a rigorous procedure for making this distinction, namely, to list the former sentence under

3.2. Montague's grammar

the heading 'meaning postulates' and not to list the latter. But it is clear that such an *ad hoc* approach to the problem of classification and characterization of elements in particular languages will be of no help to linguists, who are interested in the *general grounds* by which those elements and relations are established in each particular case, [Chomsky 55]. A second objection can be made: any attempt to enumerate these postulates is doomed to failure.

Montague has formalized the semantic relations that occur in the *PTQ* fragment of English by way of appropriate meaning postulates. We have seen a first example (*MP*1) in the preceding subsection. The dictionary of *PTQ* contains some infinitive-complement verbs, namely, **wish** and **try** (which constitute the syntactic category $IV//IV$). For this reason, it is necessary to formalize the semantic equivalence between **seek** and **try to find**, for example. This can be done by introducing the following meaning postulate:

$$MP2: \forall x \forall \mathcal{P} \Box [\ \mathbf{seek}'(x, \mathcal{P}) \equiv \mathbf{try}'(x, \char`\^[\mathbf{find}'(\mathcal{P})])].$$

This meaning postulate allows us for example to transform the translation of the *de dicto* reading of the sentence **John seeks a unicorn** (§ 3.2.12), i.e.,

1. $\mathbf{seek}'(j, \char`\^\lambda Y \exists x[\mathbf{unicorn}'(x) \land \check{} Y(x)])$,

into the equivalent expression

2. $\mathbf{try}'(j, \char`\^[\mathbf{find}'(\char`\^\lambda Y \exists x[\mathbf{unicorn}'(x) \land \check{} Y(x)])$.

We can transform this last expression so as to make its meaning more obvious, in the following way:

3. $\mathbf{try}'(j, \char`\^\lambda z[\mathbf{find}'(\char`\^\lambda Y \exists x[\mathbf{unicorn}'(x) \land$
$\check{} Y(x)])\ (z)])$ $\quad (\lambda)$
4. $\mathbf{try}'(j, \char`\^\lambda z[\mathbf{find}'(z, \char`\^\lambda Y \exists x[\mathbf{unicorn}'(x) \land$
$\check{} Y(x)])])$ $\quad (P)$
5. $\mathbf{try}'(j, \char`\^\lambda z[\check{}\check{}\lambda Y \exists x[\mathbf{unicorn}'(x) \land \check{} Y(x)]$ $\quad (MP1; § 3.2.13)$
$(\char`\^\lambda y[\ \mathbf{find}'_*(z,y)])])$
6. $\mathbf{try}'(j, \char`\^\lambda z[\exists x[\mathbf{unicorn}'(x) \land \lambda y[\mathbf{find}'_*(z,y)]$ $\quad (\lambda)$
$(x)]])$
7. $\mathbf{try}'(j, \char`\^\lambda z[\exists x[\mathbf{unicorn}'(x) \land \mathbf{find}'_*(z,x)]])$ $\quad (\lambda)$

3.2.15 Translation of sentences with modalities

The belief contexts are those that are under the scope of the verb **believe**. The syntactic category of **believe** is IV/t: this verb combines with a sentence (of syntactic category t) to produce an expression of category IV. Frege's compositionality principle forces us to combine **believe** with the constituent sentence prefixed by the intensional operator ^ rather than to combine it simply with the constituent sentence. Hence, if **sentence'** is the translation of a **sentence**, we have to write **believe'**(^**sentence'**) instead of simply **believe'**(**sentence'**) to take into account that this sentence is situated in a believe context (translation rule Tj of Subsection 3.2.7).

In Subsection 3.2.4 we have analysed the sentence **John believes that a woman talks**. Its *de dicto* translation is the following:

$$\textbf{believe'}(^\wedge\exists x[\textbf{woman'}(x)\wedge \textbf{talk'}(x)])(j).$$

After P transformation, this becomes

$$\textbf{believe'}(j,^\wedge\exists x[\textbf{woman'}(x)\wedge \textbf{talk'}(x)]).$$

Note the similarity between the last expression and the *de dicto* translation of the sentence **John seeks a unicorn**, i.e.,

$$\textbf{seek'}(j,^\wedge\lambda Y\exists x[\textbf{unicorn'}(x) \wedge \check{}Y(x)]).$$

The *de re* translation of the sentence **John believes that a woman talks** is the following:

$$\exists x[\textbf{woman'}(x)\wedge \textbf{believe'}(j,^\wedge \textbf{talk'}(x))].$$

Note again the similarity between this last expression and the *de re* translation of the sentence **John seeks a unicorn**, which is

$$\exists x[\textbf{unicorn'}(x)\wedge \textbf{seek'}_*(j,x)].$$

The analysis of the two translations of the sentence **John believes that a woman talks** is very similar to what we have seen in Subsection 3.2.12 about the sentence **John seeks a unicorn**; we shall not go into further detail.

The sentence adverb **necessarily** is treated as a member of the syntactic category t/t; this word combines with a sentence to form another sentence. Again, the Frege compositionality principle forces us to combine **necessarily** with the constituent sentence prefixed by the intensional operator ^ rather than to combine it simply with the constituent sentence. Let us analyse the sentence **Necessarily John walks**, and translate it into intensional logic.

1. he₁ walks $\Rightarrow$ **walks'**(x_1)
2. necessarily $\Rightarrow \lambda p[\Box \check{}p]$
3. necessarily he₁ walks $\Rightarrow$
    $\lambda p[\Box \check{}p](^\wedge[\textbf{walk'}(x_1)])$
4. $\Box[\textbf{walk'}(x_1)]$
5. John $\Rightarrow \lambda P[\check{}P(j)]$
6. necessarily John walks $\Rightarrow$
    $\lambda P[\check{}P(j)](^\wedge\lambda x_1\Box[\textbf{walk'}(x_1)])$
7. $\lambda x_1\Box\ [\textbf{walk'}(x_1)](j)$
8. $\Box[\textbf{walk'}(j)]$

Going from line 7 to line 8 requires the introduction of a new meaning postulate:

$$MP3: \exists x \Box (x = j).$$

This meaning postulate allows us to move the individual constant j into the scope of the modal operator $\Box$. (See [Dowty et al. 81], p.233, for a detailed discussion of this question.)

3.3 Conclusion and alternative formalisms

3.3.1 Difficulties inherent in Montague's semantics

One of the most original contributions of Montague lies in the construction of a compositional and recursive semantics which covers both intension and extension. Kripke's possible-world semantics gave the impetus. Montague perceived its fecundity at once; he exploited the notion of possible world as a means to provide the notion of intension with a rigorous status within an extended theory of reference.

Possible-world semantics of the sort used by Montague, however, is unable to account for some distinctive phenomena of natural language. The concept of intension that is obtained by widening the concept of extension with the help of the idea of a possible world is far too elementary. It is lacking in structure and sometimes fails to agree with the compositionality principle. A very simple example will bring this out. Let us consider the following two past participles: 'bought' and 'sold'. They share not only their extension but also their intension, since in all possible worlds in which the open sentence 'x has been bought' is satisfied by an object, the same object also satisfies the open sentence 'x has been sold'.

Let us now consider the two phrases 'bought by Mary' and 'sold by Mary'. We would like to be allowed to say that they are respectively obtained by combining 'bought' with 'by Mary' and 'sold' with 'by Mary'. As the intension of 'bought' is the same as the intension of 'sold' the intension of the two phrases is expected to be the same. Yet it is not. Hence a dilemma is facing us: either we have to abandon the compositionality principle or we have to renounce reducing the properties of phrases to intensions defined along Montague's lines.

Another weakness of the Montague semantics stems from the fact that two sentences which bear the same truth value across all possible worlds are treated as if they were expressing identical propositions. It follows that all tautologies become synonymous and all inconsistent sentences as well. Clearly this consequence is counter-intuitive; the relationship 'p has the same intension as q'

should be considered as a necessary condition, but not as a sufficient condition for the propositions represented by the sentences p and q to be identical.

A third objection can be made to Montague. His semantics can be blamed for presupposing that the speaker has an exhaustive knowledge of all possible worlds. One should, however, not be led into thinking that Montague postulates *factual omniscience*, by arguing that if the speaker knows the composition of all the possible worlds, the actual one included, then he knows the latter *ipso facto*. This conclusion, however, does not follow. What he is missing is the ability to tell the actual world from the other possible world, i.e., the ability to recognize the actual world as such.

Although Montague does not postulate factual omniscience, he nevertheless imputes too much knowledge to the speaker. His semantics enables us to account for the distinction between linguistic and factual knowledge, but fails to do justice to the 'dynamism' of knowledge acquisition.

3.3.2 Theory of discourse representations

Kamp's discourse representation theory is an example of a semantic theory built on the notion of *partial model* [Kamp 84]. Semantic representations are small models, called 'partial models', whose domains are finite and restricted. The representation of a sentence will contain the conditions which have to be fulfilled by the world for the sentence to be true. These conditions are represented by a partial model which must be compatible with the model which describes reality [Riche 89].

The following sentence is ambiguous: Every farmer who owns a donkey beats it. In a first reading, the quantified expression 'a donkey' falls within the scope of the quantified expression 'every farmer'; hence it can be paraphrased as: For every farmer, there is a donkey such that he owns it and beats it. In a second reading, more contrived, 'every farmer' falls within the scope of 'a donkey' and the sentence can be paraphrased as: There is a donkey such that every farmer who owns it beats it.

The formalism of the PTQ grammar cannot generate the tree that corresponds to the first reading. Nor can it, consequently, account for the meaning attached to the first paraphrase. The trouble comes from the anaphoric pronoun 'it'. To shed light on the nature of the obstacle let us try to translate the first paraphrase into predicate calculus; we obtain the formula

$$\forall x[(Farmer(x) \supset \exists y(Donkey(y) \wedge Own(x,y))) \supset Beat(x,y)].$$

Unfortunately this formalization is unacceptable. The quantification $\exists y$ cannot bind the third occurrence of the variable y, featuring in $Beat(x,y)$, as this occurrence of y does not fall within the quantifier's scope. To account for the first paraphrase within predicate calculus, we should write

3.3. CONCLUSION AND ALTERNATIVE FORMALISMS

$$\forall x[Farmer(x) \supset \forall y(Donkey(y) \land Own(x,y)) \supset Beat(x,y)].$$

But such a formal rendering does violence to the expression that we are trying to represent. We were forced to substitute the quantifier 'every' for the indefinite determiner 'a'.

The difficulty raised by sentences of that type, called 'donkey-sentences', is related to the interpretation of *anaphoric pronouns*. As the antecedent of an anaphoric pronoun can be located in a sentence which precedes the sentence containing this pronoun, the problem raised by donkey-sentences goes beyond the scope of sentence semantics; it arises at the text level or discourse level.

Let us consider the text or discourse constituted by the following sequence of sentences: A man walks. He sings. Like in the case of donkey-sentences it is possible to represent this sequence within predicate calculus, but the representation will be strongly distorted with respect to natural language. A dynamic sequence of sentences (a discourse) in which the first sentence is serving as a context for the second one would be represented by a single formula such as

$$\exists x[Man(x) \land (Walk(x) \land Sing(x))].$$

As a conclusion, neither predicate calculus, nor Montague's formalism can account for the *linguistic context*.

As far as the *situation context* is concerned, the matter is much more manageable. By adding place indices, speaker-hearer indices, and so on, to the usual time indices, it is possible to account for the way *deictics* ('here', 'I', and so on) depend upon the situation context [Gochet 80]. Once we have enriched the model by adding appropriate indices, we can express how the deictics semantic values depend upon the situation context by modifying the assignment function.

Such a method, however, cannot be extended to an anaphoric pronoun. This is all the more disturbing as the situation context which bestows a reference to a anaphoric pronoun can be replaced by a linguistic context which supplies the anaphoric pronoun replacing the deictic pronoun with an antecedent. The following example exhibits a substitution of that kind. Imagine that a shopwalker who is shown a photograph of a man who is suspected of shoplifting says: He has stolen the item *a*. This sentence in which the pronoun 'he' is deictic could be replaced — the meaning remaining unchanged — by the following sentence in which 'he' is anaphoric: Someone was taken a picture of by the camera of the shop. He has stolen *a*.

Kamp worked out his theory of discourse representations precisely to solve both the problem raised by donkey-sentences and the problem raised by anaphora across discourse [Kamp 84]. Each sentence is represented by what is called a *discourse representation structure* (*DRS*). Whenever a sentence plays the role of a context for another sentence, the *DRS* which corresponds to the latter is embedded into the *DRS* of the context-supplying sentence. Discourse representation structures are generated by an algorithm. A new account of meaning emerges: the (discursive) meaning of a sentence lies in the capability for the sentence to extend the *DRS* from which we started to a new *DRS* in which the former one is embedded. A sentence such as 'He has left' can be given an interpretation if it follows 'Someone has come in', but not so if it follows 'It is not the case that nobody came in', although the latter is logically equivalent to the former. Thus, meaning is conceived by Kamp not as a total function but only as a partial function.

Kamp's discourse theory also differs from Montague's in so far as it introduces an intermediary level between syntactic representation — the analysis tree — and interpretation in a model. The role of this intermediary representation is not similar to the role of Montague's translation of syntactic representation into intensional logic. Moreover, Kamp's discourse representation theory violates the compositionality principle to a certain extent. Kamp's treatment fails to give us a way of recovering the knowledge of the parts from the knowledge of the whole and of the mode of construction. For instance, we fail to recover the interpretation of 'farmer who owns a donkey' from 'Every farmer who owns a donkey beats it'.

3.3.3 Situation Semantics

In Barwise and Perry's situation semantics, the key notion is no longer that of possible world [Barwise and Perry 83], [Gochet 80]. These authors operate with the actual world only, i.e., with the world we are living in. This world is represented by a model $\mathcal{M}$. Situations are partial submodels of $\mathcal{M}$, i.e., fragments of the world. The key notion is now that of the *set of all situations* which support the truth of sentences, within the world represented by $\mathcal{M}$.

Situations, i.e., aggregates of objects, properties and relations, are more than predicate forms (or ordered tuples). They are objects which have properties and which stand in relation with one another as is indicated by prepositions in the examples 'the paper *on* the desk', 'the sled *in* snow'. The situation described by the phrase 'the paper on the desk' differs from the fact described by the statement 'the paper *is* on the desk'. A fact is 'a situation *polluted* by language'. In the copula-verb 'is' occurring in the statement, Barwise and Perry presumably see the trace of a mental act of judging which is missing in the phrase deprived of a verb 'the paper on the desk'.

In situation semantics, possible worlds (used in intensional logic) are left out of the picture. Their role is replaced by that of *situation*, which are complexes

3.3. CONCLUSION AND ALTERNATIVE FORMALISMS

of interrelated entities; thus, situations can be viewed as fragments of the world, i.e., as 'partial models'. Expressions are assigned denotations relative to the utterance context, to the link between the speaker who describes a situation and the situation itself, and to a 'state of affairs', i.e., a description of a part of the world which specifies how related entities hang together [Riche 89].

Some situations are perceived; Barwise and Perry dub them *scenes*. Scenes have to be brought in if we want to supply *perception* report sentences with a correct semantics. As an example, let us take the following statement in natural language:

1. Whitehead saw Russell wink.

If we try to represent this sentence within predicate logic, we shall have to rewrite it as the formula

$$Saw(W, R) \wedge Wink(R).$$

This can be paraphrased as:

2. Whitehead saw Russell and Russell winked.

Unfortunately, the latter sentence fails to capture the content of the former. Sentence 2 is not equivalent to sentence 1. Sentence 2 is implied by sentence 1 but does not imply it. Sentence 2 could be true while sentence 1 is false. This will happen if Whitehead is seeing Russell and Russell does wink, although Whitehead fails to see him wink. One might think that the problem can be solved by introducing the word 'that' and translating it formally by the intension operator. Sentence 1 would be represented in intensional logic by the formula

$$Saw(W, R) \wedge Saw(W, {\char`\^}Wink(R)).$$

This can be paraphrased as:

3. Whitehead saw Russell and Whitehead saw Russell wink.

Sentence 3, however, is not equivalent to sentence 1 either. The difference can be described in the following way. As the grammatical construction 'see + infinitive' does not carry any epistemic connotation, the logic of the verb 'see' is not *referentially opaque*, i.e., does not fall within the scope of the logic of *non-referential context* (§ 2.3.3), as opposed to the verbs 'believe' and 'know' for example. Although the logic of perception verbs is not referentially opaque in the naked infinitive constructions 'see + infinitive', 'hear + infinitive', it is nevertheless *intensional*, since it does not allow for the application of the rule of substitution of logically equivalent infinitives. For example, although the following infinitives are logically equivalent:

4. Mary enter by the back door,
5. Mary enter by the back door and Brown enter or not enter by the back door,

they are not interchangeable in context 6 in which the symbol (−) stands for a naked infinitive:

6. Fred saw (−).

They are not so because Fred might be located in such a way that Brown fails to belong to his visual field, although he is in the room which Mary and Brown are entering simultaneously. We find here the reason why Barwise and Perry isolate a special subset within the set of situations, namely, the subset made up of perceived situations, that is to say the subset of scenes.

3.3.4 Boolean semantics

Boolean semantics uses the concepts and methods of Boolean algebra, or propositional logic [Keenan and Faltz 85]. The work by Keenan and Faltz is based on the classical model-theoretical approach to semantics which could be applied directly to a fragment of the English language. In attempting to present some of the Montague insights in an elementary way, these authors observed that their task could be simplified by using Boolean algebras and Boolean homomorphisms in the models.

The fundamental relation that Keenan and Faltz try to represent is *logical implication* (or *logical consequence*), often referred to as the *entailment* relation. A sentence $S1$ entails a sentence $S2$ if and only if $S2$ is true whenever $S1$ is; we say also that $S2$ is a logical consequence of $S1$. This notion of 'logical consequence' exactly corresponds to the one that has been used by Montague in his formal approach to the semantics of natural language. Classically, the entailment relation holds between expressions of the syntactic category 'sentence'. The novelty in the Boolean semantics is that it allows a natural generalization of the entailment relation to expressions of a syntactic category other than 'sentence'. To show this, Keenan and Faltz slightly reformulate the intuition of entailment as follows: sentence $S1$ entails sentence $S2$ if and only if sentence $S1$ contains all the information in $S2$ (and possibly more). This intuitive definition tells us that sentence $S1$ claims everything that sentence $S2$ does, and possibly more. But the definition now covers expressions of other categories as well.

For example, we shall say that the verb 'work' is a logical consequence of the verb phrase 'work and walk'. Similarly, the determiner 'some' is a logical consequence of the compound determiner 'some but not all', in the sense that for any noun n, the properties which 'some but not all n's' have must all be properties which 'some n's' have. Stated otherwise, we see that 'work and

3.3. CONCLUSION AND ALTERNATIVE FORMALISMS

walk' is more informative than 'work', and that 'some but not all' is more informative than 'some'. The aim of the concept of logical consequence is precisely to formalize the intuitive notion of 'more informative'.

The basic idea of the Boolean semantics is the following. In all model-theoretic approaches to natural language, we associate with each syntactic category C of expressions a set of possible denotations, which corresponds to the *type* for C, defined in terms of the semantic primitives of the model. We have seen (Chap. 2) that in predicate extensional logic, the primitives of a model are a non-empty domain of discourse denoted S, and the set of truth values, taken as $\{1, 0\}$. The semantic primitives of predicate extensional logic, namely S and $\{1, 0\}$, have quite different statuses. Since S is an arbitrarily selected set, required only to be non-empty, it cannot be assumed to have any structure at all. We cannot, for example, assume that the elements of S bear any given relation to each other. On the other hand, the set $\{1, 0\}$ is equipped naturally with a Boolean structure. This means that the set of possible semantic values associated with sentences is a Boolean algebra. Let us remember that a Boolean algebra exhibits a partial ordering relation defined on the domain of the algebra in terms of meet and join operations (these operations are defined by the truth tables for conjunction and disjunction respectively).

This partial ordering relation is generally denoted $\leq$ and read 'less than or equal to'. Now, it turns out that the informativeness relation (entailment relation) discussed earlier is such a Boolean partial ordering relation. More precisely, for expressions $E1$ and $E2$ in any syntactic category C (one whose set of possible denotations is the domain of a Boolean algebra), expression $E1$ contains at least as much information as $E2$ if and only if for every model, the denotation of $E1$ in that model is 'less than or equal to' the denotation of $E2$ in that model. The expression $E2$ is then said to be a logical consequence of the expression $E1$.

The semantics of natural language stands surety for the following inferences: 'John sings in the garden, hence John is in the garden', 'John sees Mary in the garden, hence Mary is in the garden', but not for the following one: 'John sees Mary in the garden, hence John is in the garden'. This stands in agreement with Boolean semantics, which allows us for example to take into account the fact that the stative locative 'in' can be combined either with a transitive verb (as 'see') or with an intransitive verb (as 'sing') and that the inferences that can be made in these two cases are quite different.

We have just seen that there is motivation for regarding stative locatives (as 'in') as modifiers of transitive verbs as well as modifiers of intransitive verbs. Similar motivation exists in the case of a substantial number of other expressions. Consider for example the preposition 'from'. A phrase like 'from the bus' denotes the starting-point of a trajectory of motion executed by some entity, or else the starting-point of a trajectory describing the progress or effect of the action of some entity. The entity in question may be the one denoted by the subject or the object, and, in some cases, there is ambiguity. As in the case of stative locatives, the facts depend on the transitive verb involved. For example, if the verb is one of perception, a source phrase refers unambiguously to the subject. Thus in the sentence 'John saw Mary from the garden', John is in the garden, and the source phrase refers to the conceived trajectory of John's perception. On the other hand, in the sentence 'John knocked Mary from the roof', the source phrase seems to refer unambiguously to the trajectory of motion executed by the denotation of the object, that is, by Mary. A sentence like 'John shot Mary from the roof' seems to us ambiguous, with either John on the roof shooting Mary, or with Mary initially on the roof and falling off as a result of being shot. A formal treatment of these examples can be performed by using Boolean semantics (see [Keenan and Faltz 85], pp. 151–177).

3.3.5 Generalized phrase structure grammars

It seems that the most precise and detailed analyses of syntax/semantics interaction are those which have been developed within Montague's grammar and its extensions. Recent proposals within Montague's general framework have virtually all adopted the position that for every syntactic rule within the grammar, a corresponding semantic rule must be stated which specifies how structures of the sort analysed by that rule are to be interpreted. Gazdar and his co-authors implement this rule-to-rule hypothesis in special types of grammars, called *generalized phrase structure grammars* [Gazdar et al. 85].

In such a grammar, grammar rules are pairs consisting of:

1. a phrase structure syntactic rule,

2. a semantic rule which specifies how the constituent analysed by the phrase structure rule 1 is to be translated into an appropriate logical expression.

A grammar rule of this type will for example be represented as follows:

$\{sentence \rightarrow verb\_phrase, noun\_phrase; noun\_phrase'(\char`\^verb\_phrase')\}$.

This grammar rule can be compared with a pair of rules $\{Sj; Tj\}$ in the Montague's grammar, such as for example:

- $S4$: If $\alpha \in P_T$ and if $\beta \in P_{IV}$, then $F_4(\alpha, \beta) \in P_t$.
  The function $F_4(\alpha, \beta)$ is $\alpha\beta^*$ where β^* is the result of replacing the first verb in β by its third person singular present form.

3.3. CONCLUSION AND ALTERNATIVE FORMALISMS

- $T4$: If $\alpha \in P_T$ and if $\beta \in P_{IV}$, and if α, β translate into α', β' respectively, then $F_4(\alpha, \beta)$ translates into $\alpha'(\char`\^\beta')$.

The rule $S4$ takes a noun phrase (member of P_T) and a verb phrase (member of P_{IV}) and combines them to form a sentence (member of P_t), performing 'subject–verb agreement' in the process. The rule $S4$ simply constructs strings and does not assign any constituent structure. By contrast, the phrase structure syntactic rule admits (in conjunction with other parts of the grammar) a local syntactic tree (see chapter 5 of [Thayse 88]). While the semantic rule $T4$ shows how the strings defined by the syntactic rule $S4$ induce a translation into intensional logic, it seems reasonable to view the semantic rule of the generalized phrase structure rule as part of an inductive definition for translating the syntactic trees generated by the grammar into appropriate expressions of intensional logic.

In summary, the goal pursued by Gazdar and his co-authors is to incorporate the Montague grammar into a phrase structure grammar formalism, which itself can be translated, in a relatively easy way, into the logic programming language Prolog (see chapter 5 of [Thayse 88]).

Let us finally mention the work by Janssen, who has employed a van Wijngaarden-style two-level grammar to define a generalization of Montague's PTQ syntax [Janssen 86a], [Janssen 86b].

Chapter 4

Temporal logic

4.1 Propositional linear time temporal logic

4.1.1 Introduction

Linear time temporal logic is a language for describing sequences of situations and for reasoning about them. These sequences can be interpreted in various ways. In a purely temporal interpretation, a sequence of situations represents the evolution of the state of the world with time. We will also consider another interpretation where the sequence of situations represents the successive states a program goes through during its execution. This interpretation will allow us to use temporal logic for program specification and verification.

There are numerous versions of linear time temporal logic. We will describe one that has been widely used for the specification and verification of concurrent programs. This last subject is probably the one for which temporal logic has proved the most useful in computing science.

4.1.2 Syntax of linear time temporal logic

The language of propositional temporal logic is that of propositional calculus augmented with four symbols representing *temporal operators*. These symbols are the following.

1. $\bigcirc$ which is read 'at the next time';

2. $\square$ which is read 'always';

3. $\diamond$ which is read 'eventually';

4. $\mathcal{U}$ which is read 'until'.

All these operators are unary except $\mathcal{U}$ which is binary. We thus have the following syntax.

- All the syntactic rules of propositional logic are also syntactic rules of temporal logic.

- If A and B are formulas, then $\bigcirc A$, $\square A$, $\lozenge A$, $A\,\mathcal{U}\,B$ are formulas.

The notation used for the operators 'always' ($\square$) and 'eventually' ($\lozenge$) is identical to the one used in modal logic (§ 1.2.3). This is motivated by the fact that the temporal logic we are studying can be viewed as an extension of a particular modal logic.

4.1.3 Semantics of linear time temporal logic

A *temporal frame*[1] is a pair $\mathcal{F} = (S, R)$ where S is a finite or countable nonempty set and where R is a functional relation. A temporal frame is thus a special case of a modal frame where the accessibility relation satisfies equation 7 of Subsection 1.2.5. As the semantics given below will make clear, the relation R will be used for defining the meaning of the operator $\bigcirc$, whereas the meaning of the operators $\square$ and $\lozenge$ will be defined in terms of the reflexive and transitive closure of R.

Given a set P of atoms, a *temporal interpretation* is a triple (S, R, I) where (S, R) is a temporal frame and where I is a *temporal interpretation function*, i.e., a mapping from $S \times P$ to $\{\mathbf{T}, \mathbf{F}\}$ assigning a truth value $I(s, p)$ to each state $s \in S$ and each proposition $p \in P$.

We can now give the rules that, given a temporal interpretation $\mathcal{I} = (S, R, I)$, assign a truth value to each pair consisting of a state $s \in S$ and a formula A.

Let us first consider the case of an atomic proposition:

- $\mathcal{I}(s, p) =_{def} I(s, p)$.

The semantics of the logical connectives is defined as in propositional calculus (§ 1.2.3), for instance

- $\mathcal{I}(s, A \wedge B) =_{def} \mathcal{I}(s, A) \wedge \mathcal{I}(s, B)$,

- $\mathcal{I}(s, \neg A) =_{def} \neg \mathcal{I}(s, A)$.

In the definition of the semantics of the temporal operators, we will use the notation $R^i(s)$ to represent the ith successor of s, for $i = 0, 1, 2$, etc., that is, the $(i+1)$st element of the sequence

$$s,\ R(s),\ R(R(s)),\ R(R(R(s))),\ \ldots$$

We then have:

- $\mathcal{I}(s, \bigcirc A) =_{def} \mathcal{I}(R(s), A)$,

---

[1] We follow the terminology of [Turner 84] and [Goldblatt 87]. The expression *temporal structure* is also used for the same concept.

4.1. PROPOSITIONAL LINEAR TIME TEMPORAL LOGIC

- $\mathcal{I}(s, \Box A) = \mathbf{T}$ if and only if $\mathcal{I}(R^i(s), A) = \mathbf{T}$ for all $i \geq 0$,
- $\mathcal{I}(s, \Diamond A) = \mathbf{T}$ if and only if $\mathcal{I}(R^i(s), A) = \mathbf{T}$ for some $i \geq 0$,
- $\mathcal{I}(s, A \mathcal{U} B) = \mathbf{T}$ if and only if
  - $\mathcal{I}(R^j(s), A) = \mathbf{T}$ for all $j \geq 0$, or
  - $\mathcal{I}(R^i(s), B) = \mathbf{T}$ for some $i \geq 0$ and $\mathcal{I}(R^j(s), A) = \mathbf{T}$ for all j such that $0 \leq j < i$.

The interpretation of the formula $\Diamond A$ in the state s is the same as the interpretation of A in the state $R(s)$. The operator $\mathcal{U}$ we have just defined is often referred to as the 'weak until'. The formula $A \mathcal{U} B$ is true in two cases: when A is always true and when B becomes true (without necessarily remaining so) before A becomes false. A common variation of the operator $\mathcal{U}$, the 'strong until', is not satisfied when A is always true and B always false. Its semantic definition is the following:

- $\mathcal{I}(s, A \mathcal{U} B) = \mathbf{T}$ if and only if $\mathcal{I}(R^i(s), B) = \mathbf{T}$ for some $i \geq 0$ and $\mathcal{I}(R^j(s), A) = \mathbf{T}$ for all j such that $0 \leq j < i$.

From here on, we will only use the weak version of $\mathcal{U}$.

An interpretation $\mathcal{I}$ is a *model* of the formula A if $\mathcal{I}(s, A) = \mathbf{T}$ for every state s. A formula A is *valid* if all interpretations are models of A.

As in classical logic ([Thayse 88], § 1.1.6), the notation

$$\models A$$

means that the formula A is valid. If E is a set of formulas, the notation

$$E \models A$$

means that every model of E is also a model of A; one says that A is a *logical consequence* of E. Finally, the notation

$$\mathcal{I} \models A$$

means that the interpretation $\mathcal{I}$ is a model of A. By extension, one also uses the notation

$$\mathcal{I} \models_s A$$

to mean that the formula A is true in the state s of the interpretation $\mathcal{I}$; this notation is thus equivalent to $\mathcal{I}(s, A) = \mathbf{T}$.

The semantics of temporal logic can also be defined from a slightly different point of view. The relation R being functional and the set of states S being finite or countable, it is easy to prove that if a temporal formula has a model, then it has a model where S is the set of natural numbers $\mathbf{N}$ and where R is the successor function $(n \mapsto n+1)$ on that set. It is thus customary to give the semantics of temporal logic in terms of a mapping π from $\mathbf{N} \times P$ to $\{\mathbf{T}, \mathbf{F}\}$, the relation R being implicit. In this semantics, a temporal interpretation is thus an infinite *sequence* of assignments of truth values to the atomic propositions. The semantics of the temporal operators can be defined directly in terms of sequences, for example for $\mathcal{U}$:

- $\pi(s, A \, \mathcal{U} \, B) = \mathbf{T}$ if and only if
    - $\pi(s+j, A) = \mathbf{T}$ for all $j \geq 0$, or
    - $\pi(s+i, B) = \mathbf{T}$ for some $i \geq 0$ and $\pi(s+j, A) = \mathbf{T}$ for all j such that $0 \leq j < i$.

As before, the notation

$$\pi \vDash_s A$$

is defined to be equivalent to $\pi(s, A) = \mathbf{T}$. One says that a temporal formula A is *true for a sequence* π, and one writes $\pi \vDash A$, if A is initially true in π (true at 0). The notation

$$\pi \vDash A$$

thus has the same meaning as $\vDash_0 A$; it does not imply that $\pi \vDash_s A$ holds for every natural number s.

The two semantics (temporal frames and sequences) are equivalent in the sense that the formulas they make valid are the same. More formally, we have $\pi \vDash A$ for every sequence π if and only if $\mathcal{I} \vDash A$ holds for every temporal interpretation $\mathcal{I}$. The first semantics we introduced has the advantage of showing the link between temporal and modal logic. The second semantics makes the nature of linear temporal frames more explicit.

Remark. Several terminologies and several notations have been used for modal logic. This can sometimes lead to confusion. The same situation exists in temporal logic which can be seen as a particular modal logic (see Chapter 1). One important variation in terminology centers around the notion of model. This word is sometimes used to mean a temporal frame (S, R), or a temporal interpretation (S, R, I). In this chapter, the word 'model' has a more restrictive meaning: a *model* of a formula is an interpretation in which that formula is true.

4.1.4 Examples

1. The formula p is satisfied by all sequences in which p is true in the first state.

4.1. PROPOSITIONAL LINEAR TIME TEMPORAL LOGIC

2. The formula $\Box(p \supset \bigcirc q)$ is satisfied by all sequences in which each state where p is true is immediately followed by a state in which q is true.

3. The formula $\Box(p \supset \bigcirc(\neg q \, \mathcal{U} \, r))$ is satisfied by all sequences such that if p is true in a state, then q remains false from the next state on and until the first state where r is true.

4. The formula $\Box\Diamond p$ is satisfied by all sequences in which p is true infinitely often.

5. The formula $\Box p \wedge \Diamond \neg p$ is not satisfied by any sequence.

6. The formula $\neg p \, \mathcal{U} \, p$ is satisfied by all sequences (it is thus valid).

4.1.5 Axiomatization

To axiomatize the propositional temporal logic we have just defined, it is natural to start from the axiomatization of modal logic given in Subsection 1.2.8. Our axiomatic system will thus first include:

(A0) all the instances of valid schemata of propositional calculus.

We then need to axiomatize the operator $\bigcirc$. Temporal logic is normal (§ 1.2.8) and the operator $\bigcirc$ is functional. We should thus include the axiom schema K and the axiom characterizing the existence and uniqueness of the successor:

(A1) $\bigcirc(A \supset B) \supset (\bigcirc A \supset \bigcirc B)$,

(A2) $\neg \bigcirc A \equiv \bigcirc \neg A$.

We now turn to the operators $\Box$ and $\Diamond$. They satisfy the duality axiom and the K axiom:

(A3) $\Diamond A \equiv \neg \Box \neg A$,

(A4) $\Box(A \supset B) \supset (\Box A \supset \Box B)$.

Rather than giving axioms that characterize the properties of the operator $\Box$ (reflexivity, transitivity, linearity, etc.) it is more convenient to give axioms that describe the relation between the operators $\Box$ and $\bigcirc$, specifically:

(A5) $\Box A \supset (A \wedge \bigcirc A \wedge \bigcirc \Box A)$,

(A6) $\Box(A \supset \bigcirc A) \supset (A \supset \Box A)$.

Finally, the operator $\mathcal{U}$ is characterized by the following two axioms:

(A7) $A\,\mathcal{U}\,B \supset (B \vee (A \wedge \bigcirc(A\,\mathcal{U}\,B)))$,

(A8) $[X \wedge \Box(X \supset (B \vee (A \wedge \bigcirc X)))] \supset A\,\mathcal{U}\,B$.

In this last axiom (which, as all the other axioms, is actually an axiom schema), X stands for an arbitrary formula. To complete our axiomatic system, we have to give the inference rules. As in modal logic, these are the *Modus Ponens* rule and the *Necessitation* rule:

(MP) if $\vdash A$ and if $\vdash A \supset B$, then $\vdash B$,

(N) if $\vdash A$, then $\vdash \Box A$.

The axiomatic system for linear time propositional temporal logic is sound and complete. More precisely, we have the following result (see for example [Gabbay et al. 80], [Wolper 82], [Banieqbal and Barringer 86]).

Theorem 4.1. *A formula is provable in the axiomatic system of linear time propositional temporal logic if and only it is valid in this logic.*

To illustrate the use of this axiomatic system, let us prove that the operator $\Box$ satisfies the formula $\Box A \supset \Box\Box A$ characterizing the transitivity of the accessibility relation. To simplify the axiomatic proofs, we will not represent explicitly all purely propositional steps (i.e., those that can be performed using the axiom (A0) and the rule (MP)); we will use the notation (PR) to describe these steps.

1. $\vdash \Box A \supset \bigcirc \Box A$ (A5, PR)
2. $\vdash \Box(\Box A \supset \bigcirc \Box A)$ (1, N)
3. $\vdash \Box(\Box A \supset \bigcirc \Box A) \supset (\Box A \supset \Box\Box A)$ (A6)
4. $\vdash \Box A \supset \Box\Box A$ (2, 3, MP)

Let us now prove that the operators $\Box$ and $\bigcirc$ commute, that is, that they satisfy $\Box\bigcirc A \supset \bigcirc\Box A$ and $\bigcirc\Box A \supset \Box\bigcirc A$. For the first of these formulas the proof is the following.

1. $\vdash \Box(\bigcirc A \supset (A \supset \bigcirc A))$ (A0, N)
2. $\vdash \Box\bigcirc A \supset \Box(A \supset \bigcirc A)$ (1, A4, MP)
3. $\vdash \Box(A \supset \bigcirc A) \supset \bigcirc\Box(A \supset \bigcirc A)$ (A5, PR)
4. $\vdash \Box\bigcirc A \supset \bigcirc\Box(A \supset \bigcirc A)$ (2, 3, PR)
5. $\vdash \Box\bigcirc A \supset \bigcirc A$ (A5, PR)
6. $\vdash \Box[\Box(A \supset \bigcirc A) \supset (A \supset \Box A)]$ (A6, N)
7. $\vdash \bigcirc[\Box(A \supset \bigcirc A) \supset (A \supset \Box A)]$ (6, A5, PR)
8. $\vdash \bigcirc\Box(A \supset \bigcirc A) \supset \bigcirc(A \supset \Box A)$ (7, A1, MP)
9. $\vdash \bigcirc\Box(A \supset \bigcirc A) \supset (\bigcirc A \supset \bigcirc\Box A)$ (8, A1, PR)
10. $\vdash \Box\bigcirc A \supset \bigcirc\Box A$ (4, 5, 9, PR)

4.2. Temporal logic and finite automata

Here is the proof of the second formula.

1.	$\vdash \Box A \supset A$	(A5, PR)
2.	$\vdash \Box(\Box A \supset A)$	(1, N)
3.	$\vdash \bigcirc(\Box A \supset A)$	(2, A5, PR)
4.	$\vdash \bigcirc\Box A \supset \bigcirc A$	(3, A1, PR)
5.	$\vdash \Box\bigcirc\Box A \supset \Box\bigcirc A$	(4, N, A4)
6.	$\vdash \Box A \supset \bigcirc\Box A$	(A5, PR)
7.	$\vdash \Box(\Box A \supset \bigcirc\Box A)$	(6, N)
8.	$\vdash \bigcirc(\Box A \supset \bigcirc\Box A)$	(7, A5, PR)
9.	$\vdash \bigcirc\Box A \supset \bigcirc\bigcirc\Box A$	(8, A1, PR)
10.	$\vdash \Box(\bigcirc\Box A \supset \bigcirc\bigcirc\Box A)$	(9, N)
11.	$\vdash \Box(\bigcirc\Box A \supset \bigcirc\bigcirc\Box A) \supset$ $(\bigcirc\Box A \supset \Box\bigcirc\Box A)$	(A6)
12.	$\vdash \bigcirc\Box A \supset \Box\bigcirc\Box A$	(10, 11, MP)
13.	$\vdash \bigcirc\Box A \supset \Box\bigcirc A$	(12, 5, PR)

Finally, to show the use of the axioms for the operator $\mathcal{U}$, let us prove the formula $\neg A \, \mathcal{U} \, A$.

1.	$\vdash A \vee \neg A$	(A0)
2.	$\vdash \Box(A \vee \neg A)$	(1, N)
3.	$\vdash \bigcirc(A \vee \neg A)$	(2, A5, PR)
4.	$\vdash \bigcirc(A \vee \neg A) \supset$ $[(A \vee \neg A) \supset (A \vee (\neg A \wedge \bigcirc(A \vee \neg A)))]$	(A0)
5.	$\vdash (A \vee \neg A) \supset (A \vee (\neg A \wedge \bigcirc(A \vee \neg A)))$	(3, 4, MP)
6.	$\vdash \Box[(A \vee \neg A) \supset (A \vee (\neg A \wedge \bigcirc(A \vee \neg A)))]$	(5, N)
7.	$\vdash (A \vee \neg A) \wedge$ $\Box[(A \vee \neg A) \supset (A \vee (\neg A \wedge \bigcirc(A \vee \neg A)))]$	(1, 6, PR)
8.	$\vdash [(A \vee \neg A) \wedge \Box[(A \vee \neg A) \supset$ $(A \vee (\neg A \wedge \bigcirc(A \vee \neg A)))]] \supset (\neg A \, \mathcal{U} \, A)$	(A8)
9.	$\vdash \neg A \, \mathcal{U} \, A$	(7, 8, MP)

4.2 Temporal logic and finite automata

We have just described for temporal logic the usual elements of the definition of a logic: syntax, semantics and axiomatic system. This being done, it would seem natural to carry on with the applications of this logic or with first-order temporal logic. We will however first go over a very useful and fundamental part of the theory of temporal logic: the relation between this logic and finite automata. The main theorem we will prove establishes the existence of a translation from temporal logic to a generalization of finite automata ([Thayse 88], § 5.2.3). As we will see, this generalization enables finite automata to recognize infinite words rather than finite words. The translation will give us information on the expressiveness of temporal logic and will yield a decision procedure. Hence, the type of axiomatic proofs we have done in the preceding section can be replaced by an algorithm. The translation will also be the theoretical basis of the application of propositional temporal logic to the verification of concurrent programs.

4.2.1 Finite automata on infinite words

Given an alphabet (set of symbols) Σ, a finite word is a finite sequence of symbols of Σ. For instance, if $\Sigma = \{a, b\}$, the sequence

$$w = abaababbbba$$

is a finite word over the alphabet Σ. A more formal way of defining a finite word is to say that a finite word of length n is a function from the set $\{0, \ldots, n-1\}$ to the alphabet Σ. In the example given above, the word w is of length 12 and is characterized by the function w from the set $\{0, \ldots, 11\}$ to the set $\{a, b\}$ satisfying $w(0) = a$, $w(1) = b$, $w(2) = a$, ..., and $w(11) = a$.

An infinite word on an alphabet Σ is simply an infinite sequence of elements from Σ. More formally, it is a function from the set of natural numbers (which we will denote $\mathbf{N}$) to the alphabet Σ. For example, the function

$$w(x) = \begin{cases} a & \text{if } x \text{ is even} \\ b & \text{if } x \text{ is odd} \end{cases}$$

defines the infinite word where a and b alternate. It is of course impossible to write down an infinite word; it is nevertheless often possible to represent such a word in a finite way, for instance as we did above. To represent sets of infinite words, we will use an extension of regular expressions[2], which involves the infinite repetition operator ω. Besides ω, we will use (following the usual conventions of regular expressions) juxtaposition to represent concatenation, the operator $\cup$ to represent union and the operator $\star$ to denote finite repetition. In this language, the expression $(ab)^\omega$ represents the set of words containing only the infinite word where a and b alternate.

We can now introduce the concept of finite automata on infinite words, also known as *Büchi automata* from the name of their inventor [Büchi 62]. One should think of this type of automaton as an abstract formalism for defining sets of infinite words. Indeed, the definition of Büchi automata implies the manipulation of infinite sequences which, of course, is not actually possible.

---

[2] Called 'ω-regular expressions' (see § 4.2.2).

4.2. Temporal logic and finite automata

Formally, the definition of a non-deterministic finite automaton on infinite words includes the same items as the definition of a finite automaton on finite words. Such an automaton is defined by a 5-tuple $A = (\Sigma, S, \rho, S_0, F)$ where

- Σ is the *alphabet* (a finite set of *symbols*),

- S is a *finite set of states*,

- $\rho : S \times \Sigma \rightarrow 2^S$ is the *transition function* (for each state and each symbol of the alphabet, it gives the possible next states),

- S_0 is a set of *initial states* ($S_0 \subset S$),

- F is a set of *accepting states* ($F \subset S$).

A Büchi automaton defines a set of words that are said to be *accepted* by the automaton. This set of words is defined by using the notion of *execution* of an automaton on a word.

An execution of an automaton A on an infinite word $w = a_0 a_1 \ldots$ is an infinite sequence $\sigma = s_0 s_1 s_2 \ldots$ of elements of S with the following properties:

- $s_0 \in S_0$ (the first state of an execution is an initial state),

- $s_i \in \rho(s_{i-1}, a_{i-1})$ for every $i \geq 1$ (each state of the sequence is obtained from the previous one in agreement with the transition function).

The notion of execution we have just defined is the generalization to infinite words of the similar notion for automata on finite words ([Thayse 88], § 5.2.3). Let us note that Büchi automata are non-deterministic. This implies that there can be several executions of the automaton for a given infinite word. These different executions are obtained by different choices among the successors allowed by the transition function.

For an automaton on finite words, one says that an execution is *accepting* if it ends in an accepting state. When dealing with infinite words, this notion can no longer be used since the execution is infinite and hence does not have a last state. We will say that an execution of a Büchi automaton is *accepting* if it contains some accepting state an infinite number of times. More formally, an execution $\sigma = s_0 s_1 \ldots$ is accepting if there is some state $s \in F$ and an infinite number of integers $i \in \mathbf{N}$ such that $s_i = s$. Finally, we will say that a word is *accepted* by a Büchi automaton if there exists an accepting execution of that automaton on that word. The *language accepted* by an automaton A is the set of infinite words accepted by A. We will denote it by $L(A)$.

To represent an automaton, we will follow standard practice and use a graph whose vertices and edges respectively represent the states and transitions of the automaton. We will mark initial states with the symbol > and distinguish final states by a double circle.

Figure 4.1 shows an automaton accepting the language $(a \cup b)^\omega$, that is, the language that includes all words over the alphabet $\{a, b\}$. The automaton of Figure 4.2 accepts the language $a^*b(a \cup b)^\omega$ of all infinite words over $\{a, b\}$ that include at least one b. Figure 4.3 shows an automaton accepting the language of all words that include an infinite number of occurrences of the symbol b. This language is the one described the the ω-regular expression $(a^*bb^*)^\omega$. The automaton of Figure 4.4 accepts all words that contain the pattern ab infinitely often (or, equivalently, all words containing both a and b infinitely often).

Figure 4.1 : An automaton accepting $(a \cup b)^\omega$

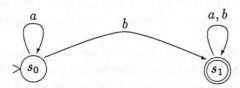

Figure 4.2 : An automaton accepting $a^*b(a \cup b)^\omega$

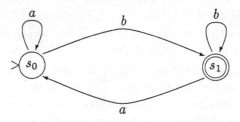

Figure 4.3 : An automaton accepting $(a^*bb^*)^\omega$

4.2. TEMPORAL LOGIC AND FINITE AUTOMATA

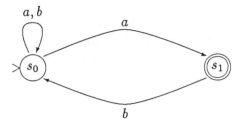

Figure 4.4 : Infinitely often ab

4.2.2 ω-Regular languages

The languages accepted by Büchi automata are usually referred to as the ω-*regular* languages. As is the case for the classical regular languages, these languages can be characterized in several ways. For instance, they correspond exactly to the languages that can be described by ω-regular expressions, i.e., expressions of the form

$$\bigcup_i \alpha_i \beta_i^\omega$$

where α_i and β_i are classical regular expressions (defined exclusively with the concatenation, union and finite repetition operators) and where $\beta_i \neq \varepsilon$.

Among the most interesting properties of the class of ω-regular languages is the fact that this class is closed under the operations of union, intersection and complementation. This means that given two Büchi automata A_1 and A_2, accepting respectively the languages $L(A_1)$ and $L(A_2)$, there are Büchi automata that accept the languages $L(A_1) \cup L(A_2)$, $L(A_1) \cap L(A_2)$ and $\Sigma^\omega \setminus L(A_1)$. We will prove the first two of these properties. The third requires a rather complex construction[3] that can be found in [Sistla et al. 87]. Another property that will be of interest to us is that it is possible to determine algorithmically if the language accepted by a Büchi automaton is empty or not.

Theorem 4.2. *Given two Büchi automata $A_1 = (\Sigma, S_1, \rho_1, S_{01}, F_1)$ and $A_2 = (\Sigma, S_2, \rho_2, S_{02}, F_2)$, one can build a Büchi automaton $A = (\Sigma, S, \rho, S_0, F)$ accepting the language $L(A_1) \cup L(A_2)$.*
Proof. The automaton A is essentially the union of the two automata A_1 and

---

[3] For finite automata on finite words, the complementation problem is usually dealt with by first determinizing the automaton using the subset construction of Rabin and Scott. To complement the resulting deterministic automaton, it is then sufficient to make accepting states non-accepting and vice versa. The subset construction is not sufficient for determinizing Büchi automata. Actually, deterministic Büchi automata define a subset of the class of ω-regular languages. It is possible to define a class of deterministic finite automata that corresponds to the ω-regular languages, but this requires the use of a more complicated condition to define accepting executions.

A_2. Indeed, as we are dealing with non-deterministic automata, we are allowed to choose either an execution of A_1 (by choosing an initial state of A_1) or an execution of A_2 (by choosing an initial state of A_2) and hence to accept the words accepted either by A_1 or by A_2. Formally, the automaton A is defined as follows (assuming that the sets of states S_1 and S_2 are disjoint):

- $S = S_1 \cup S_2, \quad S_0 = S_{01} \cup S_{02}, \quad F = F_1 \cup F_2,$

- $t \in \rho(s, a)$ if $\begin{cases} \text{either } t \in \rho_1(s, a) & \text{and } s \in S_1, \\ \text{or } t \in \rho_2(s, a) & \text{and } s \in S_2. \end{cases}$

We will leave it to the reader to check that the automaton we have constructed does indeed accept the language $L(A_1) \cup L(A_2)$. □

To prove that the ω-regular languages are closed under intersection, we will introduce a generalized form of Büchi automata and prove that these generalized automata can be reduced to standard Büchi automata. The generalization appears in the set of accepting states. A *generalized Büchi automaton* is a 5-tuple $A = (\Sigma, S, \rho, S_0, \mathcal{F})$. The first four components are identical to those of a Büchi automaton; the last component is a set of sets of states rather than a set of states. In other words, we have $\mathcal{F} = \{F_1, \ldots, F_k\}$ where each F_j, for $1 \leq j \leq k$, is a set of states (a subset of S). For this type of automata, an execution is admissible if it contains infinitely often some state of *each* F_j. Formally, an execution $\sigma = s_0 s_1 \ldots$ is admissible if for each j such that $1 \leq j \leq k$, there is some $s \in F_j$ such that, for an infinite number of $i \in \mathbf{N}$, we have $s_i = s$. Generalized Büchi automata have a more complicated definition than Büchi automata, but they sometimes allow a simpler definition of some languages. For example, the automaton of Figure 4.5 where $\mathcal{F} = \{F_1, F_2\}$ with $F_1 = \{s_0\}$ and $F_2 = \{s_1\}$ accepts the language of all words that include an infinite number of occurrences of the symbols a and b (the number j next to an accepting state indicates the set F_j to which it belongs).

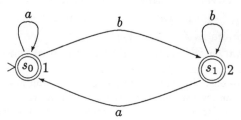

Figure 4.5 : A generalized Büchi automaton

We will now prove that any generalized Büchi automaton can be converted into a Büchi automaton accepting the same language.

4.2. TEMPORAL LOGIC AND FINITE AUTOMATA

Lemma 4.3. *Given a generalized Büchi automaton $A = (\Sigma, S, \rho, S_0, \mathcal{F})$, where $\mathcal{F} = \{F_1, \ldots, F_k\}$, it is possible to build a Büchi automaton $A' = (\Sigma, S', \rho', S'_0, F')$ accepting $L(A)$.*

Proof. We shall use a construction that creates k 'copies' of the automaton A. The execution starts in the first copy and goes form copy j to copy $(j \bmod k) + 1$ whenever a state in F_j is reached. The set of accepting states of A' is the set of states of the first copy that are in F_1. Before proving the correctness of our construction, let us define it formally. The Büchi automaton A' is defined as follows.

- $S' = S \times \{1, \ldots, k\}$. Each state in S' is a pair whose first component is a state in S and whose second component is an integer between 1 and k. This integer specifies the current 'copy' of the automaton A. For instance, the state $(s, 3)$ is the state s of the third copy of A.

- $S'_0 = S_0 \times \{1\}$. The initial states of A' are the initial states of the first copy of A. Actually, this choice is arbitrary; one could take as initial states the initial states of any copy.

- The transition relation ρ' is defined by $(t, i) \in \rho'((s, j), a)$ if

$$t \in \rho(s, a) \text{ and } \begin{cases} i = j & \text{if } s \notin F_j, \\ i = (j \bmod k) + 1 & \text{if } s \in F_j. \end{cases}$$

  When in copy j, if one is not in a state in F_j, one stays within copy j, following the transition relation of A; and if one is in a state in F_j, one moves to the next copy, while still following the transition relation of A.

- $F' = F_1 \times \{1\}$. The accepting states are the states of the first copy that are in F_1. Note that the special role of the first copy is arbitrary; one could as well define the accepting states as the states of the jth copy that are in F_j (for any value of j).

To convince oneself that the construction above has the required property, it is sufficient to note that the only way one can go infinitely often through an accepting state of A' is to go infinitely often through all the copies of A and hence to go infinitely often through at least one state of each of the sets F_j. □

Figure 4.6 shows the result of applying the construction of Lemma 4.3 to the automaton of Figure 4.5.

We can now move on to the announced theorem.

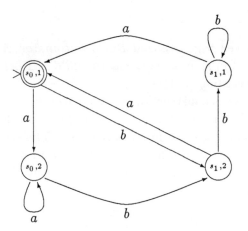

Figure 4.6 : Example for Lemma 4.3

Theorem 4.4. Given two Büchi automata $A_1 = (\Sigma, S_1, \rho_1, S_{01}, F_1)$ and $A_2 = (\Sigma, S_2, \rho_2, S_{02}, F_2)$, it is possible to build a Büchi automaton accepting the language $L(A_1) \cap L(A_2)$.

Proof. We present the construction of a generalized Büchi automaton $A = (\Sigma, S, \rho, S_0, \mathcal{F})$ that accepts $L(A_1) \cap L(A_2)$. Lemma 4.3 then makes it possible to transform this automaton into a Büchi automaton. The generalized Büchi automaton simulates the simultaneous execution of A_1 and A_2. This is done by building an automaton whose states are pairs of states of the two automata and whose transitions are those that are simultaneously possible in both automata. As for the accepting states, the set $\mathcal{F}$ of the constructed automaton has two components corresponding respectively to the sets F_1 and F_2 of the automata A_1 and A_2. Formally, the automaton A is defined by

- $S = S_1 \times S_2$, $S_0 = S_{01} \times S_{02}$,
- $\mathcal{F} = \{F_1 \times S_2, F_2 \times S_1\}$,
- $(u, v) \in \rho((s, t), a)$ if $u \in \rho_1(s, a)$ and $v \in \rho_2(t, a)$.

All that remains is to convince oneself that, for any word, there is an accepting execution of A_1 and A_2 if and only if there is an accepting execution of A. We leave this to the reader. □

Let us now turn to the problem of checking algorithmically if the language accepted by a Büchi automaton is empty or not (one often says 'checking if the automaton if empty or not'). To state the result we are interested in, we need to define the concept of accessibility between states of a Büchi automaton. Intuitively, a state s_2 is accessible from a state s_1 in a Büchi automaton if

4.2. TEMPORAL LOGIC AND FINITE AUTOMATA

there is a path from s_1 to s_2 in the graph representing the automaton. More formally, given a Büchi automaton $A = (\Sigma, S, \rho, S_0, F)$, a state s_2 is *accessible* from a state s_1 if there is some finite word $w = a_1 \ldots a_k$, with $k \geq 1$, and a sequence of states $t_1, \ldots, t_{k+1}$ such that

- $t_1 = s_1$, $t_{k+1} = s_2$,
- $t_{j+1} \in \rho(t_j, a_j)$ for $j = 1, 2, \ldots, k$.

Theorem 4.5. *A Büchi automaton $A = (\Sigma, S, \rho, S_0, F)$ is non-empty if and only if there is a state $s \in F$ accessible from a state $s_0 \in S_0$ and also accessible from s itself.*

Proof. Assume first that the accessibility condition is satisfied. An accepted word can then be obtained by following the path that goes from the state s_0 to the state s and then indefinitely following the path that goes from s to s.

To prove that the condition is necessary, assume that the automaton accepts some word w. Thus, there is an accepting execution of A on w. To be accepting, this execution must start in an initial state s_0 and must contain some state $s \in F$ an infinite number of times. The state s must then be accessible from s_0 and from itself. □

Theorem 4.5 yields an easily usable criterion for checking if a Büchi automaton is non-empty. Indeed, to check if the criterion given in the theorem is satisfied, it is sufficient to compute the strongly connected components of the graph representing the automaton. This can for instance be done with the algorithm described in [Aho et al. 74]. This algorithm is quite efficient since its running time is a linear function of the size of the graph.

Note that all the examples of Büchi automata we gave previously are non-empty. This can be checked trivially. The benefit of having an algorithm for checking if an automaton is non-empty appears when the automaton is large and cannot be inspected visually. This will be the case for the automata we will build from formulas of linear time temporal logic.

4.2.3 From temporal logic to Büchi automata

We have seen that the semantics of linear time temporal logic could be expressed by a function from $\mathbf{N} \times P$ to $\{\mathbf{T}, \mathbf{F}\}$. Equivalently, such a mapping can be described as a member of the set $(\mathbf{N} \to (P \to \{\mathbf{T}, \mathbf{F}\}))$, i.e., as an infinite sequence of choices of truth values for the atomic propositions. A temporal interpretation is thus an infinite word over the alphabet $(P \to \{\mathbf{T}, \mathbf{F}\})$, or alternatively over the alphabet 2^P. (Recall that the notation 2^P stands for the set of subsets of P. This set is isomorphic to the set of mappings from P

to $\{\mathbf{T},\mathbf{F}\}$. To an element $s \in 2^P$ corresponds the mapping $f_s : P \to \{\mathbf{T},\mathbf{F}\}$ such that for every atomic proposition $q \in P$, one has $q \in s$ if and only if $f_s(q) = \mathbf{T}$.)

From this point of view, a temporal logic formula defines a language of infinite words, namely the set of temporal interpretations that satisfy it. Let us consider an example. The formula $\Diamond p$ is true of every sequence in which p is eventually true. It thus defines the set of infinite words over the alphabet $\Sigma = \{\emptyset, \{p\}\}$ in which the symbol $\{p\}$ appears at least once. This set of words is the one accepted by the Büchi automaton of Figure 4.7.

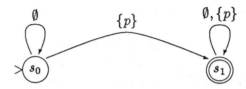

Figure 4.7 : An automaton for $\Diamond p$

The fact that the set of models of the formula $\Diamond p$ is accepted by a Büchi automaton is not serendipitous. We will prove that, for any propositional temporal logic formula, it is possible to build a Büchi automaton that accepts exactly the infinite words satisfying that formula. This result gives us a first measure of the expressiveness of propositional temporal logic: it can at most describe sets of infinite words that are ω-regular. It is natural to ask the reverse question: can every ω-regular set of infinite words be described by a propositional temporal logic formula? The answer is negative, but, as we will see (§ 4.2.10), it is possible to extend temporal logic in order to obtain an exact correspondence[4].

Before defining formally the construction that makes it possible to go from temporal logic to Büchi automata, we will attempt to explain the intuition that underlies it. Suppose one wants to build the models of a formula f. This can be done step by step by extracting from f what should be true at the initial time instant as well as what should be true from the next time instant on. For example, if f is the formula $\Box p$, one can use the identity

$$\Box p \equiv (p \land \bigcirc \Box p). \tag{4.1}$$

To check that a sequence satisfies $\Box p$, it is thus sufficient to check that p is

---

[4] The temporal logic we have defined corresponds actually to a subset of the ω-regular languages known as the counter free ω-regular languages. These languages are characterized by a restricted set of Büchi automata (see [Thomas 81] for more details).

4.2. TEMPORAL LOGIC AND FINITE AUTOMATA

true initially and that $\Box p$ is true from the next time instant on. There are identities similar to (4.1) for the other temporal operators:

$$\Diamond p \equiv (p \vee \bigcirc \Diamond p), \qquad (4.2)$$
$$p \, \mathcal{U} \, q \equiv (q \vee (p \wedge \bigcirc(p \, \mathcal{U} \, q))). \qquad (4.3)$$

To build the Büchi automaton for an arbitrary formula, one starts by using the identities (4.1–4.3) until every temporal operator is in the scope of at least one $\bigcirc$ operator. Once this is done, it is possible to determine which choices of truth values can be assigned to the atomic propositions in the initial state. These choices are those that do not make the formula identically false. For such a choice, the formula that should be true from the next time instant on is the one obtained by substituting the chosen truth values for the atomic propositions that are not in the scope of a temporal operator.

For example, the formula $\Box \Diamond p$ is transformed as follows:

$$\Box \Diamond p \equiv (\Diamond p \wedge \bigcirc \Box \Diamond p),$$
$$\equiv ((p \vee \bigcirc \Diamond p) \wedge \bigcirc \Box \Diamond p).$$

Initially, p can be **T** or **F**. If p is **T**, then $\bigcirc \Box \Diamond p$ still needs to be satisfied, i.e., $\Box \Diamond p$ needs to be satisfied from the next time instant on. If p is **F**, then $\bigcirc \Diamond p \wedge \bigcirc \Box \Diamond p$ still needs to be satisfied, i.e., $\Diamond p \wedge \Box \Diamond p$ needs to be satisfied from the next time instant on.

This examples shows that the formulas that have to be satisfied as one moves through the sequence are Boolean combinations of subformulas of the formula one started with[5]. There is only a finite number of such combinations. Thus, after a finite amount of time, one will encounter a combination that has been seen before. This means that it is possible to represent the constraints corresponding to a temporal formula as a finite automaton whose states are the Boolean combinations of the subformulas of the formula of interest. Note that it is sufficient to consider conjunctions of subformulas. Indeed, it is always possible to write a Boolean formula in disjunctive normal form ([Thayse 88], § 1.1.10) and, assuming that one builds a non-deterministic automaton, the disjuncts can be represented by separate states of the automaton. In this way, one obtains the automaton of Figure 4.8 for the formula $\Box \Diamond p$.

The automaton we have constructed is not completely defined. No set of accepting states has been specified. One might think that the set of accepting states should be the set of all states and hence that all infinite executions of the automaton are accepting. Unfortunately, this definition is not adequate

---

[5] A Boolean combination of formulas $f_1, \ldots, f_k$ is a formula obtained from $f_1, \ldots, f_k$ through the use of the logical connectives $(\wedge, \vee, \neg)$.

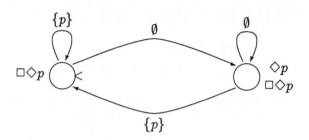

Figure 4.8 : A sketch of an automaton for $\Box\Diamond p$

for the following reason. When the formula $\Diamond p$ is decomposed, one obtains the disjunction $p \vee \bigcirc\Diamond p$. This means that either p should be true immediately (the p alternative) or that one can postpone making p true (the $\bigcirc\Diamond p$ alternative). The problem is that nothing prevents us from postponing forever the time at which p will be true and hence never making p true. Formulas of the form $\Diamond p$ are called *eventualities*. The problem we have just identified is referred to as the *realization of eventualities*.

Eventualities are formulas of the form $\Diamond f$ and, less obviously, formulas of the form $\neg(f_1 \, \mathcal{U} \, f_2)$. Indeed, when applying (4.3) to such a formula, one obtains $\neg(f_1 \, \mathcal{U} \, f_2) \equiv (\neg f_1 \wedge \neg f_2) \vee (\neg f_2 \wedge \bigcirc\neg(f_1 \, \mathcal{U} \, f_2))$. But, $\neg(f_1 \, \mathcal{U} \, f_2)$ is not true of a sequence where $(\neg f_1 \wedge \neg f_2)$ is never true (given that $(f_1 \, \mathcal{U} \, f_2)$ is true in that case). For the eventuality $\neg(f_1 \, \mathcal{U} \, f_2)$ to be realized, $(\neg f_1 \wedge \neg f_2)$ has to be eventually true. Note that this problem does not occur for a formula of the form $f_1 \, \mathcal{U} \, f_2$, given that the semantics of $\mathcal{U}$ makes $f_1 \, \mathcal{U} \, f_2$ true when f_1 is always true. (If we had used the strong definition of $\mathcal{U}$, a formula of the form $f_1 \, \mathcal{U} \, f_2$ would be an eventuality. On the other hand a formula of the form $\neg(f_1 \, \mathcal{U} \, f_2)$ would not be an eventuality.)

4.2.4 The Büchi automaton corresponding to a formula

We will now give a more formal and complete description of the construction of a Büchi automaton from a temporal logic formula[6]. As the preceding discussion indicates, the construction makes use of the subformulas of the formula under consideration. This leads us to define the notion of the *closure* of a formula f. The closure $cl(f)$ of a formula is the smallest set of formulas satisfying the following conditions:

- $f \in cl(f)$,

---

[6]The reader who so desires can skip this detailed description and resume his reading at Subsection 4.2.10 without compromising his ability to understand the rest of this chapter.

4.2. TEMPORAL LOGIC AND FINITE AUTOMATA

- $f_1 \wedge f_2 \in cl(f) \Rightarrow f_1, f_2 \in cl(f)$,
- $f_1 \vee f_2 \in cl(f) \Rightarrow f_1, f_2 \in cl(f)$,
- $f_1 \supset f_2 \in cl(f) \Rightarrow f_1, f_2 \in cl(f)$,
- $\neg f_1 \in cl(f) \Rightarrow f_1 \in cl(f)$,
- $f_1 \in cl(f) \Rightarrow \neg f_1 \in cl(f)$ if f_1 is not of the form $\neg f_1'$,
- $\bigcirc f_1 \in cl(f) \Rightarrow f_1 \in cl(f)$,
- $\square f_1 \in cl(f) \Rightarrow f_1 \in cl(f)$,
- $\Diamond f_1 \in cl(f) \Rightarrow f_1 \in cl(f)$,
- $f_1 \, \mathcal{U} \, f_2 \in cl(f) \Rightarrow f_1, f_2 \in cl(f)$.

The closure of f is thus the set consisting of the subformulas of f and their negation. For example, the closure of the formula $\Diamond p$ is the set

$$cl(\Diamond p) = \{\Diamond p, \neg \Diamond p, p, \neg p\}.$$

Let us note that the closure of a formula f contains at most $2|f|$ elements, where $|f|$ represents the *length* of f, i.e., the number of symbols (propositions, connectives and operators) appearing in the representation of f. We can now state the following theorem.

Theorem 4.6. *Given a propositional linear time temporal logic formula f, it is possible to build a Büchi automaton accepting exactly the infinite sequences satisfying f.*

We will give a commented proof of this result; this proof stretches until the end of Subsection 4.2.8. It is convenient to build the automaton in two parts: on the one hand, an automaton that checks that the sequence satisfies all conditions imposed by the formula, except the realization of eventualities; on the other hand, an automaton that checks that the eventualities are realized. The first automaton will be called the *local automaton* because the conditions it checks require only a step by step (local) check on the sequence. For obvious reasons, the second automaton will be called the *eventuality automaton*.

A slight problem arises from the fact that we build the local and eventuality automata separately. This problem is due to the fact that some communication between the two automata is necessary. Indeed, the eventuality automaton must be able to determine which eventualities have to be realized, but this information appears only in the state of the local automaton. This problem can be solved very easily. One just builds the local and eventuality automata in such a way that they operate on sequences of elements of $2^{cl(f)}$ rather than on sequences of elements of 2^P, i.e., temporal interpretations. The infinite sequences of elements of $2^{cl(f)}$ that the automata will accept are obtained from the sequences of elements of 2^P that satisfy the formula f by associating with each time instant (position in the sequence) the elements of $2^{cl(f)}$ that are true at this time.

With this approach, the information needed by the eventuality automaton lies in the sequence and thus no longer needs to be extracted from the local automaton. This allows a simple composition of the two automata. Once this composition is done, it is straightforward to modify the resulting automaton in such a way that it operates on sequences of elements of 2^P rather than on sequences of elements of $2^{cl(f)}$.

4.2.5 Building the local automaton

The local automaton associated with a formula f is the 5-tuple $L = (2^{cl(f)}, N_L, \rho_L, N_f, N_L)$ defined as follows:

- The alphabet $2^{cl(f)}$ is the set of subsets of the closure of f.

- The set of states N_L includes all subsets s of $cl(f)$ that are propositionally consistent. More precisely, the states s must satisfy the following conditions:

    - For every $f_1 \in cl(f)$, we have $f_1 \in s$ if and only if $\neg f_1 \notin s$[7].
    - For every $f_1 \wedge f_2 \in cl(f)$, we have $f_1 \wedge f_2 \in s$ if and only if $f_1 \in s$ and $f_2 \in s$.
    - For every $f_1 \vee f_2 \in cl(f)$, we have $f_1 \vee f_2 \in s$ if and only if $f_1 \in s$ or $f_2 \in s$.
    - For every $f_1 \supset f_2 \in cl(f)$, we have $f_1 \supset f_2 \in s$ if and only if $\neg f_1 \in s$ or $f_2 \in s$.

    For example, for the formula $\Diamond p$, the set N_L of states of the local automaton is

    $$N_L = \{\{\Diamond p, p\}, \{\Diamond p, \neg p\}, \{\neg \Diamond p, p\}, \{\neg \Diamond p, \neg p\}\}.$$

- The transition function ρ_L must check that the symbol being read is compatible with the state of the automaton and that the next state is

---

[7]Note that this condition implies that for every pair of formulas f_1, $\neg f_1$ in $cl(f)$, a state s includes either f_1, or $\neg f_1$.

4.2. TEMPORAL LOGIC AND FINITE AUTOMATA

compatible with the semantics of the temporal operators. To ensure the first of these properties, one requires that $\rho_L(s, a)$ be non-empty if and only if $s = a$. In other words, a transition is only possible if the symbol being read coincides with the (fictitious) label of the current state. For the second property, one also requires that a state t be element of $\rho_L(s, a)$ only if the following conditions are satisfied:

- For every $\bigcirc f_1 \in cl(f)$, we have $\bigcirc f_1 \in s$ if and only if $f_1 \in t$.
- For every $\Box f_1 \in cl(f)$, we have $\Box f_1 \in s$ if and only if $f_1 \in s$ and $\Box f_1 \in t$.
- For every $\Diamond f_1 \in cl(f)$, we have $\Diamond f_1 \in s$ if and only if either $f_1 \in s$, or $\Diamond f_1 \in t$.
- For every $f_1 \, \mathcal{U} \, f_2 \in cl(f)$, we have $f_1 \, \mathcal{U} \, f_2 \in s$ if and only if either $f_2 \in s$, or $f_1 \in s$ and $f_1 \, \mathcal{U} \, f_2 \in t$.

- The set N_f of initial states is the set of states that include the formula f. Indeed, we have to recognize the sequences whose initial state satisfies f.

- The set of accepting states is N_L, the set of all states. Every infinite execution of the automaton is thus accepting.

The local automaton for the formula $\Diamond p$ is shown in Figure 4.9. In that figure, the label of each transition is identical to the label of the state it originates from.

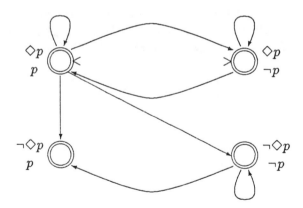

Figure 4.9 : The local automaton for $\Diamond p$

4.2.6 Building the eventuality automaton

The eventuality automaton is supposed to check that the eventualities are realized. For this, it must check that, each time a formula of the form $\Diamond f$ (or

$\neg(f_1 \, \mathcal{U} \, f_2))$ appears, there is a later time instant at which f (or $\neg f_1 \wedge \neg f_2$) appears. What makes this verification somewhat difficult is that it should be constantly repeated, given that eventualities can appear at any time instant and can reappear indefinitely. For example, in the case of the formula $\Box \Diamond p$ we need to check that the eventuality $\Diamond p$ is realized at each time instant. Fortunately, checking the realization of eventualities is quite easy in view of the following property. If an eventuality is not realized at a given time instant, it reappears at the next time instant. Indeed, $\Diamond f$ satisfies the formula $\Diamond f \equiv (f \vee \bigcirc \Diamond f)$ and similarly, $\neg(f_1 \, \mathcal{U} \, f_2)$ satisfies $\neg(f_1 \, \mathcal{U} \, f_2) \equiv (\neg f_1 \wedge \neg f_2) \vee (\neg f_2 \wedge \bigcirc \neg(f_1 \, \mathcal{U} \, f_2))$.

Thanks to this property, it is sufficient to check periodically that the eventualities are realized. This is exactly what the eventuality automaton does. It starts by finding out which eventualities have to be realized at the initial time instant, then it checks that these are realized. Once this is done, it repeats the same steps for the eventualities that have to be realized at the next time instant and so on forever.

The eventuality automaton is defined as follows. Let $ev(f)$ be the set of eventualities of a formula f, i.e., the set of elements of $cl(f)$ that are of the form $\Diamond f_1$ or of the form $\neg(f_1 \, \mathcal{U} \, f_2)$. The eventuality automaton is then $F = (2^{cl(f)}, 2^{ev(f)}, \rho_F, \{\emptyset\}, \{\emptyset\})$.

- As in the case of the local automaton, the alphabet is the set $2^{cl(f)}$ of subsets of the closure of the formula f.

- The set of states is the set $2^{ev(f)}$ of subsets of the eventualities of the formula f. Intuitively, being in a state $\{e_1, \ldots, e_k\}$ means that the eventualities $e_1, \ldots, e_k$ still have to be realized.

- The transition function ρ_F is defined by $\mathbf{t} \in \rho_F(\mathbf{s}, a)$ if and only if one of the following two conditions is satisfied:

    - $\mathbf{s} = \emptyset$ and
        * for all $\Diamond f \in a$, one has $\Diamond f \in \mathbf{t}$ if and only if $f \notin a$.
        * for all $\neg(f_1 \, \mathcal{U} \, f_2) \in a$, one has $\neg(f_1 \, \mathcal{U} \, f_2) \in \mathbf{t}$ if and only if $(\neg f_1 \wedge \neg f_2) \notin a$.

    When it is in the empty state, the automaton determines which eventualities have to be realized by looking at the sequence. It then goes to a state characterized by these eventualities.

    - $\mathbf{s} \neq \emptyset$ and
        * for all $\Diamond f \in \mathbf{s}$, one has $\Diamond f \in \mathbf{t}$ if and only if $f \notin a$.
        * for all $\neg(f_1 \, \mathcal{U} \, f_2) \in \mathbf{s}$, one has $\neg(f_1 \, \mathcal{U} \, f_2) \in \mathbf{t}$ if and only if $(\neg f_1 \wedge \neg f_2) \notin a$.

4.2. TEMPORAL LOGIC AND FINITE AUTOMATA

When it is in a non-empty state, the eventuality automaton determines, among the eventualities it has to realize, which ones can be realized at the present time. It then moves to a state containing the eventualities of the present state minus those that have just been realized.

- The only initial state is the state ∅ (the empty set of eventualities). Indeed, when it starts its execution, the eventuality automaton must first determine which eventualities have to be realized.

- The only accepting state is ∅. Indeed, when the automaton goes through this state, it has checked that all the eventualities that appeared the last time it went through this state have been realized. If it goes through this state an infinite number of times, the eventualities present at an infinite number of times in the sequence will be realized. As we have argued above, this is sufficient for all eventualities to be realized.

The eventuality automaton for the formula $\Diamond p$ is shown in Figure 4.10.

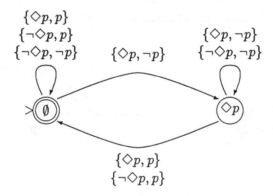

Figure 4.10 : The eventuality automaton for $\Diamond p$

4.2.7 Composing the two automata

The next step in the construction of a Büchi automaton from a temporal logic formula is to compose the local and eventuality automata to obtain the automaton that recognizes the models of the formula. We will call this automaton the *model automaton*. It accepts the intersection of the languages respectively accepted by the local and eventuality automata. It can thus be obtained by the construction of Theorem 4.4. Note however that, as the set of accepting states of the local automaton is the whole set of its states, one of the two sets of accepting states given by Theorem 4.4 is the whole set of states. Therefore, it does not impose any restriction on executions and can be ignored. One can hence directly obtain a Büchi automaton without using Lemma 4.3. Formally, this automaton is $M = (2^{cl(f)}, N_M, \rho_M, N_{M0}, F_M)$ where

- $N_M = N_L \times 2^{ev(f)}$,

- $(\mathbf{p},\mathbf{q}) \in \rho_M((\mathbf{t},\mathbf{s}),a)$ if and only if $\mathbf{p} \in \rho_L(\mathbf{s},a)$ and $\mathbf{q} \in \rho_F(\mathbf{t},a)$,

- $N_{M0} = N_f \times \{\emptyset\}$,

- $F_M = N_L \times \{\emptyset\}$.

The composition of the local and eventuality automata for the formula $\Diamond p$ is given in Figure 4.11. The labels of the transitions are not shown. By definition, the label of a transition from a state $(\mathbf{t},\mathbf{s})$ is the component $\mathbf{t}$ of that state (the one contributed by the local automaton).

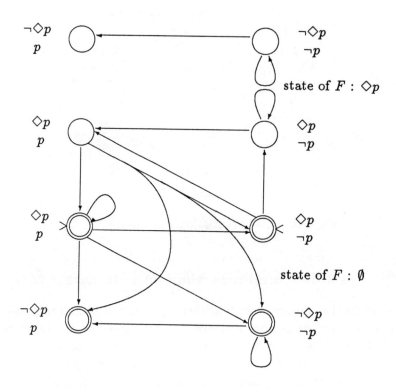

Figure 4.11 : The model automaton for $\Diamond p$

4.2.8 The automaton on the alphabet 2^P

The last step of the construction is to transform the automaton M that operates on the alphabet $2^{cl(f)}$ into an automaton M' that operates on the alphabet 2^P, where P is the set of atomic propositions of the formula f. The automaton M' is identical to M except for the alphabet and the transition relation. The alphabet is now 2^P and the transition relation is defined by $t \in \rho_{M'}(s, a)$ if and only if there is some $b \in 2^{cl(f)}$ such that $a = b \cap P$ and $t \in \rho_M(s, b)$. This means that M' is obtained from M by replacing the label of each transition by its intersection with P. Figure 4.12 shows the result of this last transformation after elimination of the states that cannot contribute to accepting executions. In this figure, the label of a transition is identical to that of the state it originates from.

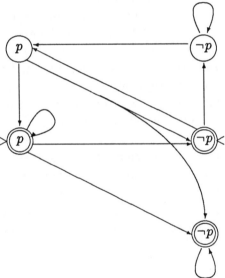

Figure 4.12 : The final automaton for $\Diamond p$

To complete the proof of Theorem 4.6, we have still to validate our construction and to prove rigorously that the sequences accepted by the constructed automaton are exactly those that satisfy the formula one started with. Such a proof is rather tedious and hardly necessary since it is easy to convince ourselves that our construction implements the semantics of temporal logic. A complete proof can be found in [Vardi and Wolper 88].

4.2.9 Algorithmic considerations

Some remarks are in order about the *size* (i.e., the number of states) of the Büchi automaton that can be constructed from a temporal logic formula. Let us first mention that the construction presented in the proof of Theorem 4.6 has been chosen for its simplicity and generality. It does not always yield a minimal automaton. This is easy to see by comparing the automata of Figures 4.12 and 4.7. By analysing the construction, one sees that the size of the constructed automaton is generally exponential and can be as large as $2^{3|f|}$. Indeed, the size of the local automaton is bounded by the size of $2^{cl(f)}$ which is at most equal to $2^{2|f|}$, and the size of the eventuality automaton is equal to the size of $2^{ev(f)}$ which is at most $2^{|f|}$ (the number of eventualities in a formula is at most equal to the length of that formula).

However, even if the construction is not optimal, the exponential growth of the size of the automaton is unavoidable. It is indeed possible to exhibit families of formulas for which the size of the Büchi automaton increases exponentially as a function of the length of the formula.

The technique we have used to prove Theorem 4.6 is quite general and can be applied to several variants of temporal and modal logic. Moreover, the connection we have established between temporal logic and Büchi automata is a fruitful source of results concerning temporal logic. For instance, in Subsection 4.4.9 we will describe an application of this connection to the verification of concurrent programs. For now, we will use it to obtain a decision procedure for temporal logic. We have the following theorem.

Theorem 4.7. *Given a propositional temporal logic formula f, it is possible to decide algorithmically if this formula is satisfiable.*
Proof. The first step is to apply the construction of Theorem 4.6 to the formula f. This yields an automaton A_f that accepts exactly the infinite words that satisfy f. To decide if such a word exists, one just has to use Theorem 4.5 to decide if the language accepted by A_f is non-empty. □

Deciding if a Büchi automaton is non-empty can be done in time proportional to the size of the automaton. The decision procedure we have just outlined thus requires an exponential amount of time to decide if a propositional temporal logic formula is satisfiable[8].

4.2.10 Extended Temporal Logic

In the preceding subsections, we have proved that any linear time temporal logic formula could be converted into a Büchi automaton. We have also mentioned that the inverse construction is not always possible: there are sets of

---

[8]One can prove that the satisfiability problem for temporal logic is complete in the class of problems solvable in polynomial space. Given what is presently known in complexity theory, it is quite unlikely that problems that are complete in this class can be solved by polynomial time algorithms.

4.2. Temporal logic and finite automata

sequences that are accepted by a Büchi automaton but cannot be described by a linear time temporal logic formula. As an example, let us consider the set of sequences in which the proposition p is true in every even state (nothing is required of odd states). This set contains, among others, the sequences

$$\begin{array}{ccccccccccccc} & 0 & 1 & 2 & 3 & 4 & 5 & 6 & 7 & 8 & 9 & 10 & \ldots \\ s_1 = & p & \neg p & p & \neg p & p & \neg p & p & \neg p & p & \neg p & p & \ldots \\ s_2 = & p & p & p & \neg p & p & \neg p & p & \neg p & p & \neg p & p & \ldots \end{array}$$

but does not contain the sequences

$$\begin{array}{ccccccccccccc} & 0 & 1 & 2 & 3 & 4 & 5 & 6 & 7 & 8 & 9 & 10 & \ldots \\ s_3 = & \neg p & \neg p & p & \neg p & p & \neg p & p & \neg p & p & \neg p & p & \ldots \\ s_4 = & p & p & \neg p & \neg p & p & \neg p & p & p & p & \neg p & p & \ldots \end{array}$$

It is rather surprising that the property defining this set of sequences (let us represent it by $even(p)$) cannot be characterized by a temporal logic formula. One might be inclined to think that one of the formulas

$$p \wedge \Box(p \supset \bigcirc \neg p) \wedge \Box(\neg p \supset \bigcirc p), \quad \text{or} \quad p \wedge \Box(p \supset \bigcirc \bigcirc p)$$

could do the job. This is not the case since these formulas are false for the sequence s_2, whereas this sequence is in the set being considered.

It is possible to prove that there is no temporal logic formula that exactly describes this set of sequences; more generally, one has the following lemma [Wolper 82], [Wolper 83], which we will not prove here.

Lemma 4.8. *Any temporal logic formula built from an atomic proposition p and including at most n operators $\bigcirc$ has the same truth value for all the sequences of the form $\sigma_k = p^k(\neg p)p^\omega$ with $k > n$.*

It is a straightforward consequence of Lemma 4.8 that the property $even(p)$ is not expressible in temporal logic. Indeed, assume it is expressed by a formula containing a given number n of operators $\bigcirc$. By the lemma, this formula has the same truth value on the sequences σ_{n+1} and σ_{n+2}, which contradicts the fact that the property $even(p)$ is true for one of these sequences and false for the other. This argument can be generalized to prove that, for any $m \geq 2$, the property stating that p must be true at every time instant that is a multiple of m is not expressible in temporal logic.

We will now present an extension of temporal logic in which it is possible to express the property $even(p)$ and, more generally, any property that can be defined by a Büchi automaton.

4.2.11 Definition of extended temporal logic

When trying to extend temporal logic in order to be able to express the property presented in the preceding subsection, one of the first observations we have to make is that a formula expressing $even(p)$ would satisfy the relation

$$even(p) \equiv p \wedge \bigcirc\bigcirc even(p). \tag{4.4}$$

This relation is very similar to the one satisfied by the formula $\Box p$, i.e.,

$$\Box p \equiv p \wedge \bigcirc \Box p. \tag{4.5}$$

The similarity between (4.4) and (4.5) suggests that it should be possible to define an operator 'even' just as we defined an operator 'always' ($\Box$). This operator, when applied to an argument p, would be true at the point i of a sequence if p is true at all points $i + 2k$ of that sequence, with $k \in \mathbf{N}$. Moreover, the relation (4.4) could be used to axiomatize this operator and to extend the construction of a Büchi automaton described in Subsections 4.2.4–4.2.8 to formulas containing the operator 'even'. It thus seems possible to extend temporal logic while preserving its most interesting properties: complete axiomatization and algorithmic construction of a Büchi automaton from a formula.

This approach is not limited to the operator 'even'. One could also use it for the operators 'multiple of 3', 'multiple of 4', and so on. What makes it possible to handle operators similar to 'even' is that they are characterized by inductive definitions similar to (4.4). These relations are analogous to the productions of a regular grammar or to the transitions of a finite automaton. From this comes the idea of a systematic way of extending temporal logic: add to temporal logic any operator definable by a finite automaton. We will call such an enriched temporal logic *extended temporal logic* and refer to it by the acronym ETL.

We still have to make this definition precise. A first question that should be answered concerns the type of finite automata to use: finite automata on finite words or Büchi automata? As the sequences considered by temporal logic are infinite, Büchi automata seem natural. We will see later that it is also possible to define extended temporal logic with a restricted class of Büchi automata or even with finite automata on finite words. Let us nevertheless start by the general definition of a temporal operator from a Büchi automaton.

Consider a Büchi automaton $A = (\Sigma, S, \rho, S_0, F)$. To define an operator from this automaton, we associate a formula with each element of the alphabet Σ. These formulas will be the arguments of our operator. Thus, if $\Sigma = \{a_1, \ldots, a_n\}$, the operator associated with the automaton will have n arguments and its application to $f_1, \ldots, f_n$ will be denoted by $A(f_1, \ldots, f_n)$. The arguments of an operator defined by an automaton can of course be arbitrary

4.2. TEMPORAL LOGIC AND FINITE AUTOMATA

formulas of extended temporal logic. The semantics of $A(f_1,\ldots,f_n)$ is defined in terms of the words accepted by the automaton A. Intuitively, a sequence satisfies $A(f_1,\ldots,f_n)$ if there is some word accepted by the automaton such that if the symbol a_i appears at the position j in that word, then f_i is true at that position. More formally, we have:

- $\mathcal{I} \models_s A(f_1,\ldots,f_n)$ if there is an infinite word $w = a_{w_0} a_{w_1} a_{w_2} \ldots$ accepted by the automaton A such that $\mathcal{I} \models_{R^j(s)} f_{w_j}$ for every $j \geq 0$.

If one defines the semantics of temporal logic in terms of a sequence $\pi : \mathbf{N} \times P \to \{\mathbf{T}, \mathbf{F}\}$, this clause becomes:

- $\pi \models_s A(f_1,\ldots,f_n)$ if there is an infinite word $w = a_{w_0} a_{w_1} a_{w_2} \ldots$ accepted by the automaton A such that $\pi \models_{s+j} f_{w_j}$ for every $j \geq 0$.

The rest of the semantics of extended temporal logic is identical to that of classical temporal logic.

As a first example, consider the automaton A_1 of Figure 4.13. The operator $A_1(f_1)$ defined by this automaton is the operator $\Box$. Indeed, this automaton accepts just one word which is a_1^ω. A sequence thus satisfies $A_1(f_1)$ if and only if f_1 is true at every time instant; this is exactly the property specified by the operator $\Box$. The other operators of temporal logic can be defined in a similar way. The operator $\Diamond$ is the dual of the operator $\Box$; one thus has $\Diamond f \equiv \neg A_1(\neg f)$. For $\mathcal{U}$, one has $f_1 \,\mathcal{U}\, f_2 \equiv A_2(f_1, f_2, \mathbf{T})$ where A_2 is the automaton of Figure 4.14. Finally, for the operator $\bigcirc$, one has $\bigcirc f \equiv A_3(\mathbf{T}, f, \mathbf{T})$, with the automaton A_3 that is defined in Figure 4.15.

The property $even(p)$ is expressed similarly with the operator defined by the automaton of Figure 4.16. This automaton accepts just one word, $(a_1 a_2)^\omega$. The formula $A_4(f_1, f_2)$ will thus be true of a given sequence if and only if f_1 is true in all the even states of that sequence and f_2 is true in all the odd states. Therefore, the property $even(p)$ can be expressed by $A_4(p, \mathbf{T})$.

Figure 4.13 : An automaton defining the operator $\Box$

In all the examples we have given so far, the automata have a special characteristic: all their states are accepting. We will call this type of automata

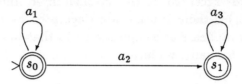

Figure 4.14 : An automaton defining the operator $\mathcal{U}$

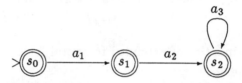

Figure 4.15 : An automaton defining the operator $\bigcirc$

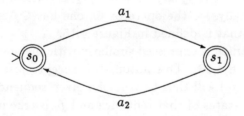

Figure 4.16 : An automaton defining the operator $even(p)$

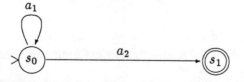

Figure 4.17 : An automaton defining the operator $\Diamond$

4.2. TEMPORAL LOGIC AND FINITE AUTOMATA

looping automata[9] and denote by ETL_l the extended logic where all operators are defined by looping automata[10]. The interesting point is that this logic is sufficient for describing any set of sequences accepted by a Büchi automaton. It is thus equivalent to the logic ETL, as far as expressive power is concerned. However, the construction that transforms an ETL formula into an ETL_l formula is quite complex [Wolper 82], [Vardi and Wolper 88]. An advantage of the logic ETL_l is that it is much easier to deal with than the logic ETL [Sistla et al. 87], [Vardi and Wolper 88]. This greater simplicity comes from the fact that it is possible to complement looping automata with the Rabin-Scott subset construction, which is not possible for Büchi automata. In particular, a complete axiomatization has been given for ETL_l [Wolper 83], [Banieqbal and Barringer 86], whereas no such axiomatization is known for ETL.

4.2.12 Automata on finite words

There is another natural restricted version of extended temporal logic which has the same expressive power as Büchi automata. In this version of the logic, the operators are defined by *finite automata on finite words* and not by finite automata on infinite words. We will denote this version of the logic by the acronym ETL_f. The semantics of an automaton operator of ETL_f is defined similarly to that of an ETL automaton operator. Consider a finite automaton on finite words $A = (\Sigma, S, \rho, S_0, F)$ where $\Sigma = \{a_1, \ldots, a_n\}$. The semantics of $A(f_1, \ldots, f_n)$ is defined in terms of the finite words accepted by the automaton A. Intuitively, an infinite sequence satisfies $A(f_1, \ldots, f_n)$ if there is some finite word of length l accepted by the automaton A such that if the symbol a_i appears in the position j of that word, with $j < l$, then the formula f_i is true in that position. Nothing is imposed on the sequence for the positions following l. More formally, we have:

- $\mathcal{I} \vDash {}_s A(f_1, \ldots, f_n)$ if there is a finite word $w = a_{w_0} a_{w_1} a_{w_2} \ldots a_{w_{l-1}}$ accepted by A such that $\mathcal{I} \vDash_{R^j(s)} f_{w_j}$ for $j = 0, 1, \ldots, l-1$.

Just as the logic ETL_l, the logic ETL_f can express any property definable by a Büchi automaton. There is a duality between these two logics similar to the one that exists in temporal logic between the operators $\Box$ and $\Diamond$. An ETL_l operator can be expressed by the negation of an ETL_f operator and vice versa. To illustrate this, consider the automaton A_5 of Figure 4.17. This automaton can be used to define the operator $\Diamond$ ($\Diamond f \equiv A_5(\mathbf{T}, f)$), as well as

---
[9] As all the states are accepting, for an execution to be accepting it just has to be infinite; it *loops* in the automaton.
[10] This logic is essentially the one that was introduced in [Wolper 82] and [Wolper 83].

the strong version of the 'until' operator ($f_1 \; \mathcal{U} \; f_2 \equiv A_5(f_1, f_2)$). In ETL$_l$, these two operators would be expressed with a negation applied to an automaton operator.

4.2.13 Conclusion

In this section, we have extended temporal logic with operators defined by finite automata. We have considered three versions of this extension:

- ETL : operators defined by Büchi automata,
- ETL$_l$: operators defined by looping automata,
- ETL$_f$: operators defined by automata on finite words.

Each of these three variants can express all the properties definable by Büchi automata. This result is obvious in the case of ETL, but requires a rather complex proof in the case of the other two logics (see [Vardi and Wolper 88]). It is possible to build a Büchi automaton from a formula of each of these logics. In each case, the size of the automaton is exponential in the size of the formula. In the case of the logics ETL$_f$ and ETL$_l$, the construction is similar to the one we have presented in Subsections 4.2.4–4.2.8. In the case of the logic ETL, it is more complex. Similarly to what was proved in Theorem 4.7 for classical temporal logic, the construction of the automaton yields a decision procedure for each of the extended logics. Finally, let us note that for ETL$_l$ and for ETL$_f$ there are some relatively simple complete axiomatizations [Wolper 83], [Banieqbal and Barringer 86], which is not the case for ETL.

The extension of temporal logic with automata not only has the advantage of increasing its expressive power but also of showing how these two seemingly distinct formalisms can be merged.

Finally, let us note that the method of extending temporal logic that we have described is not the only one possible. Two other techniques are, on the one hand, the use of fixpoint operators (see [Banieqbal and Barringer 86] and [Vardi 88]) and, on the other hand, the use of quantification on propositions [Sistla et al. 87]. It is quite interesting that the expressive power of these other extensions also coincides with that of Büchi automata.

4.3 Predicate linear temporal logic

4.3.1 Introduction

Propositional linear temporal logic can be extended by introducing variables, functions and predicates. This extension is *predicate temporal logic*. As in clas-

4.3. PREDICATE LINEAR TEMPORAL LOGIC

sical logic [Thayse 88], this extension greatly increases the expressiveness of the logic, but makes the deduction techniques more complex. Furthermore, these techniques are no longer complete. At the present time, the main application area for predicate temporal logic is the specification, design and verification of concurrent systems.

The language of predicate linear temporal logic is defined from a set of symbols denoting either 'global objects' or 'local objects'. An example of a global object is a person; the meaning of the symbol denoting this person remains the same in all states. An example of a local object is the holder of a position. Several persons can be appointed successively to this position; the symbol denoting the person holding the position will represent different persons, depending on the state being considered.

4.3.2 Syntax

The language of predicate linear temporal logic (or first-order linear temporal logic) is built from the following sets of symbols:

1. The set of *logical connectives* ($\neg$, $\wedge$, $\vee$, $\supset$ and $\equiv$),
2. The set of *temporal operators* ($\square$, $\diamond$, $\bigcirc$ and $\mathcal{U}$),
3. The set of *quantifiers* ($\forall$, $\exists$),
4. The *equality* connective (=),
5. A set P of *global predicate constants*,
6. A set F of *global function constants*,
7. A set Q of *local propositions*,
8. A set L of *local individual constants*,
9. A set V of *global variables*.

These sets are disjoint. A natural number, called the *arity*, is associated with every function constant and with every predicate constant.

The formulas of the language are obtained by construction rules, similar to those used in predicate logic, on the one hand, and in propositional temporal logic, on the other hand.

- A *term* is a variable, a local individual constant or a functional form.

- A *functional form* is the application of a function constant to an adequate number of terms. If f is a function constant of arity n and if $t_1,\ldots,t_n$ are terms, then the corresponding form is usually denoted $f(t_1,\ldots,t_n)$. If n is 0, the form is written f instead of $f(\ )$ and is called a (global) individual constant.

- A *predicate form* is the application of a predicate constant to an adequate number of terms. If p is a (global) predicate constant of arity m and if $t_1, \ldots, t_m$ are terms, then the corresponding form is usually denoted $p(t_1, \ldots, t_m)$. If m is 0, the form is written p instead of $p(\ )$ and is called a (global) propositional constant.

- An *atom* is a predicate form, a local proposition or an equality.

- An *equality* is an expression $(s = t)$, where s and t are terms.

- The concept of a *formula* is defined recursively as follows:
    - an atom is a formula;
    - if A is a formula, then $\neg A$, $\Box A$, $\Diamond A$ and $\bigcirc A$ are formulas;
    - if A and B are formulas, then $(A \wedge B), (A \vee B), (A \supset B), (A \equiv B)$ and $(A \, \mathcal{U} \, B)$ are formulas;
    - if A is a formula and if x is a variable, then $\forall x A$ and $\exists x A$ are formulas.

Let us observe that, in the logic we have defined, all symbols are global except some individual constants and some propositions which may be local. In fact, local elements of every type could be introduced, but only propositions and individual constants are considered here, since the other local elements are not used in practice.

4.3.3 Semantics

For given sets of global and local symbols as above, a *temporal interpretation* is a 5-tuple (S, R, D, I_c, I_v) with the following properties.

- (S, R) is a *linear temporal frame* (§ 4.1.3).

- D is a non-empty set, called the *interpretation domain*.

- I_c is an *interpretation function for constants*, which maps
    - each global function constant $f \in F$ of arity n to a function $I_c(f)$ from D^n into D;[11]
    - each global predicate constant $p \in P$ of arity m to a function $I_c(p)$ from D^m into the set $\{\mathbf{T}, \mathbf{F}\}$ of truth values;
    - each pair (s, l) consisting of a state $s \in S$ and a local individual constant $l \in L$ to an element $I_c(s, l) \in D$;

---

[11] By convention, a function from D^0 into B is simply an element of B.

4.3. PREDICATE LINEAR TEMPORAL LOGIC

- each pair (s,q) consisting of a state $s \in S$ and a local proposition $q \in Q$ to a truth value $I_c(s,q)$.

- I_v is an *interpretation function for variables*, which maps each variable $x \in V$ to an element $I_v(x) \in D$.

Remark. The notion of temporal predicate interpretation generalizes both the notion of classical predicate interpretation introduced in ([Thayse 88], § 1.2.5), and the notion of propositional temporal interpretation introduced in Subsection 4.1.3.

Given a temporal interpretation $\mathcal{I} = (S, R, D, I_c, I_v)$ and a state $s \in S$, interpretation rules can be stated. These rules associate a truth value $\mathcal{I}(A)$ with each formula A, and associate an element $\mathcal{I}(t)$ of D with each term t.

- If x is a variable, then $\mathcal{I}(s,x) =_{def} I_v(x)$.

- If l is a local individual constant, then $\mathcal{I}(s,l) =_{def} I_c(s,l)$.

- If f is a function constant of arity n and if $t_1, \ldots, t_n$ are terms, then $\mathcal{I}(s, f(t_1, \ldots, t_n)) =_{def} I_c(f)(\mathcal{I}(s,t_1), \ldots, \mathcal{I}(s,t_n))$.

- If p is a predicate constant of arity m and if $t_1, \ldots, t_m$ are terms, then $\mathcal{I}(s, p(t_1, \ldots, t_m)) =_{def} I_c(p)(\mathcal{I}(s,t_1), \ldots, \mathcal{I}(s,t_m))$.

- If q is a local proposition, then $\mathcal{I}(s,q) =_{def} I_c(s,q)$.

- If t_1 and t_2 are terms, then $\mathcal{I}(s, t_1 = t_2)$ is **T** if $\mathcal{I}(s,t_1) = \mathcal{I}(s,t_2)$ and is **F** otherwise.

- If A and B are formulas, then $\neg A$, $(A \wedge B)$, $(A \vee B)$, $(A \supset B)$ and $(A \equiv B)$ are interpreted as in propositional logic.

- If A and B are formulas, then $\bigcirc A$, $\square A$, $\Diamond A$ and $(A \,\mathcal{U}\, B)$ are interpreted as in propositional temporal logic.

- If A is a formula and if x is a variable, then $\mathcal{I}(s, \forall x\, A)$ is **T** if $\mathcal{I}_{x/d}(A)$ is **T** for every $d \in D$, and $\mathcal{I}(s, \exists x\, A)$ is **T** if $\mathcal{I}_{x/d}(A)$ is **T** for at least one $d \in D$. The symbol $\mathcal{I}_{x/d}$ denotes the interpretation identical to $\mathcal{I}$, except that $I_v(x) = d$.

As in the propositional framework, it is possible to limit oneself to the case where S is the set of natural numbers and where R is the successor function. A temporal interpretation can therefore be viewed as an infinite sequence of classical interpretations, in which the global symbols are interpreted uniformly.

Let us point out that the temporal aspect of the logic does not alter the role of the variables; in particular, they can be free or bound, as in classical predicate logic. If A is a formula, $\mathcal{I}$ an interpretation and π a sequence (an interpretation where S is $\mathbf{N}$ and R is the successor function), the notations $\mathcal{I} \vDash_s A$ and $\pi \vDash_s A$ are often used instead of $\mathcal{I}(s, A) = \mathbf{T}$ and of $\pi(s, A) = \mathbf{T}$ respectively. As in propositional temporal logic (§ 4.1.3), the notation $\mathcal{I} \vDash A$ means that $\mathcal{I} \vDash_s A$ holds for each state s of the model $\mathcal{I}$. Similarly, $\pi \vDash A$ stands for $\pi \vDash_0 A$. The formula A is *valid* if $\mathcal{I} \vDash A$ holds for each temporal interpretation $\mathcal{I}$, or if $\pi \vDash A$ holds for each sequence π (both conditions are equivalent).

Several terminologies and notations have been used to define the semantics of temporal predicate logic. As in the case of propositional temporal logic, the concept of model needs to be clarified. The word 'model' can refer to a mere temporal frame (S, R), or to a part of an interpretation (S, R, D, I_c), or to an interpretation (S, R, D, I_c, I_v). Let us recall that, in this chapter, a *model* of a formula A always refers to an interpretation for which A is true.

4.3.4 Axiomatic system

For first-order temporal logic to be actually used as a reasoning tool, an axiomatic system is necessary. The natural idea is to combine the axiomatic system for classical first-order logic ([Thayse 88], § 2.1.8), including the axioms and inference rules for equality ([Thayse 88], § 2.1.10), with the system developed for propositional temporal calculus (§ 4.1.5). However, one has to be careful in this matter. First, the schema

$$\forall x\, A(x) \supset A(t),$$

which is valid in classical logic, remains valid in temporal logic if and only if substituting the term t for the variable x does not introduce occurrences of local constants into the scope of a temporal operator. When this condition is satisfied, the term t is said to be *substitutable* for x in A [Manna and Pnueli 79], [Kröger 87]. A similar restriction applies to the substitutivity rules concerning equality. The rules

$$\frac{x_1 = t_1 \wedge x_2 = t_2 \wedge \cdots \wedge x_m = t_m}{p(x_1, x_2, \ldots, x_m) \equiv p(t_1, t_2, \ldots, t_m)}$$

$$\frac{x_1 = t_1 \wedge x_2 = t_2 \wedge \cdots \wedge x_n = t_n}{f(x_1, x_2, \ldots, x_n) = f(t_1, t_2, \ldots, t_n)}$$

are valid if and only if each term t_i is substitutable for the corresponding variable x_i in the formula $p(x_1, \ldots, x_m)$ and in the term $f(x_1, \ldots, x_n)$, respectively.

4.3. Predicate Linear Temporal Logic

Furthermore, the relations between the temporal operators and the quantifiers should be mentioned. They can be written as follows:

$$\forall x \lozenge A \equiv \lozenge \forall x A,$$
$$\forall x \square A \equiv \square \forall x A.$$

Several axiomatic systems have been introduced in the literature (see e.g. [Manna and Pnueli 79], [Kröger 87], [Abadi 87]). Such a system consists of a recursive set of axioms and of a finite set of inference rules. The set of axioms used in this chapter is defined by the rules listed below [Kröger 87].

- Every instance of a valid schema of propositional temporal logic is an axiom[12].

- If formula A does not contain local objects,
  then $(A \supset \lozenge A)$ is an axiom.

- If x is a global variable, if $A(x)$ is a formula and if the term t is substitutable for x in $A(x)$,
  then $(\forall x\, A(x) \supset A(t))$ is an axiom.

- If A is a formula and if x is a global variable,
  then $(\exists x \neg A \equiv \neg \forall x A)$ is an axiom.

- If A is a formula and if x is a global variable,
  then $(\forall x \lozenge A \supset \lozenge \forall x A)$ is an axiom.

The inference rules are the following:

- If A and $A \supset B$ are theorems,
  then B is a theorem (*Modus Ponens*).

- If A is a theorem,
  then $\square A$ is a theorem (*Necessitation*).

- If the global variable x does not occur free in A and if $A \supset B$ is a theorem,
  then $A \supset \forall x B$ is a theorem (*Generalization*).

An axiomatic system for temporal predicate calculus with equality is obtained by adding the reflexivity axiom for equality together with the substitutivity rules, restricted as described at the beginning of this subsection.

---

[12] As in classical logic, this rule is acceptable because propositional temporal logic is decidable (§ 4.2.4).

Unlike the classical predicate calculus, the temporal predicate calculus cannot be axiomatized completely [Abadi 87]. Nevertheless, the axiomatic system presented here is sufficient to deduce a lot of useful theorems.

As in classical logic, it is convenient to introduce *derived* inference rules, which give rise to shorter and easier proofs. Interesting derived rules and theorems are obtained from stating the relations between temporal operators and quantifiers. For instance, if A and B are formulas and if x is a global variable which does not occur free in B, then the formulas $\forall x\, (\bigcirc A \wedge B)$ and $\bigcirc \forall x\, A \wedge B$ are equivalent.

4.3.5 First-order temporal theories

Classical first-order logic can be adapted to specific domains ([Thayse 88], chapter 2). The result is a *first-order theory*. Such a theory consists of two components, a signature and a set of axioms, or postulates. The *signature* is a set of function constants and predicate constants which will be interpreted on the specific domain; no other constant will occur in the formulas of the theory. The *axioms* are formulas which restrict the possible interpretations of the elements of the signature. Only interpretations for which the axioms are true are taken into consideration.

Specific domains can be investigated in a temporal framework too; *temporal first-order theories* are used for that purpose. The signature of a temporal first-order theory can contain global constants, but also local individual constants and local propositions. An example will be presented in detail in Subsection 4.4.5.

4.4 Applications of temporal logic

4.4.1 Introduction

Temporal logic has proven useful to specify and verify concurrent systems. An interesting feature of temporal logic is the possibility of expressing properties like 'something (e.g. program termination) will eventually happen', which are not so easily expressed in other formalisms.

The problem investigated in what follows is the verification problem for a concurrent system. This problem is to determine whether a given system is correct with respect to a specification, stated as a temporal logic formula. Two approaches will be considered. The first one is based on predicate temporal logic and the second one on propositional temporal logic. The first approach is more general and gives an intuitive insight into the programs under investigation. When it can be applied, the second approach has the advantage of

4.4. Applications of Temporal Logic

allowing the automation of the proof process.

As a first step, an elementary example will be examined in order to introduce the concepts needed to specify and verify concurrent systems. This example will be studied in the framework of predicate temporal logic. Thereafter, both techniques for verifying concurrent systems will be presented in detail.

4.4.2 An introductory example

Parallel programming is concerned with computer systems consisting of *interactive* processes. The word 'interactive' means that the action performed by a process at a given time may depend on actions performed previously by other processes. Several mechanisms are used to realize an interaction between processes. *Shared memory* is of common use: a piece of information stored in shared memory can be accessed by several processes. Another mechanism is *message passing*, which allows explicit exchange of information between distinct processes. From a theoretical point of view, message passing can be considered as a special case of memory sharing. In practice, memory sharing is used especially in a *centralized* framework, when several processes share a common processing unit. On the other hand, in a *distributed* framework, processes are distant from each other and are executed on their own processing units; they communicate by message passing but do not share memory.

Let us consider a two-process system; one process produces pieces of information which are consumed by the other process. This 'Producer-Consumer' scheme occurs in many situations. An example is the production of results and their output on a printer. Some synchronization is required in the system: the amount of information produced and not yet consumed is limited by the finite capacity of the buffer interposed between the processes.

With classical notation [Owicki and Gries 76], the 'Producer' (P) and the 'Consumer' (C) can be modelled as follows:

```
       P :                              C :
while true do                    while true do
    begin                            begin
    await p − c < N do produce;      await p > c do consume;
    p := p + 1                       c := c + 1
    end                              end
```

Variables p and c are counters; they can be accessed by both processes. At every time, their values are respectively the number of items already produced and the number of items already consumed. The initial value is 0 for both of them. The statements *produce* and *consume* stand respectively for the production action of process P and for the consumption action of process C. The details of these actions do not matter here, since our purpose is only to in-

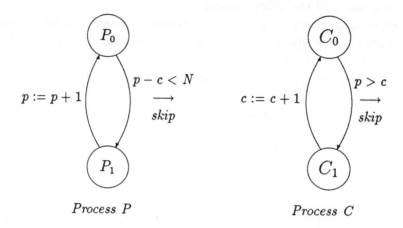

Figure 4.18 : System Producer-Consumer

vestigate the synchronization between processes P and C. Nevertheless, these actions are supposed not to alter the values of the counters p and c.

The statement 'await B do S' means 'wait until condition B is true and then execute statement S'. If condition B remains false forever, the await statement induces an infinite delay.

The processes P and C are modelled graphically in Figure 4.18. Each node has a label and each arc is associated with an atomic statement, or transition. The word 'atomic' means that the statement is executed as an indivisible action. At each time, a transition of one of the processes is executed. If, at some time, several transitions can be executed, then one of them is selected in an arbitrary way. The actions *produce* and *consume*, whose actual nature does not matter here, are modelled by *skip*. This piece of notation represents any action which does not alter the state of the memory.

When a program is executed on a computer, a sequence of machine states is generated. This sequence, or *computation*, can be viewed as a temporal interpretation. Such a correspondence between temporal interpretations and computations allows temporal logic to be used for program verification. Concurrent systems as considered here usually admit several computations, due to the fact that, at a given time, more than one statement can be selected for execution. This phenomenon is often called *non-determinism*. Obviously enough, a non-deterministic program is correct with respect to some specification if and only if *all* the computations of this program satisfy the specification.

The Producer-Consumer system or, more specifically, the abstract version of it we consider here, involves only two variables, p and c. An execution step consists in modifying the values of these variables according to a statement of

4.4. APPLICATIONS OF TEMPORAL LOGIC

one of the processes. Computations are subject to the following restriction: for each process, the statements that compose it must be executed in turn, as implied by the program of Figure 4.18.

A computation state of the Producer-Consumer system can be modelled by an instance of the 4-tuple (L_P, L_C, p, c), where L_P is either of the labels P_0 and P_1 and where L_C is either of the labels C_0 and C_1. Here p and c denote the current values of the counters p and c respectively (it is common practice to give the same name to a variable and to its current value; however, the two notions should not be confused). The occurrence of a label ℓ in the slot corresponding to a process indicates that this process is at the control point ℓ, which means that the last statement executed by this process has led to the label ℓ. In the initial state, each process is at a specified control point, denoting the beginning of the program. The first statement executed by each process starts from this control point.

The first few states of a possible computation of the Producer-Consumer system are listed below; the constant N is assumed to be 2.

	L_P	L_C	p	c
0.	P_0	C_0	0	0
1.	P_1	C_0	0	0
2.	P_0	C_0	1	0
3.	P_1	C_0	1	0
4.	P_1	C_1	1	0
5.	P_0	C_1	2	0
6.	P_0	C_0	2	1
7.	P_0	C_1	2	1
8.	P_0	C_0	2	2
9.	P_1	C_0	2	2
10.	P_0	C_0	3	2

The first two actions of any computation are necessarily due to the Producer (transition from state 0 to state 1, and then from state 1 to state 2), since the guard $p > c$ of the only transition that can be executed by the Consumer is not satisfied. In states 1 and 2, the Consumer is said to be *blocked*. Similarly, the action executed from state 5 is necessarily due to the Consumer. On the other hand, the action executed from state 6, for instance, is arbitrarily due to the Producer or to the Consumer. This choice is often assumed to be performed by a special entity, called the *scheduler*. The selection procedure applied by the scheduler does not matter here as, in principle, the relevant aspects of the behaviour of the system do not depend on the choice policy. However, there is obviously a minimal requirement: the scheduler cannot select a blocked process for the next execution step. Additional requirements are sometimes introduced; more details will be given later.

There are infinitely many distinct computations of the Producer-Consumer system; furthermore, these computations themselves are infinite. In order to verify the system, it is necessary that a finite representation of its computations and of their properties be available.

The set of program computations will be described by a set of temporal formulas, usually called the axioms of the program. These axioms define the temporal theory of the program. This theory is obtained from the text of the program by a technique that will be introduced in Subsection 4.4.5.

Properties of programs are also described by temporal formulas. There are many kinds of such properties, depending on the structure of the formulas that represent them. The most important kinds of properties will be introduced in Subsection 4.4.6. As an example, let us consider again the Producer-Consumer system. Its main property is that the two processes are adequately synchronized, provided the initial state of the computation is acceptable. This property can be modelled by the formula $(Init \supset \Box Z)$, where $Init$ and Z are

$$Init : (at\ P_0 \land at\ C_0 \land p = 0 \land c = 0),$$
$$Z : (0 \leq p - c \leq N);$$

$at\ P_0$ and $at\ Q_0$ are local propositions, and p and c are local variables.

No temporal operator occurs in the formula $Init$ since it expresses a property of the initial state only. On the other hand, the property modelled by the formula Z is common to all the states of the computation: the number of items already produced but not yet consumed remains between 0 and N.

Let us also consider the formula

$$I : \quad [at\ P_0 \supset p - c \leq N] \land [at\ P_1 \supset p - c < N]$$
$$\land\ [at\ C_0 \supset p \geq c] \land [at\ C_1 \supset p > c].$$

It is rather clear that if the formula I is satisfied in some state of the computation, then it will still be satisfied in the next state. (The technique for proving such assertions will be explained later.) As a consequence, if the formula I is true initially, it will remain true throughout the execution. A further consequence is the validity of the property modelled by the formula $(Init \supset \Box Z)$.

The problem of program verification consists in proving that a program is correct with respect to its specifications. In the framework of this section, the representation of a program is a temporal theory, and specifications are modelled by temporal formulas belonging to the language of this theory. The verification problem therefore reduces to a classical problem in logic: is this formula (modelling the specifications) a theorem of this theory (modelling the program)?

4.4.3 Program modelling

We formally define a small programming language, restricted to the usual case of systems consisting of a constant finite number of processes. These processes communicate only by sharing variables. The description of a system involves a set $\{P_1, \ldots, P_n\}$ of processes. These processes access a finite set $X = \{x_1, \ldots, x_k\}$ of variables. To avoid any confusion with logical variables, these variables are called *program variables*. Each program variable has a type. The set X is called the *memory* of the system.

Each process is represented by a *flowchart*, that is, a finite, connected and directed graph. Each node of the graph is identified by a *label*, or *control point*; there is a distinguished node called the *initial node*. The starting point and the end point of an arc may be the same node, and several arcs can share the same starting point and the same end point. A *statement* is associated with each arc. A statement S is an ordered pair (C, A) (often denoted $C \to A$) where C is a *condition*, or *guard*, and where A is an *assignment*. The condition C is an open formula of first-order logic or, more precisely, a Boolean term involving global constants and program variables. The assignment A has the form

$$(x_1, \ldots, x_k) := (t_1, \ldots, t_k),$$

where $t_1, \ldots, t_k$ are terms involving global constants and program variables. If t_j is x_j, then both occurrences of x_j are omitted in the assignment A. For instance, if the memory of the system is the set $\{x, y, z\}$, then the notation

$$(x, y) := (f(x), g(x, z))$$

models an assignment which does not alter z. The identity assignment, which modifies no program variable at all, is represented by *skip*. The expression $C\,?$ is often used instead of $C \to skip$. Similarly, A is an abbreviation for $true \to A$.

As the graphical representation is not always convenient, a lexical representation is often used instead. A directed, connected graph can be described by a list of triples of the form (starting point, arc identifier, end point). In the case of flowcharts, starting points and end points are identified by labels and arcs are identified by statements. More precisely, a flowchart is represented by the set of its transitions. The *transition* associated with the statement (C, A) is denoted

$$(\ell,\ C \longrightarrow A,\ \ell'),$$

where the label ℓ identifies the starting point of the arc associated with the statement and where ℓ' identifies the end point of the same arc. The set of labels of a process P is denoted L_P; the set of its transitions is denoted Tr_P.

The *output condition* of $\ell \in L_P$ is the disjunction $out(\ell)$ of the transitions whose starting point is labelled by ℓ (this condition is identically false if no transition originates from ℓ).

Let us now outline the operational semantics for the kind of programs we have just defined. The representation of programs in the framework of temporal logic will be based on this semantics.

We consider a system of n processes $P_1, \ldots, P_n$, with memory X. A *memory state* is a total function on the domain X, mapping each variable to a value of the corresponding type. The set of memory states is denoted Σ. A *control state* is an element of the Cartesian product $\Gamma = L_{P_1} \times \cdots \times L_{P_n}$. In order to avoid any risk of confusion, the sets L_{P_i} are assumed to be pairwise disjoint. A *system state* is an ordered pair $s = (L, \sigma)$ where L is a control state and σ is a memory state. A state $s = ((\ell_1, \ldots, \ell_n), \sigma)$ is said to be *final* if $out(\ell_i)(\sigma)$ is false, for all i.

A transition $T = (\ell, C \to A, \ell')$ of any process P_i defines a transformation of the system state. This transformation is modelled by a (partial) function, also denoted T, from the set $\Gamma \times \Sigma$ into itself. Let $s = (L, \sigma)$ and $t = (L', \tau)$ be two system states. The equality $t = T(s)$ holds if and only if the following three requirements are satisfied.

- The coordinates of L and L' are pairwise identical, except that, for some $i \in \{1, \ldots, n\}$, the ith coordinate of L is ℓ whereas the ith coordinate of L' is ℓ'.

- The memory state σ satisfies condition C.

- The memory state τ is $A(\sigma)$, that is, the memory state defined by $A(\sigma)(x_j) = t_j(\sigma)$, where $t_j(\sigma)$ denotes the value of the term t_j when the values of the program variables are given by σ. For instance, for the assignment $A : (x, y) := (f(x), g(x, z))$, the function $A(\sigma)$ is given by $A(\sigma)(x) = f(\sigma(x))$, $A(\sigma)(y) = g(\sigma(x), \sigma(z))$ and $A(\sigma)(z) = \sigma(z)$.

A *computation* of the system is a sequence $(s_0, s_1, s_2, \ldots)$ of states. The first state, $s_0 = ((\ell_1^0, \ldots, \ell_n^0), \sigma_0)$, is the *initial* state of the computation; each ℓ_i^0 is the label of the initial node of process P_i. If s_{k+1} exists, then a transition T exists such that $s_{k+1} = T(s_k)$. A computation can be finite or infinite, but the last state of a finite computation has to be a terminal state. As the sequences of situations considered in linear temporal logic are infinite, it is convenient to extend a finite execution by repeating its last state indefinitely.

4.4.4 Example: Peterson's algorithm

The problem of *mutual exclusion* is a common problem in the framework of concurrent systems. Let us assume that several processes have to access from

4.4. APPLICATIONS OF TEMPORAL LOGIC

time to time a single resource (a printer for instance). Necessarily this access can take place in mutual exclusion only: at each time, at most one process can use the resource.

Peterson's algorithm gives a simple solution to this problem, in the special case of two processes communicating by shared variables [Peterson 81]. The computation internal to each process does not matter here; we are interested only in the cooperation mechanism ensuring mutual exclusion. This mechanism involves three shared variables, namely, the Boolean variables inP and inQ, and the binary variable $turn$, which can only take the values p and q.

Let us first give the text of the algorithm, using standard notation.

P :
while $true$ do
  begin
    non-critical section;
    $inP := \mathbf{T}$;
    $turn := q$;
    await $\neg inQ \vee turn = p$ do *skip*;
    critical section;
    $inP := \mathbf{F}$
  end

Q :
while $true$ do
  begin
    non-critical section;
    $inQ := \mathbf{T}$;
    $turn := p$;
    await $\neg inP \vee turn = q$ do *skip*;
    critical section;
    $inQ := \mathbf{F}$
  end

This algorithm is symmetric: the text for process Q is obtained from the text for process P by exchanging the roles of the variables inP and inQ, and the roles of the values p and q.

The principle of the algorithm is very simple. Let us consider process P, for instance. At the beginning of the execution, the shared resource is not needed; the process is said to be in *non-critical section*. If access to the resource is required, then the *entry protocol* is executed. The variable inP is first assigned the value $\mathbf{T}$, which means that access is requested. Afterwards, the variable $turn$ is assigned the value q: process Q may take the opportunity to perform an access of its own. Finally, the access of process P is delayed until either $inQ = \mathbf{F}$ (meaning that process Q does not request access), or $turn = p$ (the variable $turn$ indicates which process has the priority, when each of them has requested access). When one of these conditions is satisfied, process P enters its *critical section* and can use the resource. When its task is completed, the *exit protocol* is executed, and process P goes back to the non-critical section.

A formal representation of Peterson's algorithm appears in Figure 4.19. Initially, both processes are in their non-critical section (nodes p_0 and q_0); the value of both variables inP and inQ is $\mathbf{F}$; the value of variable $turn$ can be arbitrarily p or q.

Remark. As some program variables are Boolean variables, logical connectives may appear in the text of the program; this is not problematic since the semantics of the logical connectives are the same in the program and in logic.

4.4.5 The temporal theory of a program

In order to write and prove program properties within temporal logic, we have to associate a temporal first-order theory with this program.

The signature of this theory will contain global and local objects. The global objects are the function constants and the predicate constants associated with the types of the program variables. An unlimited supply of (global) logical variables is also available.

The local objects will be used to describe the states of the system under analysis. A local proposition, denoted *at* ℓ, is introduced for each control point ℓ; this proposition will be true in states whose control part contains the control point ℓ. A local proposition, denoted $Next(T)$, is associated with each transition T; it will be true in a state s of a computation if the next state of this computation is the T-successor of the state s. If P is a process, $Next(P)$ is defined as the disjunction of the propositions $Next(T)$ associated with the transitions T of P; therefore, the proposition $Next(P)$ is true in some state if a transition of process P leads to the next state. Lastly, for each program variable x, a local (logical) constant, also denoted x, is introduced; the value of this constant in some system state will be the value of the corresponding program variable in this state. In the present framework, program variables are thus modelled by local constants[13].

After having defined the signature of the temporal theory associated with a concurrent system, we still have to introduce a set of axioms which allows the derivation of interesting properties of this system. There will be several kinds of axioms. A first family will consist of the axioms introduced for temporal predicate calculus (§ 4.3.4), restricted to the language defined by the signature.

A second family of axioms is concerned with the description of the data types used by the system; these axioms give the semantics of the global function constants and global predicate constants of the language. If some program variables belong to the type 'integer', for instance, the signature of the theory will contain symbols from number theory, such as the binary predicate constant $<$ and the binary function constant $+$, and an axiomatic system for number theory will be included in this family of axioms. It is often more convenient to assume that any statement which is an arithmetical truth is an axiom of the

---

[13] The representation of a program variable by a logical constant can look somewhat preposterous, but this is the only way to respect both the terminology of logic and the terminology of computing science, which give different meanings to the word 'variable'.

4.4. APPLICATIONS OF TEMPORAL LOGIC

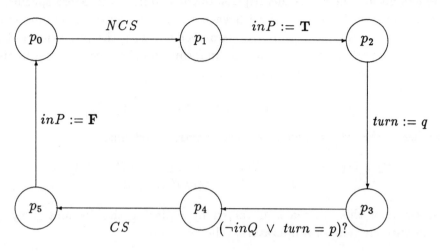

Process P

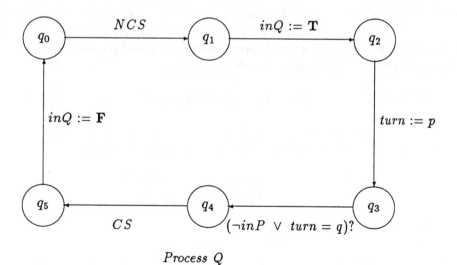

Process Q

Figure 4.19 : Formal model

theory[14].

A third family of axioms is concerned with the local objects of the language. These axioms are the formal description of the system itself. More specifically, an axiom schema is associated with each transition. Further axioms model the semantics of the local propositions associated with the processes and the control points and, finally, a special axiom formalizes the execution mechanism.

For each process P and for each transition

$$T : (\ell, C \to (x_1, \ldots x_k) := (t_1, \ldots, t_k), \ell')$$

belonging to the set Tr_P, we introduce the axiom schema

$$B_T : \Box([\mathit{Next}(T) \wedge Z(x_1/t_1, \ldots, x_k/t_k, \mathit{at}\ \ell' \leftarrow \mathbf{T})] \supset [\mathit{at}\ \ell \wedge C \wedge \bigcirc(\mathit{at}\ \ell' \wedge Z)]).$$

An instance of this schema is obtained by replacing the formula Z by any formula without temporal operators. The notation

$$Z(x_1/t_1, \ldots, x_k/t_k, \mathit{at}\ \ell' \leftarrow \mathbf{T})$$

denotes the formula Z where the program variables $x_1, \ldots, x_k$ have been replaced by the terms $t_1, \ldots, t_k$, where the local proposition $\mathit{at}\ \ell'$ has been replaced by $\mathbf{T}$, and where the other local propositions $\mathit{at}\ \ell_j$ corresponding to the process P have been replaced by $\mathbf{F}$. The axiom schema B_T indicates when the transition T can be executed and formalizes the effect of its execution.

As an illustration, let us consider the transition

$$T : (P_1,\ p := p+1,\ P_0)$$

of the Producer-Consumer system. The instance of the schema B_T for which Z is the formula I introduced in Subsection 4.4.2, is the axiom

$$\Box([\mathit{Next}(T) \wedge I(p/p+1, \mathit{at}\ P_0 \leftarrow \mathbf{T})] \supset [\mathit{at}\ P_1 \wedge \mathbf{T} \wedge \bigcirc(\mathit{at}\ P_0 \wedge I)]).$$

This axiom reduces to the formula

$$\Box([\mathit{Next}(T) \wedge p-c < N \wedge (\mathit{at}\ C_0 \supset p+1 \geq c) \wedge (\mathit{at}\ C_1 \supset p \geq c)] \supset [\mathit{at}\ P_1 \wedge \bigcirc(\mathit{at}\ P_0 \wedge p-c \leq N \wedge [\mathit{at}\ C_0 \supset p \geq c] \wedge [\mathit{at}\ C_1 \supset p > c])]).$$

The axioms about the local propositions associated with control points are better written by means of a non-classical but convenient notation, defined as

---

[14]The set of axioms is no longer a recursive set, which is generally not acceptable. However, this approach is only conceptual and amounts to studying the problem of proving properties of programs while abstracting from the problems associated with axiomatizing the data domain on which the program operates.

4.4. APPLICATIONS OF TEMPORAL LOGIC

follows. If a, b and c are propositions, then the formula $a + b + c = 1$ expresses that one and only one of these three propositions is true. More generally, if L is a set and if a proposition a_i is defined for each $i \in L$, then the formula

$$\sum_{i \in L} a_i = k$$

is true if exactly k members of the set $\{a_i : i \in L\}$ are true. (This notation becomes natural if the truth values **T** and **F** are respectively viewed as the integers 1 and 0.)

Let us now introduce the axioms about the control points and the execution mechanism. First, for each process P, the axiom

$$\Box [\sum_{p \in L_P} at\ p = 1]$$

expresses that, at each time, the process P is at a single control point. Similarly, for each process P, the axiom

$$\Box [Next(P) = \sum_{T \in Tr_P} Next(T)]$$

expresses that, at each time, at most one of the propositions associated with the transitions of P is true. If P_i and P_j are distinct processes, the axiom

$$\Box \neg [Next(P_i) \land Next(P_j)]$$

expresses that, at each time, only one process can perform an action.

The execution mechanism still has to be modelled. This is done by associating with each control state $(\ell_1, \ldots, \ell_n)$ the following axiom:

$$\Box \{(at\ \ell_1 \land \cdots \land at\ \ell_n) \supset \\ [(\bigvee_{i=1}^n Next(P_i) \land out(\ell_i)) \lor (\bigwedge_{i=1}^n \neg out(\ell_i))]\}.$$

This formula expresses that exactly one process will perform the next execution step, except if all the processes are blocked; in this case, the execution is terminated.

Some systems behave in a correct manner only when the scheduler satisfies a fairness requirement (§ 4.4.2). In this framework, the word 'fairness' may have several meanings; the most frequently used is the following: the scheduler is *fair* if every process whose execution can proceed infinitely often does actually proceed. The fairness property is formalized by adding the following axiom for each process P of the system:

$$\Box [\Diamond Next(P) \lor \bigvee_{\ell \in L_P} (at\ \ell \land \neg \Box \Diamond out(\ell))].$$

4.4.6 Program properties

Numerous properties of a program can be written in the temporal language associated with this program. The *initial conditions* are formalized into a formula without temporal operators, called *Init*. For instance, for Peterson's algorithm, the formula *Init* will be

$$at\ p_0 \land at\ q_0 \land \neg inP \land \neg inQ \land (turn = p \lor turn = q).$$

An *invariance property* is a property satisfied by all states of the execution. Such a property is modelled by a formula like

$$Init \supset \Box Z,$$

where Z contains no temporal operator.

In order to establish an invariance property, we need an *invariant*, that is, a formula I, without temporal operators, such that the three formulas listed below are valid:

$$Init \supset I, \quad \Box[I \supset \bigcirc I], \quad I \supset Z.$$

The formulas I and ($Init \supset \Box Z$) introduced in Subsection 4.4.2 for the Producer-Consumer program are respectively an example of an invariant and of an invariance property. Mutual exclusion (mentioned in connection with Peterson's algorithm) is also an invariance property; it will be formalized and proved in the next subsection.

Besides invariance properties, there is another kind of properties commonly specified for parallel systems: *liveness properties*. A liveness property asserts that some condition will eventually be true; it is formalized by a formula like

$$Init \supset \Box[Y \supset \Diamond Z],$$

where the formulas Y and Z do not contain temporal operators. Usually, the following two axiom schemata are used to establish liveness properties:

$$(A \supset \bigcirc B) \supset (A \supset \Diamond B),$$
$$((A \supset \Diamond B) \land (B \supset \Diamond C)) \supset (A \supset \Diamond C).$$

The notions of invariance and liveness properties can be generalized in such a way that any property modelled by a temporal formula becomes the conjunction of an invariance (or *safety*) property and a liveness property [Alpern and Schneider 85], [Alpern and Schneider 87]. Some kind of duality exists between invariance and liveness properties. Intuitively, an invariance property asserts that something bad will never occur, whereas a liveness property expresses that something good will eventually happen. Temporal logic is able to specify and prove both kinds of properties, but the proof techniques used in one case are quite different from those used in the other case.

4.4. APPLICATIONS OF TEMPORAL LOGIC

4.4.7 Invariance properties for Peterson's algorithm

The mutual exclusion property mentioned in Subsection 4.4.4 is modelled by the formula

$$Init \supset \Box \neg [(at\ p_4 \vee at\ p_5) \wedge (at\ q_4 \vee at\ q_5)],$$

and is, therefore, an invariance property. Another interesting invariance property is *deadlock freedom*, which means that all computations are infinite. The output conditions of all nodes are identically true, except at p_3 and q_3; however, the disjunction $[out(p_3) \vee out(q_3)]$ is identically true, which implies the desired result: there are no final states.

The invariant needed to establish mutual exclusion is defined by the formula

$$\begin{aligned} I: \quad & ((at\ p_0 \vee at\ p_1) \equiv \neg inP) \wedge ((at\ q_0 \vee at\ q_1) \equiv \neg inQ) \\ \wedge\ & (turn = p \vee turn = q) \\ \wedge\ & (at\ q_3 \vee at\ q_4 \vee at\ q_5) \supset (turn = p \vee at\ p_3) \\ \wedge\ & (at\ p_3 \vee at\ p_4 \vee at\ p_5) \supset (turn = q \vee at\ q_3). \end{aligned}$$

This formula is easily proved to be an invariant. First, the formula $[Init \supset I]$ is valid since, when $Init$ is true, I reduces to the formula

$$\begin{aligned} & ((\mathbf{T} \vee \mathbf{F}) \equiv \mathbf{T}) \wedge ((\mathbf{T} \vee \mathbf{F}) \equiv \mathbf{T}) \wedge \mathbf{T} \\ \wedge\ & (\mathbf{F} \vee \mathbf{F} \vee \mathbf{F}) \supset (turn = p \vee \mathbf{F}) \\ \wedge\ & (\mathbf{F} \vee \mathbf{F} \vee \mathbf{F}) \supset (turn = q \vee \mathbf{F}), \end{aligned}$$

which itself reduces to $\mathbf{T}$.

Second, we have to prove that the formula $\Box[I \supset \bigcirc I]$ is valid. In view of the axioms about program control, it is sufficient to verify that, for each transition $T: (\ell, C \to A, \ell')$ of each process, the formula $[I \wedge Next(T)] \supset \bigcirc I$ is valid. Let us consider the case of the transition

$$T_2^p : (p_2,\ turn := q,\ p_3).$$

We instantiate the axiom schema associated with this transition by substituting the formula I for Z. This results in the axiom

$$\Box \{[Next(T_2^p) \wedge I(turn/q, at\ p_3 \leftarrow \mathbf{T})] \supset [at\ p_2 \wedge \bigcirc (at\ p_3 \wedge I)]\}.$$

The required property will be proved if the following formula can be proved:

$$A: \quad [I \wedge Next(T_2^p)] \supset [Next(T_2^p) \wedge I(turn/q, at\ p_3 \leftarrow \mathbf{T})].$$

Taking into account the valid conditional $(Next(T_2^p) \supset at\ p_2)$, the antecedent of formula A reduces to

$$\begin{aligned}&Next(T_2^p) \wedge inP \wedge ((at\ q_0 \vee at\ q_1) \equiv \neg inQ)\\ \wedge\ &(turn = p \vee turn = q)\\ \wedge\ &(at\ q_3 \vee at\ q_4 \vee at\ q_5) \supset turn = p,\end{aligned}$$

whereas the consequent of A reduces to

$$Next(T_2^p) \wedge inP \wedge ((at\ q_0 \vee at\ q_1) \equiv \neg inQ),$$

which leads to the conclusion that A is valid. The remaining transitions are handled in a similar way.

The next step consists in verifying that the invariance of I implies mutual exclusion. To achieve that end, it is sufficient to prove the formula $[I \supset \neg((at\ p_4 \vee at\ p_5) \wedge (at\ q_4 \vee at\ q_5))]$. In fact, the negation of this formula is $[I \wedge (at\ p_4 \vee at\ p_5) \wedge (at\ q_4 \vee at\ q_5)]$, which can be rewritten as

$$\begin{aligned}&(at\ p_4 \vee at\ p_5) \wedge (at\ q_4 \vee at\ q_5)\\ \wedge\ &inP \wedge inQ \wedge (turn = p \vee turn = q)\\ \wedge\ &\mathbf{T} \supset (turn = p \vee \mathbf{F})\\ \wedge\ &\mathbf{T} \supset (turn = q \vee \mathbf{F}).\end{aligned}$$

This formula implies the inconsistent formula $[turn = p \wedge turn = q]$, and is, therefore, identically false, which proves the validity of the initial formula.

4.4.8 Fairness of Peterson's algorithm

We have just proved that Peterson's algorithm ensures mutual exclusion; furthermore, as all computations of this algorithm are infinite, no deadlock can occur. However, a last property should be established, namely, that the delay between a request for access to the critical section and the access itself is always finite. For process P, this property is formalized by the assertion

$$\Box[at\ p_3 \supset \Diamond at\ p_4],$$

or, equivalently, since formula I is an invariant of the system, by the assertion

$$\Box[(at\ p_3 \wedge I) \supset \Diamond at\ p_4].$$

The proof of this assertion is based on the seven elementary theorems listed below:

1. $\Box[(at\ p_3 \wedge at\ q_3 \wedge I \wedge turn = p) \supset \bigcirc at\ p_4]$,
2. $\Box[(at\ p_3 \wedge at\ q_3 \wedge I \wedge turn = q) \supset \bigcirc(at\ p_3 \wedge at\ q_4 \wedge I)]$,
3. $\Box[(at\ p_3 \wedge at\ q_4 \wedge I) \supset \bigcirc(at\ p_3 \wedge at\ q_5 \wedge I)]$,
4. $\Box[(at\ p_3 \wedge at\ q_5 \wedge I) \supset \bigcirc(at\ p_3 \wedge at\ q_0 \wedge I)]$,
5. $\Box[(at\ p_3 \wedge at\ q_0 \wedge I) \supset \bigcirc(at\ p_4 \vee (at\ p_3 \wedge at\ q_1 \wedge I))]$,
6. $\Box[(at\ p_3 \wedge at\ q_1 \wedge I) \supset \bigcirc(at\ p_4 \vee (at\ p_3 \wedge at\ q_2 \wedge I))]$,
7. $\Box[(at\ p_3 \wedge at\ q_2 \wedge I) \supset \bigcirc(at\ p_4 \vee (at\ p_3 \wedge at\ q_3 \wedge I \wedge turn = p))]$.

4.4. APPLICATIONS OF TEMPORAL LOGIC

The proofs of these theorems are straightforward. A proof of the assertion can now be given as follows:

8. $\Box[(at\ p_3 \wedge at\ q_3 \wedge I \wedge turn = p) \supset \Diamond at\ p_4]$ (1)
9. $\Box[(at\ p_3 \wedge at\ q_2 \wedge I) \supset \Diamond at\ p_4]$ (7,8)
10. $\Box[(at\ p_3 \wedge at\ q_1 \wedge I) \supset \Diamond at\ p_4]$ (6,9)
11. $\Box[(at\ p_3 \wedge at\ q_0 \wedge I) \supset \Diamond at\ p_4]$ (5,10)
12. $\Box[(at\ p_3 \wedge at\ q_5 \wedge I) \supset \Diamond at\ p_4]$ (4,11)
13. $\Box[(at\ p_3 \wedge at\ q_4 \wedge I) \supset \Diamond at\ p_4]$ (3,12)
14. $\Box[(at\ p_3 \wedge at\ q_3 \wedge I) \supset \Diamond at\ p_4]$ (8,2,13)
15. $\Box[(at\ p_3 \wedge I) \supset \Diamond at\ p_4]$ (9,10,11,12,13,14)

The fairness property of Peterson's algorithm has been proved without using any fairness hypothesis on the scheduler. However, such a hypothesis becomes necessary when a slightly stronger property is considered, namely, the property that is modelled by the formula

$$\Box[at\ p_0 \supset \Diamond at\ p_4].$$

In view of the result we have just obtained, it is sufficient to prove the formula $\Box[at\ p_0 \supset \Diamond at\ p_3]$, or, equivalently, the three formulas

$$\Box[at\ p_0 \supset \Diamond at\ p_1], \quad \Box[at\ p_1 \supset \Diamond at\ p_2], \quad \Box[at\ p_2 \supset \Diamond at\ p_3].$$

This is possible provided the scheduler is assumed to be fair. The fairness axiom associated with process P is

$$\Box[\Diamond Next(P) \vee (at\ p_3 \wedge \neg\Box\Diamond(\neg inQ \vee turn = p))].$$

Here is a proof of the first formula:

1. $\Box[at\ p_0 \supset \Diamond Next(P)]$ (fairness)
2. $\Box[(at\ p_0 \wedge Next(P)) \supset \bigcirc at\ p_1]$ (semantic axiom)
3. $\Box[at\ p_0 \supset \Diamond at\ p_1]$ (1,2)

The remaining two formulas can be proved in the same way.

Peterson's original paper [Peterson 81] contains an informal proof of the algorithm. A formal proof of correctness (without using temporal logic) is presented in [Dijkstra 81].

In the area of program verification, the critical point is the discovery of an appropriate invariant. Ideally, the program designer should construct the program and the invariant hand in hand. This construction is done for Peterson's algorithm in [Gribomont 85].

More complete expositions of linear temporal logic and its application to program verification can be found in [Manna and Pnueli 84] and in [Kröger 87].

4.4.9 Verification using propositional temporal logic

Proving program properties with first-order temporal logic is a general and widely usable method. However, it does require some ingenuity (in the discovery of adequate invariants, for example) and involves proofs that can be tedious.

There is an alternative method usable for systems of processes in which the number of possible states is finite. Such a system is called a *finite-state* system. This condition is satisfied if all the variables used by the processes can only take a finite number of values. Indeed, a state of the system is characterized, on the one hand, by the control state and, on the other hand, by the values of the variables. Moreover, the control state is always restricted to a finite number of possible values. Let us note that Peterson's algorithm is a finite-state system, since the variables inP and inQ can only take the values T and F, and the variable $turn$ can only take the values p and q.

We will show that any finite-state program can be represented by a Büchi automaton, in the sense that the words accepted by the automaton correspond exactly to the executions of the program. Given such a representation, there is a straightforward method for checking that the program satisfies a propositional temporal logic formula. Indeed, this problem requires us to check that all the executions of the program (i.e., all the words accepted by the automaton describing the program) satisfy the temporal logic formula. Now, Theorem 4.6 describes a construction that, given a temporal logic formula, builds an automaton accepting the sequences that satisfies this formula. It is thus possible to proceed as follows:

1. Build the automaton for the negation of the formula. (The reason for considering the negation of the formula rather than the formula itself is discussed below.)

2. Build the automaton that corresponds to the intersection of the automaton describing the program with the automaton obtained from the negation of the formula (Theorem 4.4).

3. Check that the automaton obtained in the previous step is non-empty (Theorem 4.5).

If the automaton obtained after step 2 is empty, one can conclude that there is no execution of the program that makes the negation of the formula true. In other words, all the executions of the program satisfy the given formula. Note that had we built the automaton corresponding to the formula itself rather than the one corresponding to its negation, it would have been necessary to complement this automaton to do the verification. But, complementing a Büchi automaton is an expensive operation, and we wanted to avoid it.

4.4. APPLICATIONS OF TEMPORAL LOGIC

In the next subsection, we will see in greater detail how to represent a finite-state system by a Büchi automaton and how to apply the method we have just described.

4.4.10 Modelling programs with finite automata

Rather than giving a general description of the method for representing a program by a Büchi automaton, we will show directly how it applies to Peterson's algorithm.

Our task is relatively easy given that Peterson's algorithm, as it is described in Figure 4.19, looks very much like a finite automaton. Indeed, the description of each process can be viewed as an automaton: the nodes correspond to the states of the automaton and the arcs labelled by the statements correspond to the transitions. The alphabet of the automata is, therefore, the set of statements of the program.

However, two elements are missing for this description to be an accurate representation of Peterson's algorithm by a Büchi automaton:

- The variables and their values should be represented.

- The two processes should be combined into a single automaton.

We will solve both problems simultaneously. To that end, we will define an operation on finite-state processes that represents the concurrent execution of these processes. Moreover, we will represent the variables by distinct processes. The automaton corresponding to Peterson's algorithm will then be the result of the concurrent execution of the automata representing the processes P and Q and of the automata representing the variables inP, inQ, and $turn$.

When two processes are executed concurrently, each state of the execution is the combination of the states of the two processes; going from one state to the next state is due to an action (the execution of a statement) of either of the processes. We are going to define an operation on processes that models this situation, enriched with an additional possibility: some of the actions of the two processes can be required to be identical and thus always to be executed simultaneously. This possibility will be most useful for modelling the interaction between a process and the automaton that represents a variable used by this process.

Consider two Büchi automata $A_1 = (\Sigma_1, S_1, \rho_1, S_{01}, F_1)$ and $A_2 = (\Sigma_2, S_2, \rho_2, S_{02}, F_2)$. The *partly synchronized product* of A_1 and A_2 is the generalized Büchi automaton $A = (\Sigma, S, \rho, S_0, \mathcal{F})$ defined by

- $\Sigma = \Sigma_1 \cup \Sigma_2$,

- $S = S_1 \times S_2$, $S_0 = S_{01} \times S_{02}$,

- $\mathcal{F} = \{F_1 \times S_2, F_2 \times S_1\}$,

- $(u, v) \in \rho((s, t), a)$ when

    - $a \in \Sigma_1 \cap \Sigma_2$ and $u \in \rho_1(s, a)$ and $v \in \rho_2(t, a)$,
    - $a \in \Sigma_1 \setminus \Sigma_2$ and $u \in \rho_1(s, a)$ and $v = t$,
    - $a \in \Sigma_2 \setminus \Sigma_1$ and $u = s$ and $v \in \rho_2(t, a)$.

The automaton defined by the partly synchronized product of two automata can be described informally as follows.

- If $\Sigma_1 = \Sigma_2$, the partly synchronized product of the automata A_1 and A_2 is identical to the automaton we built (Theorem 4.4) for the intersection of the languages accepted by the automata A_1 and A_2.

- If $\Sigma_1 \cap \Sigma_2 = \emptyset$, the partly synchronized product of A_1 and A_2 accepts the infinite words obtained by interleaving a word accepted by A_1 and a word accepted by A_2. Note however that the interleaving can also be of a word accepted by one of the automata and of a finite prefix of a word accepted by the other automaton.

- If $\Sigma_1 \cap \Sigma_2 \neq \emptyset$ and $\Sigma_1 \neq \Sigma_2$, the partly synchronized product of the automata A_1 and A_2 accepts the infinite words obtained by interleaving a word accepted by A_1 with a word accepted by A_2, but where the elements of $\Sigma_1 \cap \Sigma_2$ must coincide.

The partly synchronized product of Büchi automata is an associative and commutative operation. We can thus unambiguously talk about the partly synchronized product of several automata. Note that, as we have defined it, the partly synchronized product yields a generalized Büchi automaton. However, this automaton can be converted easily to a Büchi automaton by use of Lemma 4.3.

Let us now see how to model Peterson's algorithm by means of the partly synchronized product of five automata:

- an automaton for each of the processes P and Q;

- an automaton for each of the variables inP, inQ and $turn$.

4.4.11 The automata representing the processes

The automata representing the processes P and Q are almost identical to the flowcharts of Figure 4.19. They are given in Figure 4.20.

4.4. APPLICATIONS OF TEMPORAL LOGIC

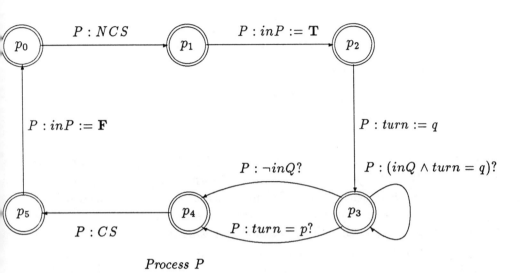

Process P

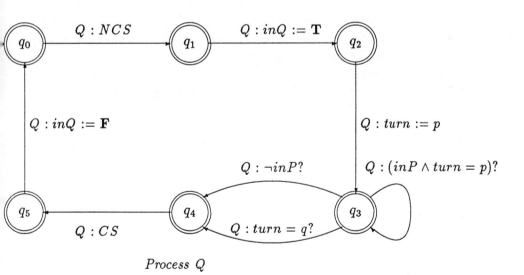

Process Q

Figure 4.20 : Automata for P and Q

Let us look at the automaton corresponding to the process P. Its set of states is $\{p_0, p_1, p_2, p_3, p_4, p_5\}$. Its alphabet is the set

$$\{P : NCS,$$
$$P : inP := \mathbf{T},$$
$$P : turn := q,$$
$$P : (inQ \wedge turn = q)?,$$
$$P : \neg inQ?,$$
$$P : turn = p?,$$
$$P : CS,$$
$$P : inP := \mathbf{F}\}$$

which consists of the statements (assignments and tests) that the process P can execute. To distinguish these statements form those of the process Q, we have prefixed them with 'P :'. The initial state of the automaton for the process P is p_0; the transitions can be read directly on Figure 4.20. The set of accepting states is the whole set of states; this means that any infinite execution is accepting.

The only difference between the transitions of the automaton of Figure 4.20 and those of the flowchart of Figure 4.19 has to do with the transitions that originate from state p_3. First, we have decomposed the transition between states p_3 and p_4 into two transitions. This will make it easier for us to model the interaction between the process and the variables, and does not change the possible executions of the process. Next, we have added a transition from p_3 to itself whose label is the negation of the exit condition for state p_3. This second modification removes the possibility of the process being blocked at state p_3. It will enable us to verify Peterson's algorithm under the assumption that all executions are infinite (this assumption is implicit in the modelling of programs by Büchi automata). If there is an execution in which the process is indeed blocked at p_3, it will be characterized by the fact that, after some time, the only action executed by the process P will be $P : (inQ \wedge turn = q)?$.

4.4.12 The automata representing the variables

The automaton representing a variable is built as follows. The number of states of this automaton is equal to the number of possible values of the variable. The transitions correspond to the actions of the processes that use the variable. An action representing a test is only possible in the states in which the test is true, and it does not modify the state of the variable. An action representing an assignment is possible in all states, and it leads to the state that gives the value of the variable after the assignment. The initial state is the state that corresponds to the initial value of the variable. The set of accepting states

4.4. APPLICATIONS OF TEMPORAL LOGIC

is the set of all states. In other words, all the infinite executions of such an automaton are accepting. The automata for the variables p, q and $turn$ are given in Figures 4.21, 4.22 and 4.23 respectively.

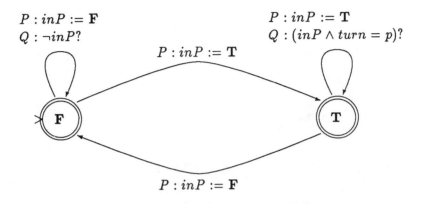

Figure 4.21 : Automaton for inP

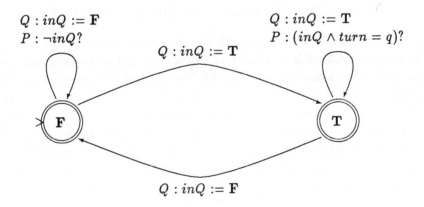

Figure 4.22 : Automaton for inQ

The automaton representing Peterson's algorithm can then be obtained as the partly synchronized product of the five automata we have described. Note that the actions $P : (inQ \wedge turn = q)?$ and $Q : (inP \wedge turn = p)?$ appear in

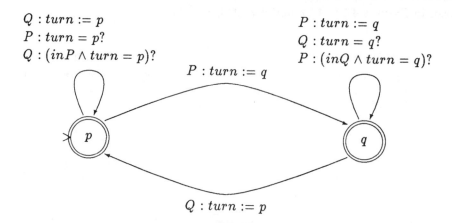

Figure 4.23 : Automaton for *turn*

three automata (respectively P, inQ and $turn$, and Q, inP and $turn$); in the partly synchronized product, these actions must be executed simultaneously by all three automata.

4.4.13 Modelling the fairness hypothesis

In all the automata we have described, the set of accepting states is the whole set of states. Therefore, all infinite executions of these automata are accepting. It is thus natural to wonder why Büchi automata are necessary for modelling concurrent programs. In this subsection, we will show that modelling concurrent programs by Büchi automata enables us to represent the fairness hypothesis that is necessary for the verification of certain liveness properties.

The fairness hypothesis we have to represent expresses the fact that any non-blocked process will eventually execute an action. Now, when modelling the processes P and Q, we took care to add the transitions that are necessary for the elimination of the possibility of blocking. This enables us to use a simpler fairness hypothesis: every process will eventually execute some action. Another way of stating this hypothesis is to say that among the infinite executions of the system, one must exclude those that do not contain an infinite number of actions of each of the processes. In the case of Peterson's algorithm, this means that the only allowable infinite executions are those in which actions of both P and Q appear infinitely often.

This restriction is easily enforced by supplementing the partly synchronized product of automata that represents Peterson's algorithm with the generalized

4.4. APPLICATIONS OF TEMPORAL LOGIC

Büchi automaton of Figure 4.24, in which $\mathcal{F} = \{\{s_P\}, \{s_Q\}\}$. One can view this automaton as a representation of the scheduler that supervises the execution of the processes P and Q. In this automaton, the transitions labelled {*action of P*} and {*action of Q*} each represent a set of transitions, respectively one for each element of the alphabet of the automaton for P and one for each element of the alphabet of the automaton for Q. The transitions labelled *choose P* and *choose Q* correspond to internal actions of the scheduler. They are used to represent correctly the fact that the choice of the next process to execute an action is made by the scheduler, independently of the processes P and Q.

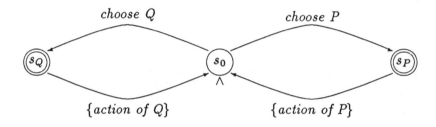

Figure 4.24 : Automaton for the fairness hypothesis

4.4.14 Specification and verification

To complete the verification of Peterson's algorithm, we still have to express the desired properties by propositional temporal logic formulas, build the Büchi automata corresponding to the negation of these formulas, and finally check that the intersection of each of these automata with the automaton representing Peterson's algorithm is empty.

In the representation of Peterson's algorithm we have just given, an execution of the algorithm is an infinite word on the alphabet that contains the actions of the process P, of the process Q and of the scheduler. Thus, we must express the properties we want to verify in terms of the elements of this alphabet. To that end, we will use the elements of the alphabet as atomic propositions in the formulas we will write. However, the automaton built from this type of formula will have as alphabet the set of subsets of the actions of processes. There are two solutions to this mismatch between alphabets. The first is to apply an obvious transformation to the automaton representing the program in order for its alphabet to become identical to that of the automaton built from the formula. The second is to modify the construction of the automaton corresponding to the formula in such a way that only one proposition is true at each time instant. The first solution has the advantage of being almost trivial, the second has the advantage of simplifying the automaton built from the formula.

We want to check two properties of Peterson's algorithm. The first is mutual exclusion. In the present context, this can be expressed by the fact that the action $P : CS$ will not be followed by the action $Q : CS$ as long as the process P has not left its critical section by executing $P : in\mathbf{F}$. This is modelled by the formula

$$\Box[(P : CS) \supset (\neg(Q : CS)\,\mathcal{U}\,(P : inP := \mathbf{F}))].$$

The symmetrical condition is of course also necessary:

$$\Box[(Q : CS) \supset (\neg(P : CS)\,\mathcal{U}\,(Q : inQ := \mathbf{F}))].$$

The second property of interest is a liveness property. This can be expressed by the fact that if process P shows its intent of entering the critical section by executing the action $P : inP := \mathbf{T}$, then it will eventually execute the action $P : CS$. This condition is modelled by the formula

$$\Box[(P : inP := \mathbf{T}) \supset \Diamond(P : CS)].$$

We will not apply the verification algorithm here. Indeed, it is not designed to be applied manually, but rather to be implemented and used on a computer. This leads us to make some comments about the complexity of the method.

A first observation is that the size of the partly synchronized product of automata can grow as the product of the sizes of the automata. Thus, if the number of concurrent processes is large, the size of the partly synchronized product can be excessive. This property is well known and is inherent in any finite-state verification method. Moreover, as we have seen, the size of an automaton generated from a temporal formula can grow exponentially as a function of the length of the formula. This certainly limits the use of the method; note, however, that specification formulas are often short and are frequently given as a conjunction, the terms of which can be treated separately.

Finally, let us mention a favourable element: once the generalized Büchi automaton has been built by taking the intersection of the automaton representing the program and of the automaton obtained from the negation of the formula, the algorithm for checking if the resulting automaton is empty or not is very efficient (its complexity is linear in the size of the automaton). For more details about the automatic verification of temporal properties of programs, the reader is referred to [Lichtenstein and Pnueli 85], [Vardi and Wolper 86], [Aggarwal et al. 87].

4.5 Propositional branching-time temporal logic

4.5.1 Introduction

In linear temporal logic, the model of time is the linearly ordered sequence of natural numbers (Sect. 4.1). Thus, each time instant has one and only one successor. It can sometimes be more convenient to consider that an instant has several possible successors. The model of time becomes then an *infinite finitary tree*, that is, a tree in which every node has a finite, non-zero number of successors. Linear time corresponds to the particular case where this number is always one. The aim of *branching-time temporal logic* is the study of infinite finitary trees of situations, or states. There exist several variants of branching-time temporal logic.

In linear temporal logic, formulas are interpreted not only on sequences but also on modal frames whose accessibility relation is functional (Sect. 4.1). The advantage of the second approach lies in the fact that modal frames can be finite. Both approaches can still be used for branching-time logic. It will be convenient to use infinite finitary trees at the informal level, but to come back to *branching-time temporal frames* at the formal level. A branching-time temporal frame is a modal frame (S, R) with an accessibility relation that satisfies the *totality* condition

$$\forall s\,(s \in S \supset \exists s'\,[s' \in S \land (s, s') \in R]). \qquad (4.6)$$

This condition asserts that every state has at least one successor.

The operators of linear temporal logic allow us to specify at which state(s) of a sequence some formulas have to be true. In branching-time temporal logic, the situation is similar, but slightly more complex. Indeed, two elements are needed to identify a node on a tree. The first one is the branch, or the *path*, containing the node, and the second element is the position of the node on that path, viewed as a sequence.

Branching-time temporal logic reflects this situation. A distinction is made between *path formulas* and *state formulas*. Path formulas essentially are formulas of linear temporal logic and are, therefore, interpreted on paths of the tree. On the other hand, state formulas indicate the paths on which the path formulas must be interpreted; they are themselves interpreted on a given state of the tree.

For example, if p is a proposition and if s is a state, the formula $\exists \Box p$ will be true for some interpretation if, for that interpretation, there exists a path, originating from s, all the states of which satisfy p. More generally, if ϕ is a path formula, then $\forall \phi$ and $\exists \phi$ are state formulas. At a fixed state s_0 of an infinite finitary tree, the formulas $\forall \phi$ and $\exists \phi$ are interpreted as follows. The first formula is true if ϕ is true for all the paths that originate from s_0; the second formula is true if ϕ is true for at least one such path.

4.5.2 Syntax

Given a set P of propositions, the syntax of branching-time temporal logic is as follows.

- A *state formula* is

    - a proposition $p \in P$,
    - an expression of the form $\neg b$, $b \wedge c$, $b \vee c$, $b \supset c$ or $b \equiv c$, where b and c are state formulas,
    - an expression of the form $\forall b$ or $\exists b$, where b is a path formula.

- A *path formula* is

    - a state formula,
    - an expression of the form $\neg b$, $b \wedge c$, $b \vee c$, $b \supset c$ or $b \equiv c$, where b and c are path formulas,
    - an expression of the form $\bigcirc b$, $\square b$, $\diamondsuit b$ or $b\,\mathcal{U}\,c$, where b and c are path formulas.

Remark. In the framework of branching-time temporal logic, the linear temporal operators $\bigcirc$, $\square$, $\diamondsuit$ and $\mathcal{U}$ are often rewritten X, G, F and U, respectively. (These symbols come from the words 'neXt', 'Generally', 'Future' (or 'Forward') and 'Until'.) Besides, the strong version of the operator 'Until' is used (§ 4.1.3). In order to keep consistency, however, the notations and definitions introduced earlier in this chapter will be maintained.

4.5.3 Semantics

A *branching-time temporal frame* is a pair $\mathcal{F} = (S, R)$ where S is a finite or denumerable non-empty set and where R is a total relation on S. A branching-time temporal frame is, therefore, a special case of modal frame (§ 1.2.3), and a linear temporal frame (§ 4.1.3) is a special case of branching-time temporal frame.

If a set P of propositions is given, then a *branching-time temporal interpretation* is a triple $\mathcal{I} = (S, R, I)$ where (S, R) is a branching-time temporal frame and where I is a function from $S \times P$ into $\{\mathbf{T}, \mathbf{F}\}$. The relation R is the accessibility relation; the function I is the interpretation function.

4.5. Propositional branching-time temporal logic

A *path* q is an infinite sequence $(s_n : n \in \mathbf{N})$ such that $s_0 \in S$ and $(s_n, s_{n+1}) \in R$, for all n. The path obtained by omitting the first i elements of q is denoted by $q^{(i)}$; thus, we have $q^{(i)} = (s_i, s_{i+1}, \ldots)$.

Given an interpretation $\mathcal{I} = (S, R, I)$, the semantic rules will associate a truth value $\mathcal{I}(s, b)$ with any state formula b in any state s, and will associate a truth value $\mathcal{I}(q, c)$ with any path formula c on any path q.

Let us first consider the case of atomic formulas.

- If p is a proposition, then $\mathcal{I}(s_0, p) =_{def} I(s_0, p)$.

The semantics of logical connectives is defined as in propositional calculus. If b_1 and b_2 are state formulas and if c_1 and c_2 are path formulas, we have e.g.

- $\mathcal{I}(s_0, b_1 \wedge b_2) =_{def} \mathcal{I}(s_0, b_1) \wedge \mathcal{I}(s_0, b_2)$,
- $\mathcal{I}(q, c_1 \wedge c_2) =_{def} \mathcal{I}(q, c_1) \wedge \mathcal{I}(q, c_2)$.

The semantics of linear temporal operators is defined as follows. Let c and d be two path formulas; the interpretation rules are

- $\mathcal{I}(q, \bigcirc c) =_{def} \mathcal{I}(q^{(1)}, c)$,
- $\mathcal{I}(q, \Box c) = \mathbf{T}$ if and only if $\mathcal{I}(q^{(i)}, c) = \mathbf{T}$ for all $i \geq 0$,
- $\mathcal{I}(q, \Diamond c) = \mathbf{T}$ if and only if $\mathcal{I}(q^{(i)}, c) = \mathbf{T}$ for some $i \geq 0$,
- $\mathcal{I}(q, c \,\mathcal{U}\, d) = \mathbf{T}$ if and only if either
    - $\mathcal{I}(q^{(j)}, c) = \mathbf{T}$ for all $j \geq 0$, or
    - $\mathcal{I}(q^{(i)}, d) = \mathbf{T}$ for some $i \geq 0$ and $\mathcal{I}(q^{(j)}, c) = \mathbf{T}$ for all j such that $0 \leq j < i$.

For the path operators ($\forall$ and $\exists$), the interpretation rules are as follows. If b is a path formula, then

- $\mathcal{I}(s_0, \forall b) = \mathbf{T}$ if and only if $\mathcal{I}(r, b) = \mathbf{T}$ for each path r that originates from s_0,

- $\mathcal{I}(s_0, \exists b) = \mathbf{T}$ if and only if $\mathcal{I}(r, b) = \mathbf{T}$ for some path r that originates from s_0.

If b is a state formula, it is also a path formula; the corresponding rule is

- $\mathcal{I}(q, b) =_{def} \mathcal{I}(s_0, b)$, where s_0 is the starting point of the path q.

This last definition shows that the semantics given here for the temporal operators is a generalization of the semantics given in Subsection 4.1.3.

By definition, the *formulas* of branching-time temporal logic are the state formulas. Path formulas are in fact auxiliary objects; they are introduced only to make easy the expression of the semantics of (state) formulas.

A formula is said to be *consistent* if it is true in some state of some interpretation; it is said to be *valid* if it is true in every state of every interpretation. A formula is said to be *inconsistent* if it is not consistent. The negation of a valid formula is inconsistent, and conversely.

Let b and c be two formulas; examples of valid formulas are

$$\neg \forall \bigcirc b \equiv \exists \bigcirc \neg b,$$
$$\neg \forall \diamond b \equiv \exists \square \neg b,$$
$$\neg \forall \square b \equiv \exists \diamond \neg b,$$
$$\forall (\neg b \, \mathcal{U} \, b),$$
$$\exists (\neg b \, \mathcal{U} \, b).$$

4.5.4 Computation tree logic

There exist several variants of branching-time temporal logic. The one that we have just introduced is denoted by the acronym CTL* [Clarke et al. 86]. It is very general; as a consequence, its expressive power is high, but its decision problem is complex [Emerson and Sistla 84], [Emerson and Halpern 86].

It proves convenient to consider a natural restriction of CTL*, in which temporal operators are applied only to state formulas (and not to arbitrary path formulas). This gives rise to *computation tree logic*, or CTL, which will now be defined formally.

The syntax of CTL is given by the rules listed below, where P is a given set of propositions. Let us emphasize that the only difference between these rules and those for CTL* concerns the definition of path formulas.

- A *state formula* is
    - a proposition $p \in P$,
    - an expression of the form $\neg b$, $b \wedge c$, $b \vee c$, $b \supset c$ or $b \equiv c$, where b and c are state formulas,
    - an expression of the form $\forall b$ or $\exists b$, where b is a path formula.

- A *path formula* is an expression of the form $\bigcirc b$, $\square b$, $\diamond b$ or $b \, \mathcal{U} \, c$, where b and c are state formulas.

The notion of branching-time temporal frame, the notion of branching-time temporal interpretation and the corresponding interpretation rules that we have introduced for CTL* in Subsection 4.5.3 hold for CTL. The definitions will not be repeated here.

4.5. PROPOSITIONAL BRANCHING-TIME TEMPORAL LOGIC

In CTL, formulas are state formulas, just as in CTL*. In fact, we can avoid introducing the notion of path formula in CTL, for the reason that any well-formed formula in CTL is obtained from non-temporal formulas by repeated applications of the *compound operators* $\exists \bigcirc$, $\forall \bigcirc$, $\exists \Diamond$, $\forall \Diamond$, $\exists \Box$, $\forall \Box$, $\exists \mathcal{U}$ and $\forall \mathcal{U}$. The first six compound operators are unary, the last two ones are binary. Formulas whose main connective is a binary compound operator can be written as $\forall (c \, \mathcal{U} \, d)$ instead of $c \, \forall \mathcal{U} \, d$, and $\exists (c \, \mathcal{U} \, d)$ instead of $c \, \exists \mathcal{U} \, d$.

The difference between CTL* and CTL is best explained with an example. The formula $\forall \Box \Diamond p$ is a CTL* formula but is not a CTL formula. Indeed, the path formula in $\forall \Box \Diamond p$ is $\Box \Diamond p$ and contains a sequence of (simple) temporal operators. On the other hand, $\forall \Box \forall \Diamond p$ is a CTL formula: a single temporal operator ($\Box$) is applied to a state formula ($\forall \Diamond p$). In the state w_0 of the branching-time temporal interpretation depicted in Figure 4.25, the CTL formula $\exists \Box \exists \Diamond p$ is true; on the other hand, the CTL* formula $\exists \Box \Diamond p$ is not true in this state. Indeed, no path originating from w_0 exists on which p would be true infinitely often. However, there exists a path (namely, the infinite loop on w_0) that enjoys the following property: each node of this path is the starting point of a path that contains a state in which p is true. The CTL formula $\forall \Box \forall \Diamond p$ and the CTL* formula $\forall \Box \Diamond p$ are both true in the state w_0 of the branching-time temporal interpretation depicted in Figure 4.26.

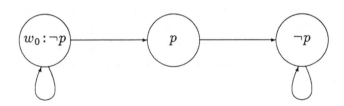

Figure 4.25 : A model of $\exists \Box \exists \Diamond p$

4.5.5 Application to program verification

We have seen in Subsection 4.4.9 how propositional linear temporal logic can be used for program verification. The method is based on the representation of programs as finite automata, followed by the verification that all infinite computations satisfy (are models of) the specification.

A similar approach is possible with branching-time temporal logic. In this framework, the automaton which represents the program can be thought of as a branching-time temporal interpretation (after some minor transformations). Verifying that all program computations satisfy a temporal specification ϕ

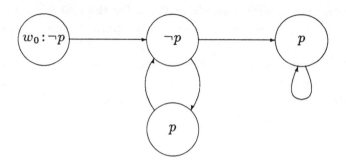

Figure 4.26 : A model of $\forall\Box\Diamond p$

amounts to verifying that the branching-time temporal interpretation which represents the program is a model of the branching-time temporal formula $\forall \phi$. In the framework of branching-time temporal logic, not only universal formulas like $\forall \phi$ but also mixed formulas like $\forall \phi \wedge \exists \psi$ can be considered. This formula specifies that all program computations satisfy ϕ and that at least one computation satisfies ψ.

Of course, the verification algorithm for properties modelled by mixed formulas is more complex than the verification algorithm for purely linear properties (modelled by universal formulas). In fact, the general algorithm consists in applying repeatedly the linear case algorithm (more details are given in [Emerson and Lei 85]). If only properties modelled by CTL formulas are considered, a very efficient verification algorithm exists [Clarke et al. 86]. Formally, the verification problem is as follows.

> Given a finite set P of propositions, an interpretation $\mathcal{I}$ and a formula a involving only elements of P, determine whether the interpretation $\mathcal{I}$ is a model of formula a.

Let us outline the description of the verification algorithm. The interpretation $\mathcal{I}$ is represented by a graph whose nodes are the states and whose arcs model the accessibility relation between states. A table is associated with each state s; it contains the truth value of the elements of P in s. Thus, the truth value of every proposition in every state can be read in the tables.

In order to decide whether $\mathcal{I}$ is a model of a, we have to complete the tables associated with the states in such a way that they contain not only the truth values attached to the atomic propositions, but also the truth values of all the subformulas of a (including a itself).

First, we consider the subformulas of a that contain a single logical connective, or a single compound temporal operator. Let us assume for instance that

4.5. Propositional branching-time temporal logic

the subformula $\forall \bigcirc p$ occurs in formula a. The truth value of this subformula in state s is **T** if and only if the value of p is **T** in all the immediate successors of s.

The truth values associated with a subformula containing $n+1$ connectives and operators can clearly be determined by looking up the tables, provided these tables already indicate the truth values associated with all subformulas containing at most n connectives and operators. This leads to a simple algorithm for checking whether $\mathcal{I}$ is a model of a or not; the construction of the tables is performed in at most n steps if a contains n connectives and operators.

As far as the graph traversals are performed efficiently, the execution time of the 'model checking algorithm' outlined above is of the order of the product of the size of formula a and of the size of interpretation $\mathcal{I}$. A detailed presentation of this algorithm can be found in [Clarke et al. 86].

Chapter 5

Formalization of revisable reasoning

5.1 Introduction

5.1.1 Revisable reasoning in artificial intelligence

Standard logic is concerned with the formalization of correct forms of reasoning. More ambitiously, artificial intelligence is also concerned with the formalization of reasonings that are just plausible or approximately correct. As we must often reason in the presence of incomplete, imperfect, uncertain or time-sensitive knowledge, our reasonings are indeed often just tentative and, therefore, revisable. For example, from our knowledge that most birds can fly and that 'Tweety' is a bird, we shall normally conclude that Tweety can fly. This inference is not rigorously correct since it does not take into account the possible exceptions to the rule asserting that birds can fly. Indeed, such an inference can be revised in the light of new information. In particular, if we discover that Tweety is an ostrich, being aware that ostriches are birds that do not fly, we must retract the assertion 'Tweety can fly'. Surprisingly, it is difficult to formalize this example of commonsense reasoning within standard logic. For this reason, it is often used as a *canonical example* in the study of the logical techniques that are developed in order to formalize revisable reasonings.

Several domains of artificial intelligence can benefit by a formal theory of revisable reasoning. On the one hand, we can mention the formalization of certain aspects of our elementary perception faculties, like vision, because they operate by means of iterated approximations and corrections [Minsky 74]. On the other hand, high-level and specialized human tasks such as diagnosis can involve forms of revisable reasoning [Reiter 87a]. The formalization of reasoning about plans and actions [Georgeff and Lansky 86], [Brown 87]; the representation of prototypal [Minsky 74] and hierarchical information that contains exceptions [Touretzky 86], [Etherington 87], and reasoning correctly about these forms of knowledge, may require formal theories of revisable reasoning. Such formal tools can be useful in the development of speech act theories too

[Perrault 87], [Appelt and Konolige 88]. Indeed, the human communication process is often based upon implicit information and conventions [McCarthy 86]. Finally, a correct treatment of incomplete, negative or implicit information both in relational or deductive databases and in logic programming systems requires the formalization of certain types of revisable reasoning.

5.1.2 Forms of revisable reasoning

From an epistemological point of view, reasoning can be revisable for various reasons. In this respect, we can distinguish between two basic forms of revisable reasoning.

On the one hand, reasoning is revisable when it is *conjectural* and simply *plausible*. Such a reasoning is uncertain because it is based on incomplete, uncertain or time-sensitive information or because of its own revisable nature. Among the numerous forms of conjectural reasoning, let us especially mention default reasoning, prototypal reasoning, and statistical reasoning.

On the other hand, reasoning is liable to revision when it enjoys an *indexical* character. Indeed, reasoning can be revised when it is based on a set of information that is supposed to be complete, but that is actually not complete or will not necessarily remain so. More generally, we often apply, in an implicit manner, revisable *conventions* and assumptions about incomplete, implicit or unknown information in order to increase our actual set of beliefs. Our reasoning can be logically correct with respect to this augmented information; however, it is often revisable because it depends on a state of beliefs that can be modified. Let us emphasize that the forms of correct deductive reasoning that standard logic aims at formalizing are revisable in this sense. Indeed, they are applied to a knowledge base that is often implicitly augmented in a simply plausible manner.

In summary, we can distinguish between two main basic families of revisable reasonings. On the one hand, our reasonings are uncertain and revisable because of their own nature or because of the information upon which they are based. On the other hand, forms of correct reasoning can be revisable when they rely on assumptions that delimit the information upon which they are based.

In this chapter, we present several logic based approaches to the formalization of revisable reasoning. Such approaches can be applied in most of the domains mentioned in Subsection 5.1.1. They involve the use of *augmenting conventions* about unknown, implicit and incomplete knowledge. The theoretical foundations of these techniques rely on a logical analysis of their associated augmenting conventions. Let us mention that the augmenting conventions will often appear as *minimization conventions* that are explored in a natural way

5.1. INTRODUCTION

with the help of a semantic analysis, and more precisely with the help of *model theory* (§ 1.2.14).

Let us emphasize that the distinction that we have made between two basic forms of revisable reasoning is valid from an epistemological point of view only. Such a distinction becomes rather superficial as soon as we undertake a logical characterization of the systems that are developed to formalize these reasonings.

5.1.3 Main properties of non-monotonic logics

Several techniques have been proposed in the literature to extend the scope of logic to the formalization of revisable reasoning. The specific properties of these techniques are reviewed in [Thayse 88]. Let us summarize them briefly.

These systems should be *non-monotonic*, which means that the set of possible conclusions can decrease when the set of premises increases. From a given set of initial information, we can sometimes build several mutually inconsistent sets of conclusions, generally called *extensions*. This property is called *pluri-extensionality*. It is often difficult to characterize the extensions in a direct and constructive way because some forms of revisable reasoning are circular: before one can infer a conclusion, one must first ensure that other formulas cannot be obtained. Two basic cases may occur. When one has to model *cautious* reasoning, the non-monotonic inference system must yield the conclusions that are present in *all* the extensions. On the contrary, when only a specific path of reasoning must be explored, the non-monotonic system must allow one to obtain the desired extension only. From a semantic point of view, conclusions that can be obtained from a revisable reasoning are often 'simply consistent' with respect to the premises. Indeed, a revisable piece of reasoning is not necessarily logically 'correct'; plausible conclusions that are not 'valid' can often be inferred.

Let us also emphasize that a formal framework for revisable reasoning must be *flexible* enough to deal with a reasoning the premises of which involve incomplete information. Such a framework must be *robust* enough to support in a convenient manner the possible evolution of the represented knowledge. It must allow the dynamical circumscription of the information that is really relevant to the problem. For example, it must permit reasoning in spite of the lack of complete information. More generally, it must support the possible evolution of both the application domain and the conceptualization thereof.

5.1.4 Reiter's, McDermott's, and Moore's logics

Three different *non-monotonic logics* have been reviewed in [Thayse 88]. Let us summarize them briefly.

- *Default logic* [Reiter 80]. In default logic, a form of revisable reasoning is captured by means of conjectural domain dependent inference rules called 'defaults'. They express statements of the form 'Most birds can fly' by means of rules of the form 'If it is consistent to infer that a particular bird can fly, then one can infer that it can fly'. These revisable rules allow us to express rules that admit exceptions, without requiring the exceptions to each rule to be listed explicitly.

- McDermott's *non-monotonic modal logics* [McDermott 82]. The aim of these logics is to provide us with a universal (i.e., domain independent) axiomatic system which allows the characterization of maximal sets of 'consistent' assertions that can be inferred from a set of premises. The formal languages used here are predicate modal logics.

- *Autoepistemic logic* [Moore 85, 88]. This logic constitutes a reconstruction of McDermott's non-monotonic logics. Autoepistemic logic involves reasoning about one's own beliefs. It is intended to model the reasoning of an ideally rational agent reflecting upon his own beliefs. This kind of reasoning is revisable since a state of beliefs can be altered.

Default logic has been proposed in order to formalize *default reasoning*. Autoepistemic logic, which is a reconstruction of modal logics of *consistency based* reasoning, formalizes an *ideally perfect introspective reasoning* that is based on the assumption that a complete state of affairs is given. Although these formalisms were intended to model different forms of reasoning, their expressive powers have been shown to be often equivalent [Konolige 87].

5.1.5 Minimization based approaches

In this chapter, we review a fourth family of logic based approaches to the formalization of revisable reasoning. These approaches involve the tools and the language of standard logic. Their fundamental principle can be explained as follows. A reasoning is non-monotonic when it is built from pieces of incomplete but time-sensitive information. A non-monotonic reasoning requires specific conventions about the unknown knowledge. These conventions often are *minimization conventions* in that they often circumscribe and restrict the positive *implicit* information that a set of incomplete knowledge may involve.

Let us introduce such conventions in an intuitive manner with the help of examples that are taken from the fields of databases and knowledge bases.

When we are dealing with relational databases that contain only elementary positive facts, we often implicitly require that any positive fact that is not represented be false. Only the information that is explicitly represented is supposed to be true, any other piece of positive information is supposed to be

5.1. INTRODUCTION

false. Let us emphasize that this assumption is often too severe; it is often excessive to assume that we are given the complete positive information.

When we are concerned with knowledge bases and with information of a more elaborate logical form, the minimization conventions become more complex; these conventions amount to minimizing the sets of individuals that satisfy particular formulas of the knowledge base.

Let us illustrate this issue with the help of the canonical example. We suppose that the only available information consists of the elementary fact 'Tweety is a bird' and the rule 'Normal birds can fly'. This information can be formalized with the help of the predicate logic formulas $Bird(Tweety)$ and $\forall x(\neg Abnormal(x) \wedge Bird(x) \supset Fly(x))$. In order to infer $Fly(Tweety)$ by *Modus Ponens*, we must be able to deduce $\neg Abnormal(Tweety)$. Yet, neither this formula nor its negation can be deduced logically from the set of initial information. By convention, we can however delimit the set of individuals that satisfy the predicate *Abnormal* to the known cases; here, we know no such individual. As we cannot deduce $Abnormal(Tweety)$ from the set of initial information, we infer $\neg Abnormal(Tweety)$ and thus become able to infer $Fly(Tweety)$. The convention we apply requires that if Tweety were abnormal, this piece of information ought to be given or deducible from the premises.

Let us emphasize that when we have *minimized* the set of individuals that satisfy the predicate *Abnormal*, we have *augmented* the verification domain of other predicates (here, of *Fly*). Indeed, we have inferred $Fly(Tweety)$ whereas this formula could not be deduced directly from the premises.

Such a reasoning is non-monotonic because it can be revised in the light of new information. If we learn that Tweety is abnormal and cannot fly, then we introduce the fact $Abnormal(Tweety)$ in the knowledge base and it is no longer possible to infer $Fly(Tweety)$.

The formalization of forms of revisable reasoning may thus rely upon *augmenting conventions* that actually enlarge the set of conclusions that can be logically inferred from a given set of premises. These conventions often minimize the implicit positive information that can be associated with the incomplete knowledge.

In databases and knowledge bases, the *closed-world assumption* (Sect. 5.3) underlies such a convention; it induces principles and techniques that are used for querying the knowledge base in a non-monotonic manner with respect to the evolution of its contents. These conventions can also take the form of axioms that must be introduced in the knowledge base (for example, *completion* (Sect. 5.4) and *circumscription* axioms (Sect. 5.6)). In logic programming systems, one uses in practice specific operators that 'approximate' to the logical negation with the help of the *negation as failure* principle (§ 5.5.1). These operators allow us to prove in a revisable manner that a statement cannot be inferred from the current state of knowledge that is contained in the logic program.

More generally, these assumptions give rise to different operations of delimitation for the verification domains of predicates and formulas and for the domains of the constants and functions of the knowledge base. These operations interact according to specific priorities and express the *specific rationality* of the agent who reasons (§ 5.7.5).

Most of these approaches to non-monotonic reasoning can be justified with the help of a *semantic* analysis. *Weakened consequence relations* can often be associated with them. These relations characterize logic systems that allow the inference of statements that are satisfied in *some specific models* of the premises.

In this chapter, we first define the logic framework of these formal approaches. We then analyse several formalization techniques of non-monotonic reasoning. Most often, we present them in the context of their specific application domains: relational and deductive databases, logic programming systems, and formalization of revisable reasoning by its own right.

5.2 Logic framework

5.2.1 Logic language

In this section, we define the logic language upon which non-monotonic systems will be based. We then summarize the main principles and tools of logic that are required for our analysis: *proof-theoretic* and *model-theoretic* considerations. (Further details on these logic domains can be found in Chapter 1. For a comprehensive presentation, the reader is referred the first two chapters of [Thayse 88].) Finally, we define several subsets of the logic language upon which non-monotonic systems of increasing complexity are built.

The formal language we make use of is a first-order language, denoted $\mathcal{L}$. The language $\mathcal{L}$ contains *formulas*; it is built in the following manner (§ 1.2.13–14).

- Let $\mathcal{V}$, $\mathcal{F}$ and $\mathcal{P}$ be finite sets of *variables*, of *function constants* and of *predicate constants* respectively. Function constants without arguments are called *individual constants*.

- The set of *terms* is defined recursively as follows. A term is either a variable in $\mathcal{V}$, or an expression $f(t_1, ..., t_n)$ where f is an n-ary function constant that belongs to $\mathcal{F}$ and where $t_1, ..., t_n$ are terms. A term that contains no variable is called a *ground term*.

- An *atomic formula* or *atom* of $\mathcal{L}$ is an expression $P(t_1, ..., t_m)$ where P is an m-ary predicate constant that belongs to $\mathcal{P}$ and where $t_1, ..., t_m$ are

5.2. LOGIC FRAMEWORK

terms. An atom or its negation is called a (positive or negative) *literal* of $\mathcal{L}$. A literal without a variable is called a *ground literal*.

- The set of *formulas* of $\mathcal{L}$ is defined recursively in the following way. A formula of $\mathcal{L}$ is either an atom of $\mathcal{L}$, either an expression $(\neg A)$ where A is a formula of $\mathcal{L}$, either an expression $(A \supset B)$ where A and B are formulas of $\mathcal{L}$, or an expression $(\forall x)A$ where A is a formula of $\mathcal{L}$ and where x is a variable that belongs to $\mathcal{V}$.

We use classical definitions for logical connectives: $(A \vee B)$ is $((\neg A) \supset B)$, $(A \wedge B)$ is $\neg((\neg A) \vee (\neg B))$, $(A \equiv B)$ is $((A \supset B) \wedge (B \supset A))$, $(\exists v)A$ is $\neg((\forall v)(\neg A))$. We omit parentheses when there is no ambiguity.

Most often in this chapter, we assume that the set of function constants reduces to the set of individual constants, which is denoted $\mathcal{C}$. In this case, terms are built from elements of the set $\mathcal{C}$ (equal to $\mathcal{F}$) of individual constants and of the set $\mathcal{V}$ of variables. By convention, the notation Δ that is used for representing a set of formulas belonging to $\mathcal{L}$, will also represent the logical conjunction of the formulas contained in that set.

5.2.2 Model-theoretic and proof-theoretic considerations

We can give the language $\mathcal{L}$ a truth based declarative semantics that is built on the concept of *interpretation* and *model* and that obeys the compositionality principle (§ 1.2.13).

Let us recall that an *interpretation* I is a triple (S, I_c, I_v), where S is a non-empty set called the *interpretation domain*; I_c is a function that maps each n-ary function constant to a function $I_c(f)$, from S^n to S, and that maps each m-ary predicate constant P to a function $I_c(P)$, from S^m to $\{\mathbf{T}, \mathbf{F}\}$; I_v is a function called *valuation* that maps each variable to an element of S.

The semantic interpretation rules obey the compositionality principle and are defined in the usual way. For a given interpretation $I = (S, I_c, I_v)$, they map each formula A that belongs to $\mathcal{L}$ to a truth value $I(A)$ that belongs to $\{\mathbf{T}, \mathbf{F}\}$.

The *extension* of an n-ary predicate P for an interpretation I is the set of n-tuples of ground terms $(t_1, ..., t_n)$ such that $I(P(t_1, ..., t_n)) = \mathbf{T}$. (Here, 'predicate' is used as an abbreviation for 'predicate constant').

Let us recall that relational database theory is centered on the concept of *relation* (Sect. 6.2). An n-ary relation R can be identified with an n-ary predicate P, in such a way that the *extension* of the relation R for an interpretation I coincides with the extension of P for I.

If a formula A is satisfied for a given interpretation $I = (S, I_c, I_v)$, i.e., if $I(A) = \mathbf{T}$, then I is called a *model* of the formula A. This is written $\models_I A$ or

$[\![A]\!]^\mathcal{I} = \mathbf{T}$. We shall use this very definition of a model in the present chapter. It does not coincide with the definition given in Subsection 1.2.14.

The *model-theoretic consequence relation* between a set Δ of formulas belonging to $\mathcal{L}$ and a formula A belonging to $\mathcal{L}$ is written $\vDash$ and is defined in the following way.

- $\Delta \vDash A$ when all models of the formulas contained in Δ are models of A.

As we have defined it, the model-theoretic consequence relation ensures the modelling of an *ideally rational reasoning*; we can only infer sure consequences of the premises. Non-monotonic logics will often allow us to formalize weakened model-theoretic consequence relations; a conclusion will be inferred when it is verified in *some specific models* of the premises but not necessarily in all models of the premises. In this chapter, we characterize these specific models for several non-monotonic logics.

Most often, a *proof-theoretic system* can be associated with this semantic characterization of the logic system. A proof-theoretic or axiomatic system is a proof system that is based on purely syntactic considerations, using *axioms* and *inference rules* in order to infer *theorems* from given premises. It defines a *proof-theoretic relation* between a set Δ of premises and a formula A. This relation is written $\vdash$.

A proof-theoretic system for the language $\mathcal{L}$ is *sound* and *complete* when the model-theoretic and proof-theoretic relations coincide (see also Subsection 1.2.2), i.e., when we have:

- $\Delta \vDash A$ if and only if $\Delta \vdash A$.

This property should be adapted to the case of non-monotonic logics that are based on an axiomatic inference system; most often it can indeed be shown that these logics induce weakened model-theoretic consequence relations. Let us emphasize that a semantic characterization is desirable even when the formalization of reasoning is not entirely developed at the *logical-object* level but also, in part or totally, at a *metatheoretical* level.

Finally, we define a *theory* to be the logical closure of a set of formulas of $\mathcal{L}$ with respect to a consequence relation. A theory is said to be *complete* when for any ground literal of the language, either this literal or its negation belongs to the theory.

5.2.3 Clausal forms

In the artificial intelligence field, it is often useful to rewrite formulas of $\mathcal{L}$ in a particular form, called the *clausal form*, because proof procedures can sometimes be simplified by this method (see the first two chapters of [Thayse 88]).

5.2. LOGIC FRAMEWORK

Recall that any formula of $\mathcal{L}$ can be written in *prenex form*, i.e., with all quantifiers in front. Eliminating the existential quantifiers in a closed prenex form, at the cost of introducing appropriate Skolem constants, results in what is called a *Skolem form*. It should be noted that this transformation preserves only the consistency of a formula, not its logical contents. Finally, a Skolem form can be written in *clausal form*, i.e., as a conjunction of clauses; here, a *clause* is a disjunction of literals the variables of which are universally quantified. (See details in [Thayse 88], § 1.2.8.)

A clause is of the general form

$$\neg P_1 \vee ... \vee \neg P_m \vee Q_1 \vee ... \vee Q_n,$$

where P_i and Q_i are positive literals, and where each variable is implicitly assumed to be universally quantified. A clause without a variable is called a *ground clause*. The clause above can be written in the logically equivalent form

$$P_1 \wedge .... \wedge P_m \supset Q_1 \vee ... \vee Q_n.$$

Depending on the number of literals that are allowed in the consequent part (the *head*) and in the antecedent part (the *body*) of this implication, we are in the presence of different subsets of $\mathcal{L}$ of variable expressiveness. The subset under consideration can coincide either with the logical equivalent of the representation language used in relational databases, either with the logic language on which Prolog is based, or with a more expressive language allowing one to express that a disjunction of positive literals can be satisfied without stating which specific literal(s) is (are) satisfied.

Depending on the families of clauses that are allowed in the logic language, one can conceive several non-monotonic systems, with variable complexity. The next subsection introduces some concepts that are relevant to a classification of these systems.

5.2.4 Different sublanguages

DISJUNCTIVE INFORMATION

If the number n of positive literals in a clause is strictly greater than 1, then this clause represents a *piece of disjunctive information*, which may have a conditional character.

More particularly, if $m = 0$ and $n > 1$, a clause only contains positive literals and is written $Q_1 \vee ... \vee Q_n$. These kinds of clauses allow us to represent pieces of incomplete information. For example, *Fly(Tweety)* $\vee$ *Fly(Coco)* is a clause the intended meaning of which is 'at least one of the two birds Tweety and

Coco flies' (but it is not specified whether only Tweety flies, whether only Coco flies, or whether both birds fly).

If $m \geq 1$ and $n > 1$, then a clause is of the very general syntactical form $P_1 \wedge ... \wedge P_m \supset Q_1 \vee ... \vee Q_n$ and represents a piece of conditional disjunctive information.

We shall see that the formalization of revisable reasoning is more complex when disjunctive information is dealt with. A disjunctive piece of information allows us to state incomplete knowledge about the world. When such a type of knowledge is present, it is difficult to formalize rational forms of reasoning. To avoid conceptual problems and to favour the efficiency of automatic proof systems, disjunctive clauses are generally forbidden in the basic logic programming systems and in the basic deductive databases.

HORN CLAUSES

In logic programming languages, such as Prolog, one generally allows *Horn clauses* only; these are clauses that contain at most one positive literal. Increasing the expressiveness of the language generally weakens the efficiency of the proof system.

DEFINITE HORN CLAUSES

A Horn clause is said to be *definite* when it contains exactly one positive literal. In particular, for $m \geq 1$ and $n = 1$, we obtain a definite Horn clause of the form $P_1 \wedge ... \wedge P_m \supset Q$. Such a clause allows the representation of conditional information, such as a *rule* of a Prolog program. The clause $Bird(x) \wedge Free(x) \supset Fly(x)$ gives an example of such a conditional rule.

On the other hand, when the antecedent part of the implication contains no literal, then we have a piece of non-conditional information. Such a clause consists of a positive literal Q, called a *positive unit clause*. We can distinguish between two cases.

- If Q does not contain any variable, then Q is a ground positive literal, called *elementary fact*, like $Bird(Tweety)$ or $Fly(Tweety)$. Such elementary facts are, for example, the instances of relation schemas of a relational database or the facts without variables of a Prolog program.

- If Q contains at least one variable, then we are faced with a more general piece of information that applies to all terms of the language. Indeed, the variables that occur in a clause are universally quantified. For example, $Fly(x)$ is a clause the intended meaning of which is 'all objects fly'.

5.2. Logic framework

Negative Horn clauses

A clause that contains no positive literal is called a *negative clause*. If the negative clause contains exactly one negative literal, i.e., if it reduces to $\neg P$, then it represents an elementary piece of negative information.

- If $\neg P$ does not contain any variable, then this clause represents a negative fact. Negative elementary facts are rarely represented in relational databases. They are implicitly taken into account by means of the *closed-world assumption* (Sect. 5.3) when their positive counterparts are not present in the database. An example of an elementary negative fact is provided by the ground negative literal $\neg Fly(Tweety)$.

- If $\neg P$ contains one or more variables, then the clause represents a piece of negative information that applies to all individual constants of the language. In databases, that kind of information is treated as an integrity constraint (Chap. 6). For example, the intended meaning of the negative clause $\neg Fly(x)$ is 'nothing flies'.

If the clause contains more than one negative literal, then it represents a more complex piece of negative information which can be written in the form $\neg(P_1 \wedge ... \wedge P_m)$. For example, the intended meaning of the clause $\neg(Fly(x) \wedge Submarine(x))$ is 'no object can fly and be a submarine object at the same time'.

In databases, such a kind of information is treated as an integrity constraint. In logic programming, a negative Horn clause is treated as a *query* that is put to the proof system. Most often, the positive information it denies is proved with the help of a refutation based procedure.

The empty clause

If $m = n = 0$, then the clause is called the *empty clause*. It denotes the truth value *false* and is never represented in a database. In logic programming, refutation based systems attempt to establish a contradiction by yielding the empty clause.

It is rather easy to develop formalization techniques for revisable reasoning when the sublanguage of $\mathcal{L}$ deals only with ground literals, Horn clauses, or, as shown below, *stratified sets of clauses*. Using a more expressive logic language often gives rise to more complex systems.

Let us emphasize that logic programming techniques generally make use of a language that is more expressive than the language of Horn clauses. One often permits the use of an operator called *negation as finite failure* (§ 5.5.1) allowing one to express negative literals in the conditional parts of the clauses. Such a negative literal will be interpreted as *true* when the corresponding positive literal cannot be 'proved' from the logic program.

Accordingly, it is often proposed to replace the language of Horn clauses by a more expressive extended language in both fields of logic programming and deductive databases. Generally speaking, *extended clauses* are allowed in such systems. The consequent of an extended clause is a positive literal while its antecedent consists of the conjunction of positive and negative literals. A set of extended clauses is said to be *stratified* when it can be organized in the following way. Each predicate is assigned a number. The number of each predicate that occurs in a positive literal of the antecedent of a clause A has to be less than or equal to the number of the predicate of the literal of the consequent of A. Moreover, the number of each predicate that occurs in a negative literal of the antecedent of A has to be (strictly) less than the number of the predicate in the literal of the consequent of A. In this chapter, we shall simply mention the main results about the formalization of forms of reasoning that can be associated with such extended languages [Apt et al. 88], [Bidoit and Froidevaux 87], [Lifschitz 88], [Przymusinski 88a] and [van Gelder 88]. A detailed treatment of that question is contained in Chapter 6.

5.2.5 Canonical Herbrand models

To characterize the semantics of relational and deductive databases, it is often natural to take into account *Herbrand models* only. Let us recall that a *Herbrand interpretation* and a *Herbrand model* are, respectively, an interpretation and a model whose interpretation domain is the set of ground terms of the language and that associate each ground term with *itself* ([Thayse 88], § 1.2.9). A Herbrand model is completely characterized by the set of ground atoms that it verifies. We can thus associate a Herbrand model with the set of ground atoms it verifies and compare Herbrand models with the help of the set-theoretic inclusion.

In both fields of logic programming and deductive databases, it is usual to select a specific model among the Herbrand models of the set Δ. This Herbrand model will be representative of every ground atom that can be inferred from Δ. This specific model is called the *canonical (Herbrand) model* of Δ. A ground atom A can be inferred only when A is true in the canonical model.

When the logic language consists of Horn clauses only, then the canonical model is the *unique minimal Herbrand model* (§ 5.5.2 and Chap. 6) of Δ, i.e., the intersection of all Herbrand models of Δ, each of these models being represented as the set of ground atoms that it verifies. When defined in this way, the canonical model is representative of the ground atoms that are verified in all the Herbrand models of Δ and that can be inferred according to the definition of the model-theoretic consequence relation.

When more expressive subsets of $\mathcal{L}$ are considered, the question of determining a canonical model is more involved. Indeed, a set of formulas of such a logic language often has several minimal Herbrand models, i.e., several Herbrand models whose sets of verified ground atoms do not possess any proper subset that can be associated with another Herbrand model. Two basic cases of non-monotonic reasoning may occur. In the first case, the reasoning agent is *brave* and explores a particular extension; *one* minimal model must be distinguished. In the second case, we wish to keep the *pluri-extensionality* property of non-monotonic systems (§ 5.1.3), and we have to select *several* specific models, allowing us to represent what can be inferred from Δ.

This approach to the semantic characterization of non-monotonic systems will be mainly carried out in Chapter 6 in the context of deductive databases.

5.3 The closed-world assumption

5.3.1 Introduction

Consider a database Δ listing the flights that connect different towns:

Flight(Paris, Brussels)
Flight(Tokyo, Paris)
Flight(London, Brussels)
Flight(Paris, Amsterdam)

The database Δ is *elementary* in the sense that it contains ground atoms only. These atoms are instances of the relation schema *Flight(x, y)*.

Is there any flight connecting London to Amsterdam? Although the atom *Flight(London, Amsterdam)* cannot be deduced or refuted logically from Δ, we are inclined to give the answer *no* to this question. In fact, a natural way to query a database requires us to deny any positive fact that is not explicitly represented in the database. This principle is not directly included in the deductive systems of standard logic. It is called the *(relational) closed-world assumption*: positive facts that we do not know to be true can be interpreted as false. Under this convention, we can avoid the explicit representation of negative information. The involved reasoning is non-monotonic because the introduction of new facts in the database can force us to revise previous conclusions. This assumption is often applied in elementary databases.

5.3.2 Formal definition

Let us give a more precise definition of the closed-world assumption. Recall that a theory $T(\Delta)$ associated with a set Δ of formulas of $\mathcal{L}$ is the set of all logical consequences of Δ.

Let Δ be a set of formulas of the first-order language $\mathcal{L}$. Let us represent by Δ' the set of negative ground literals that correspond to ground positive literals that do not belong to the theory $T(\Delta)$. The closed-world assumption *completes* (§ 5.2.2) the theory $T(\Delta)$ by defining a new theory $T(\Delta \cup \Delta')$, denoted $CWA(\Delta)$. Thus, $CWA(\Delta)$ is the logical closure of the union of the database Δ and the set Δ'. The theory $CWA(\Delta)$ represents the extended set of information that can be inferred from Δ under the closed-world assumption.

Checking whether a formula A of $\mathcal{L}$ can be *deduced from a database Δ under the closed-world assumption* consists in checking if A belongs to $CWA(\Delta)$.

5.3.3 Properties and limitations

The closed-world assumption can be applied to databases that contain ground positive literals only. In general, it cannot be applied when the database contains information that has to be written in a more complex logic language. For example, let us consider the following database Δ:

$Man(x) \lor Woman(x)$
$Professor(Smith)$

This database contains a piece of disjunctive information: any individual is either a man or a woman. The ground atoms $Man(Smith)$ and $Woman(Smith)$ cannot be deduced logically from this database. The theory $CWA(\Delta)$ will thus contain all the logical consequences of the union of the database Δ and of the set that consists of the ground negative literals $\neg Man(Smith)$ and $\neg Woman(Smith)$. The set $CWA(\Delta)$ is, however, *inconsistent* because it contains the three mutually inconsistent formulas $Man(x) \lor Woman(x)$, $\neg Man(Smith)$ and $\neg Woman(Smith)$.

There exists a theorem that states the condition under which consistency is preserved when the closed-world assumption is applied.

- Let Δ be a consistent set of formulas of $\mathcal{L}$. The theory $CWA(\Delta)$ is consistent if and only if any clause $P_1 \lor ... \lor P_n$ that is formed of positive ground literals P_i and that can be deduced from Δ, contains at least one positive ground literal P_i that can be deduced from Δ [Shepherdson 84].

The 'only if' statement is obvious. Let us give a proof *a contrario* of the 'if' statement. Assume that $CWA(\Delta)$ is inconsistent. Suppose that any clause

5.3. THE CLOSED-WORLD ASSUMPTION

that is formed of positive ground literals and that can be deduced from Δ, contains at least one literal that can be deduced from Δ too. As $CWA(\Delta)$ is the logical closure of $\Delta \cup \Delta'$, the theory $CWA(\Delta)$ is inconsistent if and only if $\Delta \cup \Delta'$ is inconsistent. According to the compactness theorem ([Thayse 88], § 1.2.9), there must exist a finite subset of Δ' that contradicts Δ. Let $\{\neg P_1, ..., \neg P_n\}$ be such a subset. It must be possible to deduce the negation of the logical conjunction of the elements of this set, i.e., $\Delta \vDash P_1 \vee ... \vee P_n$. As each $\neg P_i$ belongs to Δ', no P_i can be deduced from Δ. This result contradicts the assumption and proves the theorem.

The theorem above states the conditions under which disjunctive information can give rise to mutually inconsistent conclusions when the closed-world assumption is applied. The following corollary is of a great practical interest.

- If the clausal form of Δ is consistent and contains Horn clauses only, then the theory $CWA(\Delta)$ is consistent [Reiter 78].

We can thus apply the closed-world assumption to simple databases containing positive ground literals or Horn clauses only. In these cases, consistency is preserved. Let us however emphasize that dealing with Horn clauses is a sufficient condition but not a necessary condition to preserve consistency under the closed-world assumption.

In order to illustrate this point, let us examine the example above. The clause $Man(x) \vee Woman(x)$ in Δ is not a Horn clause. If we add the ground literal $Man(Smith)$ to the database Δ, then the theory $CWA(\Delta)$ becomes consistent. Indeed, there does not exist any clause $P_1 \vee ... \vee P_n$ that is formed of positive ground literals P_i and that can be deduced from Δ while none of the P_i's can be deduced from Δ (if the only individual constant of $\mathcal{L}$ is $Smith$).

In Section 5.7, we shall show that the closed-world assumption requires us to minimize the extensions of all the predicates occurring in the database. To preserve consistency under the closed-world assumption, the existence of a unique *minimal model* is necessary. A proof-theoretic constraint that is equivalent to this model-theoretic restriction has been established for the propositional case [Papalaskari and Weinstein 87]. This constraint is a direct consequence of the theorem given above. A subset of $\mathcal{L}$ will be said to be *subconditional* when it is equivalent to a set $\mathcal{P}_1$ of the following type. The set $\mathcal{P}_1$ is the union of a set $\mathcal{P}_2$ that consists of positive literals and of a set $\mathcal{P}_3$ that consists of disjunctions of literals containing at least one negative literal whose positive form does not appear in $\mathcal{P}_2$. Only consistent subconditional sets have a unique minimal model; their consistency is preserved under the closed-world assumption.

Let us also note that the closed-world assumption may allow undesirable inferences when *null values* (Chap. 6) or *Skolem constants* ([Thayse 88], § 1.2.8) are used. The closed-world assumption interprets these pieces of information as simple individual constants; one could thus infer undesired constraints about them.

Let us finally emphasize that the closed-world assumption takes into account the syntactical form of formulas of Δ, as a ground *negative* literal is inferred each time a *positive* ground literal cannot be deduced from Δ. To replace all occurrences of the predicate P_i by $\neg Q_i$ in Δ (with the definition $P_i \equiv \neg Q_i$) yields a new database Δ_1 that is logically equivalent to Δ, but that may have a different behaviour under the closed-world assumption. In the conceptualization of a domain, one must see to it that the number of positive facts is less than the number of negative facts that one wants to represent in an implicit manner.

5.3.4 Additional conventions

The closed-world assumption is often applied together with a second convention, called the *domain-closure assumption*. This assumption restricts the interpretation domain to the set of objects that can be defined by means of the function constants occurring in the knowledge base. Under such a convention, provided there are no function constants except individual constants, we can only instantiate a variable to an individual constant. In this case, the *domain-closure assumption* can be represented by the following axiom:

$$\forall x \left((x = c_1) \vee \ldots \vee (x = c_n)\right)$$

where $c_1, ..., c_n$ are all the individual constants of the language $\mathcal{L}$.

In the example above, the domain-closure assumption requires us to restrict the domain of interpretation of the knowledge base Δ to the one-element set $\{Smith\}$, as the individual constant $Smith$ is the only function constant that occurs in Δ.

A third convention is obtained by applying the closed-world assumption to the equality predicate. Ground terms are viewed as being equal only when they can be proved to be equal: different names represent different objects. This assumption is called the *unique-name assumption*. When there are no function constants of non-zero arity, this assumption can be represented by the set of formulas $\neg(c_i = c_j)$ corresponding to all pairs of integers (i, j) with $1 \leq i < j \leq n$, where $c_1, ..., c_n$ are the individual constants of $\mathcal{L}$. This assumption will be applied systematically in the remainder of this chapter.

In order fully to axiomatize the reasoning that is involved in the manipulation of the knowledge base, we must introduce the axioms that represent the properties of the equality predicate, since this predicate occurs in the conventions. When the only function constants are individual constants, these axioms are the following ones:

5.3. THE CLOSED-WORLD ASSUMPTION

- $\forall x\, (x = x)$: *reflexivity property*;

- $\forall x\, \forall y\, ((x = y) \supset (y = x))$: *symmetry property*;

- $\forall x\, \forall y\, \forall z\, ((x = y) \wedge (y = z) \supset (x = z))$: *transitivity property*;

- $\forall x_1...\forall x_n\, \forall y_1...\forall y_n\, (P(x_1, ..., x_n) \wedge (x_1 = y_1) \wedge .... \wedge (x_n = y_n) \supset P(y_1, ..., y_n))$: *principle of substitution of equal terms*.

The last axiom is stated for each n-ary predicate P occurring in the knowledge base.

In this way, we have formalized the different principles and assumptions discussed above with the help of additional axioms (also called *particularization axioms*) that are to be added to the initial set of information. Thus, we have characterized the reasoning formally by means of a classical deductive system that operates, in a proof-theoretic manner, over an extended knowledge base.

Let us emphasize that, in practice, it would be totally inefficient to introduce such axioms in the knowledge base and then let a theorem prover to operate. In practice, these axioms are replaced by conventions and operational techniques play an 'equivalent role'.

From a semantic point of view, the domain-closure assumption implies considering *Herbrand models* only. The role of the axioms of equality and of the unique-name assumption is to allow us to consider models the interpretation domain of which stands in bijection with the set of individual constants of $\mathcal{L}$. Let us note that the unique-name assumption is sometimes too severe, since it implies that the database can contain no aliases. The particularization axioms will be discussed more thoroughly in Chapter 6.

Let us note that the closed-world assumption together with its associated additional conventions yield a *unique model* for the theory $CWA(\Delta)$ when this theory is consistent. Indeed, the interpretation domain contains only ground terms, and the truth value of every ground literal is determined [Shepherdson 88b].

5.3.5 Restricted closed-world assumption

It is possible to weaken the closed-world assumption by only applying it to a restricted subset of predicates of the knowledge base. Indeed, it happens that our knowledge is complete with respect to some but not all relations occurring in the knowledge base.

For example, let us consider the following knowledge base Δ:

$Professor(x) \vee Student(x)$
$Professor(Smith)$
$Student(Davies)$
$Student(Johnson)$
$Woman(McKay)$

Assume that Δ contains the complete list of professors but does not necessarily contain the list of all the students. Let us query Δ under the closed-world assumption that is restricted to the predicate *Professor*. In this case, we may infer in a consistent manner that *McKay* is a student. Indeed, as the information $Professor(McKay)$ cannot be deduced from Δ, we infer $\neg Professor(McKay)$. The first formula of Δ allows us to infer $Student(McKay)$. Let us note that under the unrestricted closed-world assumption, we would have inferred that *McKay* is neither a professor, nor a student; this would have been inconsistent with Δ.

5.4 Predicate completion

5.4.1 Introduction

To define a concept consists in giving the necessary and sufficient conditions for this concept to be verified. In everyday conversation, we often interpret as being the definition of a concept what is actually a mere sufficient condition for the concept to be verified.

Let us for example consider the following statement: 'If I arrive after nine o'clock, then I am late'. Formally, this sentence gives only a particular sufficient condition for me to be late; there may exist some other ones. However, it seems natural to interpret this specific sufficient condition as being the only possible one, and thus as being the necessary and sufficient condition to be late. In other words, we often identify this sentence as the following definition: 'I am late if and only if I arrive after nine o'clock'. This is clearly a conjecture; for example, it is possible that I am also late if I arrive between eight and nine o'clock. Such an approach, which restricts the sufficient conditions about the verification of a concept to the only conditions that have been stated, can clearly give rise to revisable reasoning. Indeed, it may be necessary to retract an inference when a new sufficient condition is taken into account.

Predicate completion is based upon this principle. A set Δ of predicate logic formulas that are written *about* some predicates P_i gives sufficient conditions about the verification of these predicates. It is sometimes natural to consider that the formulas contained in Δ are representing all the sufficient conditions about the verification of the predicates P_i. By convention, these conditions

5.4. PREDICATE COMPLETION

are thus interpreted as necessary and sufficient conditions for the verification of the predicates P_i. It is possible to express this convention formally with the help of additional axioms stating that a predicate is satisfied *only* when one of the sufficient conditions is satisfied. This technique is called *predicate completion*. Let us introduce it intuitively with the help of two elementary examples.

Suppose that Δ contains only the ground literal $P(a)$. This formula is logically equivalent to the following implication formula:

$$\forall x \, ((x = a) \supset P(x)).$$

The formula above states a sufficient condition for the verification of predicate P. We shall say that it is *written about* P. In order to transform the sufficient condition into a definition, we introduce the following 'dual' implication formula into the knowledge base:

$$\forall x \, (P(x) \supset (x = a)).$$

This formula is called the *completion formula of P in Δ*. By introducing the completion formula of predicate P into the knowledge base Δ, we obtain the *knowledge base completed with respect to predicate P*. This new knowledge base is denoted $\mathcal{C}(\Delta, P)$. We see that, in our example, $\mathcal{C}(\Delta, P)$ is equivalent to the knowledge base containing the unique formula

$$\forall x \, (P(x) \equiv (x = a)).$$

In this elementary example, to complete a knowledge base with respect to P consists in transforming the implication formulas that have P as consequent into equivalence formulas.

Let us consider a second example, where Δ is a knowledge base containing only the two positive ground literals $P(a)$ and $P(b)$. The completion formula of P in Δ is

$$\forall x \, (P(x) \supset (x = a) \vee (x = b)),$$

and $\mathcal{C}(\Delta, P)$ contains the unique formula

$$\forall x \, (P(x) \equiv (x = a) \vee (x = b)).$$

To *complete a knowledge base* Δ consists in completing Δ with respect to every predicate P for which there exists at least one formula in Δ that is written about P. The completed knowledge base is denoted $\mathcal{C}(\Delta)$.

5.4.2 Formal definition

Let Δ be a set of *clauses* of the form

$$P_1 \wedge ... \wedge P_n \supset P,$$

where P is a positive literal and where $P_1, ..., P_n$ are positive or negative literals. Such a clause is said *to be written about* P (in short, it is *clause about* P). This clause is logically equivalent to the following formula:

$$\forall \mathbf{y} \, (P_1 \wedge ... \wedge P_n \supset P(\mathbf{t})).$$

Here, $\mathbf{t}$ is a tuple of terms $t_1, ..., t_k$, the literals P_i and $P(\mathbf{t})$ mention some elements of the tuple $\mathbf{y}$ of variables $y_1, ..., y_r$, and $\forall \mathbf{y}$ is an abbreviation for $\forall y_1....\forall y_r$. This formula is logically equivalent to

$$\forall \mathbf{y} \, \forall \mathbf{x} \, ((\mathbf{x} = \mathbf{t}) \wedge P_1 \wedge .... \wedge P_n \supset P(\mathbf{x})),$$

where $\mathbf{x}$ is a tuple of variables $x_1, ..., x_k$ that do not occur in the literals P_i and $P(\mathbf{t})$, and where $(\mathbf{x} = \mathbf{t})$ is an abbreviation for $(x_1 = t_1) \wedge ... \wedge (x_k = t_k)$. As variables of the tuple $\mathbf{y}$ only appear in the antecedent of this implication, the formula above is equivalent to either of the following formulas:

$$\forall \mathbf{x} \, \forall \mathbf{y} \, (\neg((\mathbf{x} = \mathbf{t}) \wedge P_1 \wedge .... \wedge P_n) \vee P(\mathbf{x})),$$

$$\forall \mathbf{x} \, (\exists \mathbf{y}((\mathbf{x} = \mathbf{t}) \wedge P_1 \wedge ... \wedge P_n) \supset P(\mathbf{x})).$$

This last formula is called the *normal form* of the clause. When the knowledge base Δ contains j clauses about P, then the j associated normal forms are

$$\forall \mathbf{x} \, (E_1 \supset P(\mathbf{x}))$$
$$\vdots$$
$$\forall \mathbf{x} \, (E_j \supset P(\mathbf{x}))$$

i.e., $\forall \mathbf{x} \, (E_1 \vee ... \vee E_j \supset P(\mathbf{x}))$, where E_i is a conjunction of literals that are existentially quantified. The *completion formula of* P *in* Δ is defined to be

$$\forall \mathbf{x} \, (P(\mathbf{x}) \supset E_1 \vee ... \vee E_j).$$

The *knowledge base* Δ *completed with respect to predicate* P is denoted $\mathcal{C}(\Delta, P)$; it is obtained by the introduction into Δ of the completion formula of P in Δ.

5.4. PREDICATE COMPLETION

We obtain the *completed knowledge base*, denoted $C(\Delta)$, when we introduce into Δ the completion formulas of all predicates about which at least one formula is written in Δ.

When Δ contains no clause about P, then the completion formula of P in Δ is $\forall x \, (P(x) \supset \mathbf{F})$. This formula is equivalent to $\forall x \, (\neg P(x))$. As Δ tells us *nothing* about the verification of P, we infer that P is never satisfied.

Let us also emphasize that predicate completion has been introduced by Clark with the objective of giving a formal semantics to the form of negation used in Prolog (§ 5.5.1) [Clark 78].

5.4.3 Example

Before presenting the main properties and drawbacks of predicate completion, let us illustrate this technique by means of the canonical example. Consider the knowledge base

$$\Delta = \{Bird(x) \supset Fly(x), Fly(Icarus), \neg Bird(Bobby)\}.$$

We wish to complete Δ with respect to the predicate Fly. The sufficient condition given in Δ about flying is either to be a bird, or to be Icarus. Therefore, the completion formula of the predicate Fly in Δ is

$$\forall x \, (Fly(x) \supset Bird(x) \vee (x = Icarus)).$$

The knowledge base Δ completed with respect to the predicate Fly, i.e., the knowledge base $C(\Delta, Fly)$, is the set

$$\{\forall x \, (Fly(x) \equiv Bird(x) \vee (x = Icarus)), \neg Bird(Bobby)\}.$$

The sufficient condition about the predicate Fly has now become a necessary and sufficient condition. We can query the knowledge base completed with respect to the predicate Fly and, for instance, infer $\neg Fly(Bobby)$ (under the unique-name assumption).

This inference can be revised when a new piece of information is introduced in Δ. For example, if we introduce the ground atom $Fly(Bobby)$ into Δ, then the completion formula of Fly in Δ becomes

$$\forall x \, (Fly(x) \supset Bird(x) \vee (x = Icarus) \vee (x = Bobby)).$$

In this case, the negative ground literal $\neg Fly(Bobby)$ can no longer be inferred from the knowledge base completed with respect to Fly.

5.4.4 Properties and drawbacks of predicate completion

Conclusions that can be inferred from a completed knowledge base $\mathcal{C}(\Delta)$ may be revised when new information is added to Δ. A knowledge base that is completed with respect to a predicate P restricts the extension of P to the sufficient conditions about the verification of P that are given by Δ. The introduction into Δ of new formulas written about P will modify the completed knowledge base and the conclusions it allows one to infer.

Let us briefly describe the main properties and drawbacks of predicate completion. First, it is important to bear in mind that this technique has been defined only when a particular set of clauses of $\mathcal{L}$ is involved (namely, clauses with at least one positive literal). This observation exhibits a first limitation of predicate completion. Next, consistency is not always preserved by predicate completion. For these reasons, it is necessary to limit severely the application domain of this technique. Let us consider for example the knowledge base Δ that only contains the formula $\neg P \supset P$. This formula is consistent; it is logically equivalent to the atom P. However, $\mathcal{C}(\Delta)$ contains the inconsistent formula $\neg P \equiv P$ whereas the knowledge base $\{P\}$ completed with respect to P is consistent. This example shows that predicate completion does not only depend on the logical contents of the formulas of the knowledge base, but also on their syntactical form.

Let us also point out the following problem. Clauses written about P may mention occurrences of P in their antecedents. As clauses written about P are interpreted as giving the definition of P, there is an obvious danger of circular definition (Chap. 6). To avoid this problem and to preserve consistency, we have to limit the application domain of predicate completion. For example, a knowledge base will be completed with respect to a predicate P only when all clauses that are written about P are *solitary* in P, i.e., when all the clauses that contain a positive occurrence of P contain exactly one occurrence of P [Genesereth and Nilsson 87]. It can be shown that the completion of a consistent set of solitary clauses with respect to P preserves consistency. Note however that restricting to clauses solitary in P is sufficient but not necessary to preserve consistency. In particular, completing a knowledge base with respect to a predicate preserves consistency in the cases of a set of definite Horn clauses [Shepherdson 88b] and of a stratified set of clauses [Apt et al. 88].

Let us also mention that it is possible to define a technique, analogous to predicate completion, the semantics of which is given within a three-valued logic and which always preserves consistency [Kunen 87]. Let us finally emphasize that circumscription and its variants, which are described in Section 5.6, extend and generalize predicate completion in several respects.

5.4.5 Closed-world assumption and predicate completion

The closed-world assumption and the predicate completion technique are superficially similar. Both techniques involve the following revisable reasoning. When a piece of positive information cannot be 'proved' from Δ, then its value is conjectured as *false*. However, these techniques often give rise to different results [Shepherdson 84, 85, 88b].

The closed-world assumption may preserve consistency whereas predicate completion does not. We have seen that the completion of $\Delta = \{\neg P \supset P\}$ with respect to P is inconsistent whereas the closed-world assumption would have given a consistent result (as $\neg P \supset P$ is equivalent to the atom P). Reciprocally, predicate completion can preserve consistency whereas the closed-world assumption does not. For example, if $\Delta = \{\neg P \supset Q\}$, then $CWA(\Delta)$ contains the three mutually inconsistent formulas $\neg P$, $\neg Q$ and $\neg P \supset Q$. Finally, both techniques might preserve consistency while giving incompatible results.

Let us examine an example where the closed-world assumption restricts more severely than predicate completion the positive implicit information that can be associated with a knowledge base. Define $\Delta = \{P(a) \supset P(b), P(b) \supset P(a)\}$. While we can infer $\neg P(a)$ and $\neg P(b)$ from $CWA(\Delta)$, the completed knowledge base $\mathcal{C}(\Delta)$ admits a model where $P(a)$ and $P(b)$ are both true. Indeed, we have $\mathcal{C}(\Delta) = \{\forall x \, (P(x) \equiv ((x = b) \wedge P(a)) \vee ((x = a) \wedge P(b)))\}$. In fact, as pointed out by Apt, Blair and Walker, the closed-world assumption is often too strong as it resolves any uncertainty in a negative way [Apt et al. 88].

5.5 Specific application domains

5.5.1 Logic programming

Logic programming is a generic name for problem solving methods that apply deductive techniques of logic to knowledge represented in a declarative way.

Classical deductive systems do not directly allow the formalization of forms of revisable reasoning. However, specific operators in logic programming languages allow us to arrange that some inferences depend on the evolution of the contents of the knowledge base given by the logic program.

In languages such as Prolog, these operators allow the refutation of a positive ground literal P when there is no finite proof of P from the current state of the knowledge base. When a new piece of information is added to the knowledge base, it may become possible to infer the atom P. Let us briefly recall the main principles of logic programming (for more details, the reader may consult chapter 6 of [Thayse 88]). Usual systems of logic programming, such as pure Prolog, make use of Horn clauses, i.e., clauses that contain at most one positive literal.

A Prolog program is made up of a sequence of *definite Horn clauses* (§ 5.2.4), i.e., clauses that contain exactly one literal in their consequent. A clause without an antecedent is called a *unit clause*; when it contains no variable, it represents an *elementary fact*. A clause with an antecedent represents a *conditional rule*. A Prolog interpreter is a refutation based automatic proof system that uses only the *resolution rule* as inference rule. To prove that the formula

$$\exists y_1 ... \exists y_r (R_1 \wedge ... \wedge R_n)$$

is a 'logical consequence' of the Prolog program, the interpreter attempts to prove that introducing the negation of this formula into the knowledge base (which is here the Prolog program) leads to a contradiction, i.e., that introducing the Horn clause with consequent $\neg R_1 \vee ... \vee \neg R_n$ (which is equivalent to $\neg(R_1 \wedge ... \wedge R_n)$) leads to the empty clause.

The Prolog algorithm is explained in detail in [Thayse 88]. Let us simply recall here that it performs a depth-first search, from left to right. Although the logic system which Prolog is based upon is sound and complete, the Prolog algorithm is not complete.

To increase the expressive power of Prolog, 'negative' expressions are allowed in the antecedents of the rules. They involve the special operator *not*. This operator is a *predefined predicate* which is based upon the *negation as finite failure principle*. In its simplest form, this principle is defined as follows.

Given a ground positive literal G, the predicate not(G) succeeds and takes the value **T** *if we cannot find a finite proof for G from the information contained in the system; otherwise, not(G) takes the value* **F**.

Let us give an example of a Prolog program that makes use of the predefined predicate *not*.

```
fly(X) :- bird(X),not(abnormal(X)).
abnormal(X) :- ostrich(X).
bird(Tweety).

?- abnormal(tweety).
---> no
?- fly(tweety).
---> yes
```

It was not possible to infer the conclusion abnormal(tweety) from the current set of information; the predicate not(abnormal(tweety)) does succeed and yields the value **T**. When we add the piece of information ostrich(tweety),

5.5. SPECIFIC APPLICATION DOMAINS

the predicate not(abnormal(tweety)) fails because abnormal(tweety) succeeds. The assertion fly(tweety) is then refuted.

```
ostrich(tweety).
?- abnormal(tweety).
---> yes
?- fly(tweety).
---> no
```

The predefined predicate *not* is based on a principle which is analogous to the closed-world assumption (Sect. 5.3). The idea they have in common is the following one. If a positive ground literal cannot be 'proved', then its negative form can be interpreted as 'true'. There is, however, a major difference. The closed-world assumption is based on the concept of 'pure' logical proof, whereas negation as failure relies upon the concept of proof that underlies the Prolog algorithm. Clark has suggested that negation as failure could be interpreted as the result of deductive operations upon the *completed knowledge base* of the Prolog program [Clark 78]. However, Shepherdson has shown that neither predicate completion, nor the closed-world assumption can provide an exact characterization of negation as failure [Shepherdson 84, 85]. Indeed, the *not* predicate of Prolog is based upon the specific algorithm of Prolog and not upon a pure logical definition.

The reader will find in [Shepherdson 88b] and in [Kunen 87] a deeper study of the treatment of negation in logic programming. Several approaches have been proposed to characterize the semantics of different classes of logic programs (with the negation as failure principle, or without it). Among the most recent ones, let us mention a *three-valued semantics* [Fitting and Ben-Jacob 88], the *stable model semantics* [Gelfond and Lifschitz 88a], the *perfect model semantics* [Przymusinski 88b], and the *weakly perfect model semantics* [Przymusinski and Przymusinska 88]. A more traditional approach, based on the *fixpoint* technique [Apt and van Emden 82], [Apt et al. 88], [Lloyd 87], is introduced in Subsection 6.3.6.

Let us mention that a particular form of the circumscription technique (that will be introduced in the next section), allows one to characterize the semantics of *stratified logic programs* [Lifschitz 88], i.e., of logic programs that consist of stratified sets of clauses. Stratified sets of clauses enjoy several interesting properties that are discussed in Chapter 6. Default logic has also been shown capable of providing a semantics for classes of logic programs [Bidoit and Froidevaux 87]. In fact, it has been shown possible to make use of fragments of several non-monotonic logics to give stratified logic programs a declarative semantics [Przymusinski 88c].

5.5.2 Deductive databases

Deductive databases form a natural and advanced application domain for techniques that formalize revisable reasoning. When analysed from the point of view of logic, deductive databases extend simple relational databases in a natural and elegant way. Knowledge representation that was restricted to elementary logical facts (ground atoms) is extended to more complex logical formulas. Querying a database that was consisting in proving or in refuting elementary facts, becomes a deduction process that deals with more elaborate formulas. Chapter 6 is devoted to the theory of deductive databases; in this subsection, we introduce, in an intuitive manner, a technique of formalization of revisable reasoning that is specific to deductive databases.

In the introduction, we have emphasized that reasoning is often revisable when it is performed in the context of disjunctive, incomplete, negative or implicit knowledge. The closed-world assumption is a satisfactory convention when the represented knowledge consists of elementary facts only and when the querying process is restricted to positive ground facts. When one wants to represent or infer disjunctive information, one must make use of more elaborate techniques.

In this subsection, we intuitively introduce an extension of the closed-world assumption that has been proposed to deal with disjunctive information; this technique is called the *generalized closed-world assumption* [Minker 82]. It is based on the following semantic considerations.

In the field of deductive databases, we only take into account Herbrand models of the databases. We can represent a Herbrand model $\mathcal{M}$ of a database Δ by means of the set of ground atoms that are verified in $\mathcal{M}$ (§ 5.2.5).

By definition, *Horn databases* (i.e., databases that contain Horn clauses only) do not contain disjunctive information. Viewed as sets of logic formulas, these databases enjoy the *model intersection* property. This property can be stated in the following way.

- The intersection of all Herbrand models of a finite consistent Horn database is a model of this database. This model is called a *least model* of the database ([Thayse 88], § 1.1.16).

Intuitively, a least model denotes the sure ground positive facts that can be associated with the database. In this model, any uncertain information is interpreted as false. In fact, to query a Horn database Δ under the closed-world assumption consists in evaluating the query with respect to the least model. The least Herbrand model is the canonical Herbrand model (§ 5.2.5).

Indefinite databases, i.e., databases that contain disjunctive information, do not enjoy the model intersection property. For example, let us consider the database $\Delta = \{P(a) \vee P(b)\}$. This database admits three Herbrand models.

5.5. SPECIFIC APPLICATION DOMAINS

In the first one, the only ground positive literal that is true is $P(a)$; in the second one, only $P(b)$ is true; in the last one, both $P(a)$ and $P(b)$ are true. The intersection of these three models is the empty set, which is not a model of Δ.

To cope with indefinite databases, Minker proposes the following method [Minker 82]. Define a *minimal model* of a database Δ to be a Herbrand model of Δ that does not contain any model of Δ as a proper subset. According to this definition, an indefinite database may admit several minimal models. In the example above, $\{P(a)\}$ and $\{P(b)\}$ are minimal models of Δ, whereas $\{P(a), P(b)\}$ is not. Intuitively, a formula that is true in all minimal models of Δ is true with respect to Δ, and a formula that is true in no minimal model of Δ is false with respect to Δ. The truth value of a formula that is true in some but not all minimal models of Δ is unknown.

Let us consider the set of all minimal models of a database Δ. The intersection of these minimal models is a set of ground atoms that we denote $\Gamma(\Delta)$. By definition, $\Gamma(\Delta)$ contains the ground atoms that are true in all the minimal models of Δ. In a dual way, for each minimal model of Δ, let us consider the set of ground atoms that are not verified in this model. The intersection of all these sets is denoted $\Gamma(\Delta)'$. By definition, $\Gamma(\Delta)'$ contains the ground atoms that are false in all minimal models of Δ. A ground atom will be interpreted as true, false, or undetermined, according to whether it belongs to $\Gamma(\Delta)$, to $\Gamma(\Delta)'$, or to none of these sets.

Under the *generalized closed-world assumption*, a ground atom can be inferred from a database Δ if and only if it belongs to $\Gamma(\Delta)$. One will be allowed to infer the negation of a ground atom when this ground atom belongs to the set $\Gamma(\Delta)'$. A deeper and more formal presentation of the generalized closed-world assumption is given in Subsection 6.5.3.

Several algorithms have been proposed to query a deductive database under the generalized closed-world assumption (see e.g. [Grant and Minker 86]). There exist several variants of the generalized closed-world assumption. In particular, let us mention a technique of applying negation as failure that is based on a generalized closed-world assumption restricted to specific predicates of the database [Gelfond and Przymusinska 86]. Unfortunately, the corresponding algorithms often have exponential complexity.

Various extensions to the generalized closed-world assumption have been proposed in the literature. Their goal is mainly to allow the inference of more elaborate logical formulas. For example, the *extended generalized closed-world assumption* in [Yahya and Henschen 85] allows one to decide whether a disjunctive ground formula of the form $\neg P_1 \vee ... \vee \neg P_n$ can be inferred. It should be stressed that the associated algorithms have exponential complexity.

An alternative technique is *subimplication* [Bossu and Siegel 85]. This technique can be interpreted as being a decidable form of the generalized closed-world assumption. Subimplication enforces specific syntactical constraints to the representation and query languages of the knowledge base.

5.6 Circumscription

5.6.1 Introduction

Circumscription is a technique of formalization of revisable reasoning that generalizes predicate completion and the conventions based on suitable versions of the closed-world assumption. Proposed originally by McCarthy [McCarthy 80, 86], circumscription is currently one of the most active areas of research in the field of non-monotonic logics. Extensive research work is being devoted to this attractive technique, with the aim of determining its properties, extending its effective scope, and limiting its drawbacks. Let us especially mention the work of Lifschitz [Lifschitz 85a, 85b, 86a, 86b, 87b, 88] and the work of Besnard's team [Besnard et al. 87].

Circumscription is based upon the following principle. To circumscribe a property in a knowledge base Δ consists in limiting the verification domain of this property to what is required by Δ. The circumscription of a property in a knowledge base Δ is obtained by introducing an additional axiom into Δ. This axiom is called a *circumscription axiom*. In general, it is a second-order formula. It expresses, in a proof-theoretic way, the minimization of the verification domain of the circumscribed property.

There exist numerous forms of circumscription. In this section, we shall deal only with *formula circumscription* [McCarthy 86]. Furthermore, we shall restrict ourselves to knowledge bases that consist of formulas of the first-order language $\mathcal{L}$.

As circumscription is based on the introduction and the use of an additional axiom, this approach to non-monotonic reasoning is primarily of a proof-theoretic nature. However, for didactic reasons, we shall first examine the semantic aspects of circumscription.

5.6.2 Minimal models

From a semantic point of view, circumscription consists in *minimizing* the verification domain of selected formulas contained in a knowledge base. In Section 5.7, we shall introduce a formal framework that allows one to characterize different approaches to non-monotonic reasoning that are based on model-theoretic minimization techniques. In this subsection, we simply give an intuitive account of the semantic role of circumscription.

5.6. CIRCUMSCRIPTION

Let Δ be a knowledge base consisting of first-order formulas that belong to $\mathcal{L}$. Let us classify the predicate constants occurring in Δ into three disjoint families, denoted $\mathcal{P}_1$, $\mathcal{P}_2$ and $\mathcal{P}_3$. The family $\mathcal{P}_1$ contains the circumscribed predicates, i.e., the predicates whose extensions must be minimized. The family $\mathcal{P}_2$ contains the predicates whose extensions are allowed to vary in the circumscription process. The family $\mathcal{P}_3$ contains all the other predicates that occur in Δ. The knowledge base Δ is written $\Delta(\mathcal{P}_1; \mathcal{P}_2)$.

Recall that a *model* of a set Δ of first-order formulas is an *interpretation* that satisfies all formulas in Δ (§ 5.2.2). Let $\mathcal{M}_1$ and $\mathcal{M}_2$ be two models of Δ. The model $\mathcal{M}_1$ will be said to be *inferior* to the model $\mathcal{M}_2$ *with respect to the set $\mathcal{P}_1$ of circumscribed predicates and to the set $\mathcal{P}_2$ of variable predicates* when the following three conditions are satisfied.

1. $\mathcal{M}_1$ and $\mathcal{M}_2$ share the same domain of interpretation; they interpret in the same manner both the function constants and the variables;

2. $\mathcal{M}_1$ and $\mathcal{M}_2$ interpret in the same manner the predicate constants belonging to the set $\mathcal{P}_3$;

3. For each predicate P belonging to the set $\mathcal{P}_1$ of circumscribed predicates, the extension of P in $\mathcal{M}_1$ is a subset of the extension of P in $\mathcal{M}_2$.

These conditions can be paraphrased as follows. Models $\mathcal{M}_1$ and $\mathcal{M}_2$ interpret 'almost identically' the formulas of Δ. Model $\mathcal{M}_1$ interprets the predicates in $\mathcal{P}_1$ to *true* 'less often' than model $\mathcal{M}_2$ does. Moreover, the behaviour of $\mathcal{M}_1$ and $\mathcal{M}_2$ may differ with respect to the predicates belonging to $\mathcal{P}_2$.

When one circumscribes the set $\mathcal{P}_1$ of predicates in the knowledge base Δ, allowing the predicates of the set $\mathcal{P}_2$ to vary, one is interested in the models of Δ that are minimal with respect to the comparison relation between models we have just introduced. *A formula can be inferred from a circumscribed knowledge base when it is satisfied in all minimal models.* In other words, a formula that can be inferred, within circumscription, from $\Delta(\mathcal{P}_1; \mathcal{P}_2)$ obeys a minimality condition with respect to the extensions of the circumscribed predicates, namely, the predicates in $\mathcal{P}_1$.

A knowledge base Δ can be circumscribed only if Δ admits minimal models. This necessary condition is not always satisfied. In fact, the existence of minimal models is only ensured when the formulas of Δ obey some specific constraints. In Subsection 5.6.5, we examine sufficient conditions for circumscription to preserve consistency.

5.6.3 Formal definition

Circumscription is usually introduced from a proof-theoretic point of view. A second-order formula, called the *circumscription axiom*, is introduced in the knowledge base.

The circumscription of the knowledge base $\Delta(\mathcal{P}_1; \mathcal{P}_2)$, which consists in minimizing the predicates that belong to $\mathcal{P}_1$ while the predicates of $\mathcal{P}_2$ are allowed to vary, is defined with the help of the following conjunctive formula:

$$\Delta(\mathcal{P}_1; \mathcal{P}_2) \wedge ((\forall \mathcal{P}'_1 \, \forall \mathcal{P}'_2) \, \neg(\Delta(\mathcal{P}'_1; \mathcal{P}'_2) \wedge \mathcal{P}'_1 < \mathcal{P}_1)).$$

Intuitively, the second conjunct of this formula prevents the extensions of the predicates in $\mathcal{P}_1$ from being made smaller, even when the predicates in $\mathcal{P}_2$ are allowed to vary. In other words, this formula enforces that the extensions of the predicates belonging to $\mathcal{P}_1$ are minimized, while the extensions of the predicates belonging to $\mathcal{P}_2$ are allowed to vary.

The expression $\forall \mathcal{Q}$ represents a second-order universal quantification over the set $\mathcal{Q}$ of predicates. If P' and P are predicates of the same arity, then $P' < P$ represents the formula $\forall \mathbf{x} \, (P'(\mathbf{x}) \supset P(\mathbf{x})) \wedge \neg(\forall \mathbf{x} \, (P(\mathbf{x}) \supset P'(\mathbf{x})))$. If $P' \leq P$ is defined to be the formula $\forall \mathbf{x} \, (P'(\mathbf{x}) \supset P(\mathbf{x}))$, then $P' < P$ is logically equivalent to the formula $P' \leq P \wedge \neg(P \leq P')$. If $\mathcal{P}'$ and $\mathcal{P}$ represent n-tuples $(P'_1, ..., P'_n)$ and $(P_1, ..., P_n)$ of predicates P'_i and P_i of the same arity, then $\mathcal{P}' < \mathcal{P}$ is defined to be the formula $P'_1 \leq P_1 \wedge ... \wedge P'_n \leq P_n \wedge \neg(P_1 \leq P'_1 \wedge ... \wedge P_n \leq P'_n)$, and $\mathcal{P}' \leq \mathcal{P}$ is defined to be the formula $P'_1 \leq P_1 \wedge ... \wedge P'_n \leq P_n$.

5.6.4 Example

Let us illustrate circumscription with the help of the canonical example [McCarthy 86]. To circumscribe a knowledge base requires us to make several decisions. We must select the predicates that have to be circumscribed and the predicates the extensions of which are allowed to vary. Moreover, we must select suitable predicates to instantiate the circumscription formula and to compute effectively, within first-order logic, what can be inferred from the circumscribed knowledge base.

As to the choice of the predicates that will be circumscribed (or minimized), we can apply a rather simple and systematic policy. In the conceptualization of the domain of discourse, it is often natural to make use of *general rules with exceptions*; these rules describe the normal state of affairs, whereas the abnormal cases are represented by specific unary predicates. These precisely are the predicates that will be minimized: they will be denoted $Abnormal_i$.

Let us give predicate logic formulas that model the following information. A normal thing does not fly; a bird is a thing that is an exception to this first rule; a normal bird flies; and an ostrich is a bird that does not fly:

- $\forall x \, (Thing(x) \wedge \neg Abnormal_1(x) \supset \neg Fly(x))$

5.6. CIRCUMSCRIPTION

- $\forall x \, (Bird(x) \supset Thing(x) \wedge Abnormal_1(x))$
- $\forall x \, (Bird(x) \wedge \neg Abnormal_2(x) \supset Fly(x))$
- $\forall x \, (Ostrich(x) \supset Bird(x) \wedge \neg Fly(x))$

Our knowledge base Δ is the set containing the four formulas above. We circumscribe the predicates $Abnormal_1$ and $Abnormal_2$ in Δ. In this way, we minimize the exceptions to the rules that are expressed by the first and third formulas. Abnormal birds and abnormal things should be deducible from Δ. The predicate Fly is allowed to vary during the circumscription process.

The circumscription axiom is the following formula:

$$(\forall \, Abnormal'_1 \, \forall \, Abnormal'_2 \, \forall \, Fly') \\ \neg(\Delta(\{Abnormal'_1, Abnormal'_2\}; \{Fly'\}) \wedge \\ (Abnormal'_1 \leq Abnormal_1) \wedge (Abnormal'_2 \leq Abnormal_2) \wedge \\ \neg((Abnormal_1 \leq Abnormal'_1) \wedge (Abnormal_2 \leq Abnormal'_2))).$$

This is a second-order formula, since a universal quantification is applied to the predicates $Abnormal'_1$, $Abnormal'_2$ and Fly'.

To make an effective use of the circumscription axiom, we have to instantiate each predicate variable. The main difficulty encountered in the practical use of circumscription lies precisely in the choice of the instantiation predicates.

In our example, intuition leads to the following instantiations:

- $Abnormal'_1(x) \equiv Fly(x)$
- $Abnormal'_2(x) \equiv Ostrich(x)$
- $Fly'(x) \equiv Bird(x) \wedge \neg Ostrich(x)$

In this case, simple but tedious first-order deductive operations allow us to infer the following formulas from Δ and from the instantiated circumscription axiom:

- $\forall x \, (Abnormal_1(x) \equiv Bird(x))$
- $\forall x \, (Abnormal_2(x) \equiv Ostrich(x))$
- $\forall x \, (Thing(x) \wedge \neg Bird(x) \supset \neg Fly(x))$
- $\forall x \, (Bird(x) \wedge \neg Ostrich(x) \supset Fly(x))$

Birds are the only exceptions to the rule stating that things do not fly. The only birds that do not fly are ostriches. The things that are not birds do not fly. Birds that are not ostriches fly.

5.6.5 Properties

In this concluding subsection we give, without proof, some properties and results about circumscription. The circumscription axiom is a second-order formula, which makes the effective use of circumscription rather difficult. Recall that the set of formulas that can be deduced from a knowledge base containing second-order formulas is generally not recursively enumerable.

The choice of a suitable instance for the circumscription axiom must often be made with the help of intuition, as the example of Subsection 5.6.4 has shown. The circumscription axiom has the character of a (second-order) induction axiom and can be interpreted as such. As Nilsson emphasizes (see [Levesque 87], p. 203), we might be required to be aware of the answer in order to apply circumscription.

For specific classes of formulas, the circumscription axiom can collapse to a first-order formula. In this case, systematic techniques have been proposed to compute circumscription. Let us consider the elementary situation where only one predicate must be circumscribed and where no predicate is allowed to vary. A formula A of $\mathcal{L}$ is said to contain a *positive occurrence* of the predicate P when P occurs positively in the clausal form of A. If all occurrences of the predicate that must be circumscribed are positive in Δ, then the circumscription axiom is collapsible to a first-order formula.

It is also possible to extend the concept of solitary clause (§ 5.4.4) that was defined for predicate completion so as to obtain the concept of solitary formula. For a given predicate constant P, a formula is said to be *solitary in* P when it can be written in the *normal form* $N(P) \wedge (E \leq P)$ where $N(P)$ is a formula that contains no positive occurrence of P, and where E is a formula that contains no occurrence of P. The circumscription of the predicate P in a knowledge base that can be rewritten as a formula solitary in P, given by the normal form $N(P) \wedge (E \leq P)$, is the first-order formula $N(E) \wedge (E = P)$, where $N(E)$ is obtained by replacing in $N(P)$ all occurrences of P by E and where $E = P$ is an abbreviation for $E \leq P \wedge P \leq E$. In particular, to circumscribe the predicate P in a set of clauses that are solitary in P amounts to completing this set with respect to P.

More generally, let us consider formulas that are *separable with respect to a predicate P* [Lifschitz 85a], i.e., formulas that

- either do not contain any positive occurrence of P,
- either are of the form $E \leq P$, where E contains no occurrence of P,
- or are composed of conjunctions and disjunctions of separable formulas.

Note that formulas that are solitary in P are separable with respect to P. A lot of formulas can be rewritten in the form of separable formulas. The cir-

5.6. CIRCUMSCRIPTION

cumscription axiom for separable formulas can collapse to a first-order formula. More precisely, let us just mention the following result [Lifschitz 85a].

If Δ is separable with respect to predicate P, then Δ can be rewritten in the associated normal form $\bigvee_i(N_i(P) \wedge (E_i \leq P))$ where $N_i(P)$ is a formula that contains no positive occurrence of P and where E_i contains no occurrence of P. The circumscription of predicate P in Δ is equivalent to the formula $\bigvee_i(D_i \wedge (P = E_i))$, where D_i is defined to be the formula $N_i(E_i) \wedge \bigwedge_{j \neq i} \neg(N_j(E_j) \wedge (E_j < E_i))$, where $N_k(E)$ is obtained from $N_k(P)$ by replacing all occurrences of P by E.

Furthermore, there exist several results allowing one to compute, in a rather systematic manner, the circumscription of an *ordered set of predicates* for some specific classes of formulas [Genesereth and Nilsson 87].

It is sometimes possible to reduce circumscription with predicates allowed to vary to circumscription without predicates allowed to vary. Let $\mathcal{P}_1$ and $\mathcal{P}_2$ be two sets of predicates. If Δ can be written in the form $N(\mathcal{P}_2) \wedge (\mathcal{E} \leq \mathcal{P}_2)$ where $N(\mathcal{P}_2)$ does not contain any positive occurrence of predicates belonging to $\mathcal{P}_2$ and $\mathcal{E}$ does not contain any occurrence of predicates belonging to $\mathcal{P}_2$, then the circumscription in Δ of the set $\mathcal{P}_1$ of predicates, with the set $\mathcal{P}_2$ of predicates allowed to vary, can be reduced to a circumscription without predicates allowed to vary. This circumscription is equivalent to the logical conjunction of the circumscription in $N(\mathcal{E})$ of the set $\mathcal{P}_1$ of predicates and the formula $N(\mathcal{P}_2) \wedge (\mathcal{E} \leq \mathcal{P}_2)$.

Just as we have given sufficient conditions under which the predicate completion technique and the closed-world assumption technique preserve consistency, we can state such conditions for the circumscription technique. Some consistent sets of formulas can become inconsistent when circumscription is applied. Let us mention a sufficient condition for consistency to be preserved under circumscription of a set of predicates [Lifschitz 86a].

- If Δ is consistent and is formed with formulas that are almost universal with respect to a predicate P, then the circumscription of P in Δ is consistent.

(A formula is said to be *almost universal with respect to a predicate P* when it is of the form $\forall \mathbf{x} \, (\phi)$ where $\mathbf{x}$ is a tuple of variables and where ϕ does not contain any positive occurrence of P in the scope of quantifiers.) In particular, formulas that are separable with respect to a predicate are almost universal with respect to this predicate.

The result above cannot be extended to the case where some predicates are allowed to vary in the circumscription process. However, if Δ is a consistent set of *universal* formulas (i.e., if Δ is a set of clauses or if the conjunctive normal form of each formula in Δ does not contain any Skolem function), then the circumscription in Δ of a set of predicates with some predicates allowed to vary is consistent [Lifschitz 86a].

Let us end this short review of properties of circumscription by mentioning that a concept of *completeness* can be defined for circumscription. Intuitively, completeness of circumscription is obtained when it is possible to infer from a circumscribed knowledge base every formula that is true in all minimal models of the premises. Besnard, Houdebine and Rolland have given general conditions of completeness and incompleteness for circumscription [Besnard et al. 87].

In the framework of an introductory book, it is not possible to study the numerous forms of circumscription that have been proposed in the literature. Among the more elaborate forms of circumscription, let us just mention *pointwise circumscription* [Lifschitz 86b] and *forms of prioritized circumscription* [McCarthy 86]. An excellent synthetic presentation of the properties of the various forms of circumscription is given in [Etherington 88]. Numerous detailed examples of application of the theorems mentioned in this subsection can be found in [Genesereth and Nilsson 87].

5.7 The preferential-model approach

5.7.1 Introduction

In standard logic, a conclusion can be inferred when it is true in *all* the models of the premises. In a non-monotonic logic, a formula can be inferred when it is true in *some specific* models of the premises. We can mutually compare most of the non-monotonic logics we have studied in this chapter by means of the following *semantic analysis*. For a given logic, we characterize the relevant set of models of the premises from which an inference can be drawn [Bossu and Siegel 85], [Shoham 87, 88], [Besnard and Siegel 88]. These specific models are called *preferred models* [Besnard and Siegel 88]. They underlie the specific analysis of the premises that is undertaken by the agent who draws the revisable inferences. By examining the sets of preferred models of different logics, one performs a comparison which has the interesting characteristic of fitting into the classical framework of *model theory*.

The minimization conventions we have been studying in this chapter underlie preferred models in which some specific predicates have a minimal extension.

By definition, the preferred models determine the specific *rationality* of the agent who conducts the reasoning. This rationality does not always correspond to an ideal and 'correct' logical approach. In fact, there exist different kinds of rationality, which direct different forms of revisable reasoning. Before we present them and before we analyse several fundamental issues concerning the very nature of the system that aims at formalizing them, let us briefly introduce

5.7. THE PREFERENTIAL-MODEL APPROACH

the preferential-model technique, which allows one to compare different non-monotonic logics.

5.7.2 Preferred models

Various techniques based on model theory have been proposed to compare non-monotonic logics. Here, we present the preferential-model approach [Besnard and Siegel 88]. This comparison technique is remarkable for its simplicity. Moreover, it can easily be extended in several directions (§ 5.7.4).

The *preferential-model approach* is based on the following idea. A non-monotonic reasoning underlies inferences in a specific set of models of the premises. The selection of such models depends on the specific rationality of the agent. This rationality is most often concerned with minimization conventions about the implicit positive information that can be associated with the premises. In semantic terms, this is expressed by a preference for the models of the premises in which the extensions of a first family of predicates are minimized, the extensions of a second family of predicates are maximized, and the extensions of a third family are preserved; no specific constraints are enforced on the extensions of the remaining predicates.

To represent a scale of satisfaction for these constraints, we shall define a *preorder* relation between the models of the knowledge base Δ. Recall that a preorder relation is a reflexive and transitive relation. Let us emphasize that if a model $\mathcal{M}$ is 'inferior' to a model $\mathcal{N}$ and conversely, then $\mathcal{M}$ and $\mathcal{N}$ are not necessarily identical.

By definition, the models that are minimal with respect to this preorder relation are the *preferred models* of the knowledge base, i.e., the models that justify the inferences allowed by the system. The characterization of the preferred models of a non-monotonic logic gives rise to a weakened model-theoretic consequence relation between a set of premises and one (or several) set(s) of conclusions that can be inferred from these premises.

A preorder relation between the models of a knowledge base Δ is defined formally as follows [Besnard and Siegel 88]. Let $\mathcal{L}$ be a first-order language based upon a set $\mathcal{P}$ of predicate constants (§ 5.2.1). We shall consider a four-class partition $\Gamma =_{def} (\mathcal{P}_=, \mathcal{P}_+, \mathcal{P}_-, \mathcal{P}_.)$ of the set $\mathcal{P}$. Here, $\mathcal{P}_=$ represents the set of predicates whose extensions must be the same in all models, $\mathcal{P}_+$ represents the set of predicates whose extensions must be maximized, $\mathcal{P}_-$ represents the set of predicates whose extensions must be minimized, and $\mathcal{P}_.$ represents the set of predicates whose extensions are allowed to be different for different models.

We are now in a position to introduce the preorder relation, denoted $\leq_\Gamma$, between the models of a knowledge base Δ, with respect to the chosen partition Γ. Let $\mathcal{M}$ and $\mathcal{N}$ be two models of Δ that have the same interpretation domain and that interpret in the same way the function constants as well as the variables. For any predicate $P \in \mathcal{P}$, let us denote $[\![P]\!]^{\mathcal{M}}$ and $[\![P]\!]^{\mathcal{N}}$ the extensions of P in the models $\mathcal{M}$ and $\mathcal{N}$ respectively. The model $\mathcal{N}$ is said to be *preferable* to the model $\mathcal{M}$ with respect to the partition Γ of $\mathcal{P}$, which is written as $\mathcal{N} \leq_\Gamma \mathcal{M}$, when the following conditions are satisfied:

- $[\![P]\!]^{\mathcal{N}} = [\![P]\!]^{\mathcal{M}}$ for all $P \in \mathcal{P}_=$,
- $[\![P]\!]^{\mathcal{N}} \supset [\![P]\!]^{\mathcal{M}}$ for all $P \in \mathcal{P}_+$,
- $[\![P]\!]^{\mathcal{N}} \subset [\![P]\!]^{\mathcal{M}}$ for all $P \in \mathcal{P}_-$.

Having defined a preorder relation between the models of a knowledge base, we can define the corresponding concept of a minimal model. A model $\mathcal{M}$ of a knowledge base Δ is said to be *minimal* with respect to the preorder relation $\leq_\Gamma$ when for each model $\mathcal{N}$ of Δ that satisfies $\mathcal{N} \leq_\Gamma \mathcal{M}$, one has $\mathcal{M} \leq_\Gamma \mathcal{N}$. The minimal models of Δ are called the *preferred models* of Δ.

5.7.3 Application to non-monotonic logics

It is possible to characterize most of the non-monotonic logics we have presented in this chapter with the help of the preferential-model approach. The set $\mathcal{P}$ of predicates used by each of these logics is partitioned into a quadruple $\Gamma = (\mathcal{P}_=, \mathcal{P}_+, \mathcal{P}_-, \mathcal{P}_.)$. A given formula A of $\mathcal{L}$ can be inferred, in a revisable manner, from a set Δ of formulas of $\mathcal{L}$ when A is true in all the preferred models of Δ with respect to the partition Γ.

For example, let us determine the nature of the preferred models that the closed-world assumption underlies. We shall restrict our attention to knowledge bases that contain only ground atoms. Recall that the domain of interpretation of a Herbrand model is the set of ground terms of the logic language. In such a model, each ground term is interpreted as itself (§ 5.2.5). The result is as follows [Lifschitz 85b], [Besnard and Siegel 88].

- A ground literal A can be inferred from the knowledge base Δ under the closed-world assumption if and only if A is true in all Herbrand models of Δ that are preferred with respect to the partition $\Gamma = (\emptyset, \emptyset, \mathcal{P}, \emptyset)$ of $\mathcal{P}$.

In other words, the closed-world assumption requires us to minimize the extension of every predicate.

The preferential-model approach generally does not allow the characterization of the predicate completion technique. Indeed, as we have emphasized in Subsection 5.4.4, this technique does not only depend on the logical contents of a formula, but also on its syntactical form.

Circumscription can be interpreted, with respect to several issues, as a generalization and an improvement of predicate completion. Circumscription can

5.7. THE PREFERENTIAL-MODEL APPROACH

be described quite naturally in the preferential-model approach. Indeed, circumscription could be redefined as follows.

- A formula A of $\mathcal{L}$ can be inferred from the knowledge base $\Delta \subset \mathcal{L}$ that is circumscribed with respect to predicate P and with predicates $P_1, ..., P_n$ allowed to vary, when A is satisfied in all preferred models with respect to the partition $\Gamma = (\mathcal{P}_=, \emptyset, \{P\}, \{P_1, ..., P_n\})$ of the set $\mathcal{P}$, where $\mathcal{P}_=$ is the complement set of $\{P, P_1, ..., P_n\}$ in $\mathcal{P}$.

Consider for example the knowledge base $\Delta = \{Bird(x) \land \neg Abnormal(x) \supset Fly(x), Ostrich(x) \supset Abnormal(x)\}$. When we circumscribe the predicate $Abnormal$ in Δ with the predicate Fly allowed to vary, we obtain the circumscribed knowledge base $\Delta' = \Delta \land (\forall x(Abnormal(x) \equiv Ostrich(x)))$. In terms of the preferential-model approach, we have partitioned the set of predicates into the quadruple $(\{Bird, Ostrich\}, \emptyset, \{Abnormal\}, \{Fly\})$. Thus, the preferred models are the minimal models with respect to the preorder relation for which the predicate $Abnormal$ is minimized, the predicate Fly is allowed to vary, and the extensions of the predicate $Bird$ and of the predicate $Ostrich$ are the same. To circumscribe a knowledge base Δ consists in introducing a circumscription axiom into Δ. This is equivalent to defining a new knowledge base the models of which are the preferred models of Δ.

Let us give another example of application of the preferential-model approach. We shall examine the knowledge base Δ that contains the following information:

$Man(Smith)$
$Man(Johnson)$
$\forall x \ (Man(x) \land Normal(x) \supset \neg(Fortune\,Teller(x)))$
$\forall x \ (Soldier(x) \supset Man(x))$

Let us consider the partition $\Gamma = (\mathcal{P}_=, \mathcal{P}_+, \mathcal{P}_-, \mathcal{P}_.)$ of the set of predicates $\mathcal{P} = \{Man, Normal, Fortune\,Teller, Soldier\}$ given by $\mathcal{P}_= = \{Man\}$, $\mathcal{P}_+ = \{Normal\}$, $\mathcal{P}_- = \{Fortune\,Teller\}$, $\mathcal{P}_. = \{Soldier\}$. We choose this partition because it seems natural to prefer models that maximize the extension of the predicate $Normal$ and that minimize the extension of the predicate $Fortune\,Teller$, while giving the same extension for the predicate Man and allowing the predicate $Soldier$ to vary.

We shall compare several models of Δ and, first, the two models $\mathcal{M}$ and $\mathcal{M}'$ defined as follows:

$[\![Man]\!]^\mathcal{M} = \{Smith, Johnson\}$ $\quad [\![Man]\!]^{\mathcal{M}'} = \{Smith, Johnson\}$
$[\![Normal]\!]^\mathcal{M} = \emptyset$ $\quad [\![Normal]\!]^{\mathcal{M}'} = \{Smith\}$
$[\![Fortune\,Teller]\!]^\mathcal{M} = \{Johnson\}$ $\quad [\![Fortune\,Teller]\!]^{\mathcal{M}'} = \emptyset$
$[\![Soldier]\!]^\mathcal{M} = \{Johnson\}$ $\quad [\![Soldier]\!]^{\mathcal{M}'} = \{Smith\}$

The model $\mathcal{M}'$ is preferable to the model $\mathcal{M}$ with respect to the partition Γ of $\mathcal{P}$ (i.e., $\mathcal{M}' \leq_\Gamma \mathcal{M}$). However, there exist other models that are preferable to both these models, with respect to the same partition. Consider the model $\mathcal{M}''$ defined as follows:

$$[Man]^{\mathcal{M}''} = \{Smith, Johnson\}$$
$$[Normal]^{\mathcal{M}''} = \{Smith, Johnson\}$$
$$[Fortune\ Teller]^{\mathcal{M}''} = \emptyset$$
$$[Soldier]^{\mathcal{M}''} = \{Smith\}$$

This model is a preferred model of Δ with respect to the partition Γ of $\mathcal{P}$; there does not exist any model of Δ that would be preferable to it with respect to this partition.

5.7.4 Possible generalizations

The preferential-model approach to the foundations of non-monotonic logics is interesting in many respects. This very simple approach can be generalized and extended in several ways. For example, it is possible to take into account possible priorities between the predicates that are to be minimized or maximized. In particular, the semantics of *forms of prioritized circumscription* [McCarthy 86], [Lifschitz 86b] can be characterized by means of preferred models that result from an ordered minimization of predicates [Lifschitz 88]. More generally, the preferential-model approach may allow the characterization of forms of rationality different from the ones we have described so far. Let us recall that these ones are concerned with agents who minimize or possibly maximize specific implicit information. Although such a form of rationality underlies a lot of reasonings, it may be interesting to consider other kinds of rationality.

An agent who reasons in a revisable way may infer conclusions that are simply consistent with respect to his basic set of beliefs. He selects one or several preferred models of these premises. This choice is made in accordance with the *specific rationality* of the agent. The preferred models represent the point of view of the agent about the modelled domain. They derive from specific partitions of the sets of formulas, predicates, functions, and variables of the language, and from specific operations concerning their extensions, values, and domains. These partitions and these operations express the specific rationality of the agent. In the next subsection, we briefly examine the possible natures of the rationality of an agent who reasons in a revisable way.

We must, however, emphasize that the 'correct' reasoning that an external agent could hold with respect to the effective state of beliefs of an agent whose reasoning is revisable, must remain a strict logical calculus, or at least a careful

5.7. THE PREFERENTIAL-MODEL APPROACH

operation. Therefore, this reasoning must enforce a form of *minimality constraint* to what can be inferred with respect to the state of beliefs of the agent, even when this agent does not apply some specific minimization operations.

5.7.5 The rationalities of revisable reasoning

There exist several classes of rationality that direct forms of revisable reasoning. A logic that aims at modelling the reasoning of a particular agent must respect the specific rationality of this agent. The nature of this rationality often depends on the considered application. It influences the design of the logic system at several levels and in several respects. In this subsection, we briefly examine different rationalities of revisable reasoning. We shall distinguish between two components of these rationalities: the *logical component* and the *subjective component*. The logical component allows one to determine the nature and the range of reasonings that are theoretically accessible to the rational agent. The subjective component dictates a specific choice among the alternatives within this range of reasonings for a specific application.

THE LOGICAL COMPONENT OF THE RATIONALITY

In this book, we are interested in forms of reasoning that preserve consistency. Thus, we are considering only agents who avoid inferring, at the same time, several mutually inconsistent conclusions from an initial consistent set of premises. *Consistency preservation* is a criterion that limits the application domain of non-monotonic logics.

Generally, an agent who obeys this only constraint is an agent whose view of the world is simply consistent with respect to his initial set of beliefs. The subjective component of the rationality of the agent will allow him to select one particular extension among the possible alternatives. Such a selection is retractable in its own right; for example, when the agent is given complementary premises. This form of reasoning has an obvious *conjectural* character.

On the other hand, the agent can be *ideally rational*. Such an agent infers only tautological consequences of his basic set of beliefs. Let us emphasize that the qualifier 'ideally rational' does not refer to an absolute and unique reality. Although it is usually meant for purely tautological reasoning, it is sometimes applied to other forms of reasoning (for example, intuitionistic reasoning, which is a mode of reasoning that denies the law of the excluded middle). An ideally rational agent obeys the laws of standard logic. Therefore, an ideally rational reasoning is unretractable by nature. However, it may become revisable when it is based on augmenting conventions about incomplete or unknown information (§ 5.1.2).

Indeed, the reasoning of an ideally rational agent can be introspective and can consider the limits of beliefs and knowledge of the agent himself. Such a reasoning requires appropriate conventions regarding the treatment of unknown or incomplete information. Such conventions have either a cautious character or a conjectural character. For example, Minker's generalized closed-world assumption (§ 5.5.2) can be qualified as being cautious because only the information that is verified in all the minimal models of the premises is interpreted as true while uncertain information is considered as such. Nevertheless, most of the approaches that are based on a form of minimization of the implicit information have a conjectural character, since they involve revisable conventions to cope with unknown information.

The 'ideally rational' qualifier is sometimes considered to be still relevant when the modelled reasoning remains 'logically correct' with respect to the premises, provided the augmenting convention that states 'completeness of positive information' is taken into account (§ 5.1.2). An ideally rational reasoning can thus be retracted when it is based on principles and conventions about the implicit information an agent is given. In fact, the formalization of an ideally introspective reasoning can be viewed as the definition of a new syllogism. This syllogism consists in the application of a strict logical reasoning to a state of knowledge that is supposed to be *complete* but that is not necessarily so.

To sum up, there exist several forms of *logical components* in the rationalities that direct forms of revisable reasoning. This aspect of the rationality obeys the consistency preservation constraint of non-monotonic logics. Two main cases should, however, be distinguished. On the one hand, the reasoning can be purely conjectural. In this case, the agent will infer simply consistent conclusions according to the subjective component of his rationality. On the other hand, the reasoning can be ideally rational. Such a reasoning is performed in an introspective way and takes into account specific conventions to deal with incomplete or unknown information. Non-monotonicity is caused by these conventions, which generally are minimization conventions. In both cases, it can happen that mutually incompatible reasonings are accessible to the agent.

Let us emphasize that such a distinction, which opposes a simply conjectural and consistency based reasoning to a form of deductive reasoning that is performed on an augmented knowledge base, is only valid in the context of an epistemological analysis. In fact, some specific logical tools that formalize the two types of reasoning can be shown to have the same expressive power [Konolige 87]. A perfect deductive reasoning that is performed on an augmented knowledge base can coincide with a purely conjectural reasoning that is performed on the initial knowledge base.

5.7. THE PREFERENTIAL-MODEL APPROACH

THE SUBJECTIVE COMPONENT OF THE RATIONALITY, AND THE SELECTION OF AN EXTENSION

The *subjective* component of the rationality dictates the behaviour of the agent with respect to the range of reasonings and extensions (§ 5.1.3) that the logical component of his rationality has characterized as possible. For example, a *cautious* agent will infer what is verified in all the possible extensions, and nothing else. In default logic, the modelled reasoning is often *conjectural* since the agent may infer simply consistent conclusions.

Furthermore, the different possible subjective components of the rationality can be expressed at different levels with respect to the logic system. For some authors, the subjective component should be expressed at the logical-object level, i.e., within the logic system itself. Other authors have proposed describing it at a more abstract level, inside a metalanguage [Weyrauch 80], [Zadrożny 87]. Let us analyse this issue for two kinds of non-monotonic systems: default logic and autoepistemic logic.

In default logic, primitives belonging to the metalanguage are used to express the conjecture rules. The subjective component of the rationality of the agent is expressed at two different levels. A first part of the subjective component is expressed in the default rules themselves; these rules determine a first subset of possible extensions of the premises. The reasoning is conjectural only with respect to these rules. The second part of the subjective component of the rationality of the agent is expressed by the way in which the agent actually applies the default rules; this strategy determines a specific extension. Let us emphasize that the system seldom is provided with a language for clearly expressing this rationality in the use of the default rules. Indeed, the selecting problem of a specific extension is not always directly addressed in default logic. Several authors have, however, proposed organizing the knowledge in such a way that a specific path of reasoning is favoured (see [Froidevaux 86], [Loui 87], [Lukaszewicz 84], [Moinard 87], [Poole 85], [Touretzky 84], [Zadrożny 87]).

In autoepistemic logic, the modelled reasoning is said to be 'ideally rational'. Several extensions, called *stable expansions*, are, however, sometimes possible. The selection of a specific stable expansion is treated in none of the basic versions of autoepistemic logic [Moore 85, 88]. The subjective component of the rationality of the agent is not expressed within the logic system. It is sometimes possible to constrain the set of initial beliefs of the introspective agent in such a way that a unique stable expansion can be obtained. In particular, Gelfond has proposed a class of autoepistemic theories called *stratified autoepistemic theories* that yield unique stable expansions [Gelfond 87].

We have mentioned that most non-monotonic logics enjoy the property of *pluri-extensionality*, which means that several extensions can be obtained from the same set of premises. In practice, an agent will select a specific extension. Most of the proposed non-monotonic logics do not allow a clear and direct expression of the subjective component of the rationality of the agent, i.e., of the component that dictates the choice of the agent's reasoning [Hanks and McDermott 86].

There exist, however, several techniques to enforce a system to select a specific path of reasoning and a specific extension when it is faced with several possibilities. As we have mentioned, in several systems, as for example in default logic, this selection can be implemented directly within the formulation of the conjecture rules [Reiter and Criscuolo 81]. One can also organize knowledge according to its specificity and its syntactical complexity [Loui 87], [Moinard 87], [Touretzky 86]. More generally, a non-monotonic system can take into account a specific order between the available information, the order of the rules it applies, or a specific ordered selection of the predicates that must be minimized. For example, a *circumscription policy* that determines the predicates to be minimized and the predicates allowed to vary, together with the priorities between these predicates, can be expressed either in a metalanguage, or, at the logical-object level, in the form of axioms [Lifschitz 87b]. As a second example, let us consider the inference system of Prolog. This system takes into account the order of the clauses of the given logic program. Note that this contradicts the fundamental principle of logic programming that asks for a strict independence between knowledge and reasoning. The clause order *should not* influence the answer to queries that are put to the system. Ideally, when this order has to be taken into account, either it should be expressed and used as a metaknowledge that is given to the system in the form of clauses, or it should be expressed and used in a metalanguage.

When stratified sets of clauses (viewed as logic programs or deductive databases) are under consideration, the problem of selecting the desired extension is resolved in an implicit manner. The stratification condition gives rise to a unique natural extension which can be characterized by means of a unique preferred model (i.e., a *perfect Herbrand model* in the terminology of the semantics of [Przymusinski 88b]) that corresponds to the specific interpretation that we want to associate with the knowledge base [Apt et al. 88], [Lifschitz 88], [Przymusinski 88c]. In particular, fragments of default logic [Bidoit and Froidevaux 87], of prioritized circumscription [Przymusinski 88c], [Gelfond and Lifschitz 88b], and of autoepistemic logic [Gelfond 87] can be translated into stratified sets of clauses and, therefore, do enjoy this 'unique extension' property. These results have been shown useful in allowing one to give *skeptical inheritance theories* a semantics in terms of the major non-monotonic logics [Grégoire 89].

In a number of application fields, specific domain dependent factors will dictate the choice of the specific extension. For example, when one reasons about plans and actions, *time* and *evolution* factors play an essential role. Bringing such factors explicitly into play in the represented knowledge, and making the causality theories that are involved in the represented phenomena quite explicit will sometimes allow one to determine the judicious choice of the extension [Kautz 86], [Lifschitz 87a], [Shoham 88].

In this subsection, we have briefly analysed different kinds of rationality that direct forms of revisable reasoning. Let us finally emphasize that a *natural* rationality of revisable reasoning can differ in several other respects from the kind of rationality that underlies classical deductive reasoning. For example, transitivity between inferences is not always desired in default reasoning [Reiter 87b].

5.8 Conclusion

The formalization of revisable reasoning is a major area of research in artificial intelligence. Whereas the formalization of 'strictly correct' reasoning can be performed directly within standard logic, the formalization of revisable reasoning cannot, and requires the development of more elaborate frameworks.

The formalization of revisable reasoning must take into account the *logical component* of the rationality of the modelled agent. This component determines the range of mutually incompatible reasonings that seems to be accessible to the agent. The *subjective component* of the rationality of the agent will determine his effective behaviour with respect to these incompatible alternatives. For example, the agent will only consider what is shared by all possible reasonings. On the other hand, he may prefer considering just one specific alternative of reasoning. The different possible rationalities of revisable reasoning are often based on implicit conventions that are not directly included in logical deductive systems.

In this chapter, we have analysed several variants of a *natural* rationality that underlies most of our reasonings and that directs our usual way of processing knowledge. This rationality corresponds to different minimization operations about the positive implicit information that we can associate with an initial information set.

As far as logic is concerned, a reasoning is an inference. In this respect, we have shown how different forms of revisable reasoning can be modelled with the help of specific inference systems. From a semantic point of view, taking into account the rules and conventions that direct forms of revisable reasoning has led us to define *weakened model-theoretic consequence relations*. In such logical systems, conclusions can be inferred when they are true in some specific models of the premises, called *preferred models*.

Whereas standard logic in essence is the tool that allows the propagation of truth from statement to statement, non-monotonic logics take part in a more ambitious project, namely the development of languages and inference systems whose scope includes the expression and the formalization of commonsense reasoning.

Chapter 6

Logic and databases

6.1 Introduction

This chapter presents two ways of describing, with the help of logic, the semantics of relational databases and of some of their extensions.

A first approach finds its origin in the traditional semantics of relational databases, that is, their management by relational systems from research and industry.

From a logical point of view, this traditional approach amounts to (1) associating with every database a first-order language, and (2) describing the database as a Herbrand model of some formulas of the logic language. Evaluating a query then amounts to selecting the ground instances of the query that are true in the model in question.

More recently, as approaches to enhance the relational model with deduction were gaining interest (e.g., for integrating Prolog and relational databases), another presentation was proposed that describes a database not as a model but as a theory expressed in a first-order language. Evaluating a query then amounts to selecting the ground instances of the query that are theorems in the theory.

The traditional approach is usually called 'model-theoretic', while the approach which views a database as a theory is called 'proof-theoretic', after Reiter [Reiter 84b], who first clearly formulated the alternative.

Section 6.2 is devoted to the usual relational model. The basic relational jargon has become classical, but, in practice, it is not always used very consistently; our logical approach must be able to rely upon a precise definition of the main relational concepts (§ 6.2.2). Definitions are presented for the relational algebra (§ 6.2.3), for a logic based relational language (§ 6.2.4), and for their correspondence (§ 6.2.6).

The relational language of Subsection 6.2.4 is directly based on first-order logic, while being very similar to existing languages of the relational model. The definition of its semantics is a bridge between the relational model and the logical view of databases, presented in Section 6.3 and studied throughout the rest of this chapter.

A logical definition is given for the concept of database (§ 6.3.1); then the model-theoretic (§ 6.3.2) and proof-theoretic (§ 6.3.3) approaches are introduced for characterizing the information actually modelled in a relational database.

In fact, for classical relational databases, both approaches directly correspond to each other and are clearly equivalent (§ 6.3.4): the model chosen in the model-theoretic approach to characterize the information in the database is the unique model of the theories adopted in the proof-theoretic approach.

The correspondence is less immediate, and thus more interesting, for deductive databases, treated in Sections 6.4 and 6.5, which generalize relational databases. Deductive databases comprise an extensional part (a classical relational database) and an intensional part (made of virtual relations, or predicates, defined by deduction rules in terms of ordinary relations and other virtual relations).

Section 6.4 is devoted to definite deductive databases, for which recursion is allowed in deduction rules but negation is not so. Then, a database admits a least Herbrand model and query evaluation concerns efficient methods for computing that model. The proof-theoretic approach is only really practical in the absence of recursion, as recursive rules considerably complicate the building of theories.

Section 6.5 considers deductive databases where not only recursion but also negation are involved in defining virtual relations. Then, the model-theoretic approach must explicitly consider several minimal models, each representing an equally possible state of the world. Query evaluation in the model-theoretic approach must, one way or another, examine the different models. There is, however, a subclass of deductive databases whose semantics is expressed naturally with a single model; it consists of the stratified databases (§ 6.5.4).

Section 6.6 deals with databases that represent an incomplete knowledge about the world. Here again, the model-theoretic approach is confronted with the existence of several models. In certain cases, the incomplete knowledge can exactly be represented in a first-order theory of the proof-theoretic approach, which then gives its full measure. It enables in effect a uniform and natural representation of all the knowledge available about an incompletely known reality. Still, the manipulation of incomplete information remains a vast and complex topic, which we barely skim in this chapter to illustrate how some of the existing approaches can be described with a logical formalism.

The logical modelling of databases is currently a vast and active research area, close in some aspects to logic programming. This chapter is a systematic introduction to the topic. It tries to integrate in a unifying framework a number of results scattered in various forms in the literature.

For the sake of completeness, it should be noted that logic is also used as a formal basis for describing and studying several other aspects of databases, including the design of database schemas (e.g., functional dependencies), the design of user-oriented database languages, and the formulation and management of database integrity (e.g., transaction consistency). Those topics are not addressed in this chapter.

Several papers have presented an introduction and survey of the logical modelling of databases. Let us mention three of them, with a word of comment. [Gallaire et al. 84] is the oldest of the three; the paper is less detailed than our presentation and, thus, of necessity, it does not contain the most recent results. [Gardarin and Simon 87] mainly deals with the architecture of systems and is less directly based on logic. From that point of view, our presentation is closer to that of [Chisholm et al. 87], but our treatment is more unified.

6.2 Relational model

6.2.1 Importance of the relational model

A *data model*[1] is a general approach, and also a collection of formalisms and of more or less formal methods for describing databases and their manipulation.

The *relational model* is relatively recent [Codd 70]. Its advent was particularly important as it marked the transformation of the database field from an empirical discipline into a scientific one.

In addition to representing a step forward for both practitioners and theoreticians, the relational model provided a coherent synthesis between those two worlds.

For practitioners, a relation is a remarkably simple data structure, compared with those that prevailed before. A relation constitutes a simple abstraction of the concept of flat file, independent of physical structures and access methods. Such a simplicity made possible the definition of data manipulation languages whose conciseness contrasts in spectacular ways with the complexity of programs accessing databases of previous generations.

For theoreticians, the success of the relational model results from the possibility of developing a mathematical and logical basis for a relational theory [Maier 83]. A series of concepts concerning databases and their management systems could thus receive, for the first time, a precise definition. A relation may, in effect, be viewed quite naturally as a logical predicate. Many data manipulation languages relate, more or less directly, to an applied predicate calculus, that is, a predicate calculus whose interpretations are tied to the database [Pirotte 78]. Furthermore, as we will see, generalizations of relational

---
[1] This usage of 'model' is different from that of logic.

databases are quite naturally presented in a logical formalism.

6.2.2 Fundamental relational concepts

The basic concepts of the relational model have become classical, even if there does not exist a unique reference definition [Date 81], [Ullman 82]. Differences concern, for example, the degree of formalization of concepts, the importance of the role given to domains, and certain specific constraints considered, or not, as essential in a basic definition. Such distinctions are not discussed any further here.

A *domain* is a finite (non-empty) set of atomic values.

A *relation* is defined by a schema and a value, themselves defined as follows.

The *value* or *extension* of a relation is a subset of a Cartesian product of domains, that is, a set of n-tuples $(c_1, \ldots, c_n)$ such that $c_i \in D_i$, where the D_i's are domains (not necessarily distinct). The arguments in the Cartesian product are distinguished through indices i ($1 \leq i \leq n$) called *attributes* of the relation.

It is convenient to be able to distinguish among the arguments of a relation not through their position, but through a symbolic name identifying them. The value of a relation is then defined with the help of a generalized or indexed Cartesian product, which does not depend on the order of the set arguments [Pirotte 82]. In this chapter, both representations are used interchangeably (attribute identification by position or by name).

The value of a relation is sometimes visualized as a table of values, where each column has a name (an attribute) and values appearing in any given column all belong to the same database domain.

Example 6.1. The *father*, *mother*, and *spouses* relations can be defined as follows[2]:

father	Father	Child
	Dior	Elwing
	Tuor	Eärendil
	Eärendil	Elrond
	Eärendil	Elros

mother	Mother	Child
	Nimloth	Elwing
	Idril	Eärendil
	Elwing	Elrond
	Elwing	Elros

---

[2]The examples are inspired from the fantastic universe of J. R. R. Tolkien [Tolkien 77].

6.2. RELATIONAL MODEL

spouses	Husband	Wife	#Children
	Finarfin	Eärwen	5
	Elrond	Celebrian	3
	Eärendil	Elwing	2
	Dior	Nimloth	1
	Tuor	Idril	1

The database domains are as follows:

$$D_{woman} = \{E\ddot{a}rwen, Celebrian, Elwing, Nimloth, Idril\},$$
$$D_{man} = \{Finarfin, Elrond, Elros, E\ddot{a}rendil, Dior, Tuor\},$$
$$D_{human} = \{E\ddot{a}rwen, Celebrian, Elwing, Nimloth, Idril,$$
$$\qquad\qquad Finarfin, Elrond, Elros, E\ddot{a}rendil, Dior, Tuor\},$$
$$D_{ten} = \{0, 1, 2, 3, 4, 5, 6, 7, 8, 9, 10\}.$$

Thus, D_{ten} is the domain of the '#Children' attribute (number of children) of the *spouses* relation and D_{woman} is the domain of the attributes 'Wife' of the *spouses* relation and 'Mother' of the *mother* relation.

The *schema of a relation* describes the information that is independent of the value of the relation, i.e., the name of the relation and a set of attribute/domain pairs. The schema is also said to represent the *intensional* definition of the relation. The schema of a relation r is denoted

$$r(A_1/D_1, \ldots, A_n/D_n)$$

or, if domains are understood,

$$r(A_1, \ldots, A_n) \text{ or } r(\boldsymbol{A})$$

with $\boldsymbol{A} = (A_1, \ldots, A_n)$, or still, simply, r if the attributes A_i are numerical indices comprised between 1 and n.

Example 6.2. The schemas of the relations of Example 6.1 are

$father(\text{Father}/D_{man}, \text{Child}/D_{human}),$
$mother(\text{Mother}/D_{woman}, \text{Child}/D_{human}),$
$spouses(\text{Husband}/D_{man}, \text{Wife}/D_{woman}, \#\text{Children}/D_{ten}).$

Given a relation schema, there exist many possible values, compatible with the given schema, for the relation in question. Conversely, speaking of the value of a relation normally implies the existence of a schema for that relation.

A *database* comprises domains, relations (schema + value), and integrity constraints.

The *extension of the database* comprises the value of relations.

The *schema of a database* comprises the domains, the relation schemas, and the integrity constraints.

An *integrity constraint* expresses a condition on the extension of the database.

Example 6.3. The following is an integrity constraint on the relations of Example 6.1:

> *If two individuals are, respectively, the father and the mother of another individual, then both parents appear together in a tuple of the spouses relation.*

The above definitions are summarized in Figure 6.1.

Relation	= relation schema
	+ relation value or extension
Relation schema	= relation name
	+ attribute/domain pairs
Relation value	= set of tuples
Database	= domains
	+ relations
	+ constraints
Database schema	= domains
	+ relation schemas
	+ constraints
Database extension	= relation values

Figure 6.1 : Basic relational concepts

Just as for relations, there exist, for a database, many extensions compatible with a given schema; but speaking of the extension of a database presupposes a schema for that database.

A database evolves with time through *updates* of the schema (creation, suppression of relation schemas, of domains, or of constraints) or of the extension (addition, deletion, modification of tuples in the relations).

In the usual mode of operation, updates of the database schema are much less frequent than updates of the extension. In other words, the schema of a database describes information that is more stable with time than the extension.

Integrity constraints, which, remember, are part of the database schema, have both a static and a dynamic aspect.

6.2. RELATIONAL MODEL

In its dynamic aspect, a constraint expresses a condition that updates of the database extension must satisfy.

Example 6.4. For Example 6.1 with the constraint of Example 6.3, the information '*a is the father of c*' and '*b is the mother of c*' cannot be introduced into the *father* and *mother* relations without also introducing in the *spouses* relation, if it is not yet there, the information '*a is the spouse of b with n children*' for some *n*.

The rest of this chapter contains little about updates and, therefore, about the dynamic aspect of constraints.

In its static aspect, a constraint summarizes, in the database schema, some information from the database extension. Strictly speaking, such information is redundant with respect to the extension (thus, the constraint of Example 6.3 can be deduced from the relations of Example 6.1), but that information is expressed by constraints at a more abstract level, valid for a number of particular extensions. This question will be discussed further in Subsection 6.4.1.

Several *data manipulation languages* have been associated with the relational model. *Queries* in a data manipulation language specify an answer relation in terms of other relations, or they request the verification of a condition, the answer then being either *yes* or *no*.

Example 6.5. The following is a query of the first kind, addressed to the relations of Example 6.1:

Who are the grandfathers (that is, the fathers of the father or of the mother) of Elros ?

For the relations of Example 6.1, the answer is a one-attribute relation whose value is the set $\{Tuor, Dior\}$.

Example 6.6. A yes–no query is as follows:

Is it true that all the children of Eärendil have a spouse ?

For the relations of Example 6.1, the answer is *no*.

Data manipulation languages of the relational model comprise the *relational algebra*, and the *domain* and *tuple* relational calculi.

All three languages have approximately the same expressive power. A definition of the main operations of the relational algebra is given in Subsection 6.2.3. The domain calculus is directly based on the relational language defined in Subsection 6.2.4.

The evaluation of queries and integrity constraints takes into account hypotheses that clearly specify the universe of reference and make explicit a treatment of the negative information. Those hypotheses are most often implicit in definitions of the relational model, but they are actually implemented in relational database management systems.

- The *closed-world assumption* or *CWA* says that all information that is not true in the database (that is, not explicitly present in the extension of relations) is considered as false (see also Section 5.3).

    Example 6.7. For the database of Example 6.1, the CWA leads to the conclusion that *Tuor* is not the father of *Elwing*.

- The *unique-name assumption* or *UNA* says that two distinct constants (in the database or in a data manipulation language) necessarily designate two different objects.

    Example 6.8. For the database of Example 6.1, the UNA says that *Tuor* and *Elwing* are distinct individuals.

- The *domain-closure assumption* or *DCA* says that there are no other objects in the universe than those designated by constants of the database.

    Example 6.9. For the database of Example 6.1, the DCA says that the only individuals whose existence is known by the database belong to the set $D_{woman} \cup D_{man} \cup D_{human} \cup D_{ten}$.

6.2.3 Relational algebra

The formulas of the relational algebra are constructed by composition from a set of primitive operations. A query addressed to the database is a formula of the algebra and its answer is obtained by evaluating the formula against the database. The primitive algebraic operators have one or two relation operands and one relation result. Usual definitions are limited to a description of the extension of the relation result [Ullman 82] and that is what is done in this section. A more complete definition, clearly specifying the schema of the relation result, is somewhat more complex [Pirotte 82].

The principal primitive algebraic operations are now defined. Several syntaxes are possible to express their combinations. In general, an operator is necessary for renaming relation attributes.

- Set-theoretic operators.

    As the value of a relation is a set of tuples, the usual set-theoretic operators are included in the algebra (union, intersection, difference, Cartesian

6.2. Relational model

product), with compatibility conditions, reasonably obvious in principle at least, on the argument relations.

- Projection.
  For a relation $r(A, B)$ with attributes (A, B), the projection of r on attributes A is

  $$\pi_A(r(A,B)) = \{a \mid (a,b) \in r \text{ for some } b\}.$$

  If the relation is visualized as a table, the operation results in only retaining the columns corresponding to the A attributes with a single instance of any identical tuples thus produced.

  Example 6.10. The *wives* of the *spouses* relation constitute the set
  $$\pi_{\text{Wife}}(spouses) = \pi_2(spouses)$$
  $$= \{\text{Eärwen, Celebrian, Elwing, Nimloth, Idril}\}.$$

- Restriction or selection.
  The following expressions are selections in relation $r(A)$, with $A = (A_1, \ldots, A_i, \ldots, A_j, \ldots, A_n)$:

  $$\sigma_{A_i \theta A_j}(r(A)) = \{a \in r \mid a_i \; \theta \; a_j\},$$

  $$\sigma_{A_i \theta c}(r(A)) = \{a \in r \mid a_i \; \theta \; c\},$$

  where a is an n-tuple of constants $(a_1, \ldots, a_i, \ldots, a_j, \ldots, a_n)$, c is a constant, and θ is a comparison operator ($=$ or $\neq$ and, for numerical or character string attributes, $<$, $\leq$, $>$ or $\geq$).
  The operation selects in r the n-tuples $(a_1, \ldots, a_i, \ldots, a_j, \ldots, a_n)$ for which the condition $a_i \theta a_j$ or $a_i \theta c$, respectively, is true.
  The definition extends to each expression $\sigma_{F(A)}(r(A))$, where $F(A)$ is a formula combining comparisons of the form $A_i \theta A_j$ and $A_i \theta c$ with the $\wedge$ and $\vee$ logical connectives. The formula $F(A)$ is called the *qualification* of the selection $\sigma_{F(A)}$.

  Example 6.11. The tuples of the *father* relation with *Eärendil* as father are as follows:

  $$\sigma_{2 \; = \; \text{Eärendil}}(father) = \{(\text{Eärendil, Elros}), (\text{Eärendil, Elrond})\}.$$

- Join.
  This operator specifies the tuples of the result relation on the basis of the equality of certain values in tuples of the two operand relations. Given

two relations $r_1(A,C)$ and $r_2(C,B)$ such that $A \cap B = \emptyset$, the join of r_1 and r_2 is

$$r_1 \bowtie r_2 = \{(a,c,b) \mid (a,c) \in r_1 \text{ and } (c,b) \in r_2\}.$$

The operation constructs a tuple (a,c,b) of the result from each pair of tuples $(a,c) \in r_1$ and $(c,b) \in r_2$ with the same constants c. If relations r_1 et r_2 do not share the desired attributes, then either some attributes have to be renamed or, as in Example 6.12 below, the operation itself must explicitly indicate the corresponding attributes in the relation arguments.

Example 6.12. The grandfathers, their children, and their grandchildren are obtained by a combination of join and union as follows:

$$father \underset{2=1}{\bowtie} (father \cup mother)$$
$$= \{(Tuor, E\ddot{a}rendil, Elrond), (Tuor, E\ddot{a}rendil, Elros),$$
$$(Dior, Elwing, Elrond), (Dior, Elwing, Elros)\}.$$

- Division.
  This operator expresses a restricted form of universal quantification. The division of relation $r_1(A,B)$ by $r_2(B)$ is defined as follows:

  $$r_1(A,B) \div r_2(B) = \{a \mid (a,b) \in r_1 \text{ for all } b \in r_2\}$$

  or

  $$r_1(A,B) \div r_2(B) = \{a \mid r_2 \subset \pi_B(\sigma_{A=a}(r_1))\}.$$

Example 6.13. The query of Example 6.6 (*Do all children of Eärendil have a spouse ?*) is expressed as follows:

$$(\pi_1(spouses) \cup \pi_2(spouses)) \div (\pi_2(\sigma_{1\ =\ E\ddot{a}rendil}(father))) = \{\}.$$

That query is a yes–no query and its answer $\{\}$ must be interpreted as no. The answer $\{()\}$ would have been interpreted as yes.

6.2.4 Relational language

This subsection defines a first-order language very similar to the domain calculus of the usual relational model described in [Lacroix and Pirotte 77] and [Ullman 82]. This language will be called *relational language* as in [Reiter 84b]. It is associated with the schema of a database or, more precisely, as the constraints do not intervene in its definition, with the domains and the relation schemas in the schema of a database.

6.2. RELATIONAL MODEL

SYNTAX

The relational language $\mathcal{L}$ is obtained by adapting the syntax of the first-order predicate calculus (see chapter 1 of [Thayse 88]) to a database schema S in the following manner [Pirotte 78].
- The individual constants of $\mathcal{L}$ are finite in number; the set of constants is the union of the domains of S.
- There are no functions (except for individual constants, viewed as functions without arguments).
- The number of predicates in $\mathcal{L}$ is finite.
- Among the predicates of $\mathcal{L}$, there exists a distinguished binary predicate, equality (denoted =).
- Every predicate in $\mathcal{L}$ other than equality is associated with a relation schema in S and conversely. Each attribute in the relation schema corresponds with an argument of the associated predicate. Such a predicate is called a *relation predicate*. For simplicity, the same name is given to both a relation predicate and the associated relation.
- The formulas of the relational language are constructed like those of predicate calculus.

The relational language $\mathcal{L}$ serves as a basis for defining a query language. A *query* is an expression of the form:

$$< \boldsymbol{x} \mid Q(\boldsymbol{x}) >$$

where
- $\boldsymbol{x} = (x_1, \ldots, x_n)$ is an n-tuple of variables called *target variables* or *query variables*.
- $Q(\boldsymbol{x})$ is a formula of the language where the only free variables belong to the n-tuple $\boldsymbol{x}$.

In particular, yes–no queries are of the form

$$< \mid Q >$$

where Q is a closed formula (without free variable).

Example 6.14. The query of Example 6.5 is expressed as follows in the relational language:

$$< x \mid \exists y \ (father(x, y) \land (father(y, Elros) \lor mother(y, Elros))) > .$$

Integrity constraints are any closed formulas of $\mathcal{L}$.

Example 6.15. The constraint of Example 6.3 is expressed as follows in the relational language:

$$\forall f\ \forall m\ \forall c((father(f,c) \land mother(m,c)) \supset \exists n(spouses(f,m,n))).$$

From a logical point of view, the absence of functions and the fact that constants are finite in number are considerable simplifications. In particular, the Herbrand universe (or Herbrand domain) of every set of formulas of $\mathcal{L}$ reduces to those constants occurring in the formulas. Remember that, in general, the *Herbrand universe* $\mathcal{U}_\mathcal{F}$ of a set $\mathcal{F}$ of formulas comprises the terms obtained by applying, in all possible ways, the function symbols appearing in the formulas to the constants of the formulas and to other terms of the Herbrand universe (see chapter 1 of [Thayse 88]).

Similarly, the Herbrand base of a set of formulas of the language $\mathcal{L}$ is finite. Remember that the *Herbrand base* $\mathcal{B}_\mathcal{F}$ of a set $\mathcal{F}$ of formulas is the set of atoms obtained by applying, in all possible ways, the predicate symbols appearing in $\mathcal{F}$ to tuples of constants of the Herbrand universe $\mathcal{U}_\mathcal{F}$.

SEMANTICS

As in predicate calculus, the semantics of the relational language $\mathcal{L}$ involves the definition of interpretations and models for the formulas of $\mathcal{L}$.

An *interpretation* of $\mathcal{L}$ is a 3-tuple $I = <\mathcal{D}, I_c, I_v>$ where
- $\mathcal{D}$ is a non-empty set of constants,
- I_c is a function associating an element of $\mathcal{D}$ with every constant of $\mathcal{L}$ and a mapping from $\mathcal{D}^n$ into $\{true, false\}$ with every n-ary predicate of $\mathcal{L}$ other than equality.
- I_v is a function associating an element of $\mathcal{D}$ with every variable of $\mathcal{L}$.

Equality is interpreted in $\mathcal{D}$ as identity, just as in the predicate calculus with equality.

Every formula F of $\mathcal{L}$ has, in the usual manner, a truth value in each interpretation I of $\mathcal{L}$. If a formula is true in a given interpretation, then this interpretation is said to be a *model* of the formula.

A *Herbrand interpretation* of a set $\mathcal{F}$ of formulas is the assignment of a truth value to each formula of the Herbrand base $\mathcal{B}_\mathcal{F}$. By extension, the subset of atoms in $\mathcal{B}_\mathcal{F}$ with truth value *true* in the interpretation is also called Herbrand interpretation of $\mathcal{F}$.

In practice, among all interpretations of $\mathcal{L}$, the most interesting ones are the Herbrand interpretations that are in direct correspondence with database extensions whose schema S is associated with $\mathcal{L}$. In particular, as will be seen in Section 6.3, the query $<x \mid Q(x)>$ characterizes a set of tuples c of

6.2. RELATIONAL MODEL

constants for which $Q(c)$ evaluates to true in a certain interpretation of $\mathcal{L}$ or, equivalently, for which $Q(c)$ is a theorem in a certain theory of $\mathcal{L}$.

6.2.5 Well-formedness of queries

According to the definition given above, a query is essentially any formula of $\mathcal{L}$. Such a definition is too wide, as shown in the following example.

Example 6.16. The query '*Who is not a father of Elwing ?*' is expressed in the relational language as

$$<x \mid \neg father(x, Elwing)>.$$

Because of the closed-world assumption (§ 6.2.2), the answer is $D_{man} \backslash \{Dior\} \cup D_{ten} \cup D_{woman}$, that is, all the constants in the database except $Dior$. It is clear that the presence in the answer of the set D_{ten}, and perhaps also of D_{woman}, is not satisfactory.

A natural solution, which is classical for programming languages, involves associating *types* with variables and constants of the query language. For databases, types are naturally defined from database domains [Pirotte 78], [Lacroix and Pirotte 81], [Reiter 84b].

Example 6.16 above no longer raises a problem if variable x has a type corresponding to the domain D_{human} (or D_{man}, depending on the exact intent of the query).

Another solution is that of restricting the form of acceptable queries so as to eliminate those queries producing unsatisfactory answers [Demolombe 79], [Ullman 82].

The principal forms of unreasonable queries are as follows:

- $<x \mid \neg F(x)>$ is not a reasonable query, as was shown by Example 6.16. On the contrary, the query $<x \mid T(x) \wedge \neg F(x)>$ is acceptable provided that T suitably restricts the domain of x.

- $<y \mid \forall x\ F(x,y)>$ is acceptable only if F is an implication suitably restricting the quantification on x. In that case, the query has in fact the form $<y \mid \forall x\ (T(x) \supset F'(x,y))>$.

- $<x,y,z \mid F_1(x,y) \vee F_2(y,z)>$ is not a reasonable query if F_1 does not contain variable z or if F_2 does not contain variable x, as its answer would contain 3-tuples (a,b,c) where (a,b) is in the answer to $<x,y \mid F_1(x,y)>$ and c is any constant in the database (as well as similar 3-tuples for F_2). The query $<x,y,z \mid F(x,z) \wedge (F_1(x,y) \vee F_2(y,z))>$ raises no problem if $F(x,z)$ suitably restricts the domains of x and of z.

From now on, queries will be supposed to be suitable formulas of a relational language, either adapted by introducing types or restricted by syntactic constraints. It can be shown that, when doing so, nothing essential is lost concerning the expressive power of the query language.

6.2.6 From relational language to relational algebra

Query evaluation by database management systems typically is performed by evaluating primitive operations of the relational algebra. Compared with the tuple and domain relational calculi, the relational algebra permits a more procedural query formulation, in that it suggests more directly an order of evaluation.

There exists, however, a very direct correspondence (which expresses the equivalence between algebra and domain calculus) between algebraic operations on the one hand, and logical connectives and quantifiers on the other.

This subsection shows the principle of a translation (it is not the only one possible; see for example [Louis and Pirotte 82], [Ullman 82]) of queries in the relational language of Subsection 6.2.4 into operations of the relational algebra. This translation does not pretend to be efficient. In particular, it does not take into account desirable restrictions like those shown in Subsection 6.2.5. Most often, the resulting algebraic queries are far from optimal. Query optimization is an essential function of database management systems [Jarke and Koch 84], [Kim et al. 85], but this topic is outside the scope of our presentation.

The basic idea consists in associating with every query $<x \mid F(x)>$ in the relational language a relation whose attributes are precisely the variables x. The extension of that relation will be noted $|<x \mid F(x)>|$ and it will be called the *algebraic correspondent* of the query.

Translation rules progressively define the algebraic correspondent of formulas of $\mathcal{L}$. In the rules, D denotes the set of constants of $\mathcal{L}$.

1. **Relation predicates**

    Remember that to each relation corresponds a relation predicate with the same name in the relational language. With a query of the form $<x \mid r(x)>$, where r is a relation predicate, is associated the relation r with attributes x (which are in fact new names for the original attributes A of r in the database schema). The following translation rule holds[3]:

    $$|<x \mid r(x)>| = r.$$

    For a query of the form $<x \mid r(x, a)>$, where r is a relation predicate and a is a tuple of constants, the algebraic correspondent is obtained through

---

[3] Here and in the sequel, symbol r also denotes the extension of relation r.

6.2. Relational model

a combination of selection (σ) and of projection (π) on the base relation r with attributes (x, y) as follows:

$$|<x \mid r(x,a)>| = \pi_x(\sigma_{y=a}(r)).$$

2. **Addition of attributes**
   If y does not occur free in $F(x)$, then the algebraic correspondent of the query $<y, x \mid F(x)>$ is a relation with attributes (y, x) whose extension is defined by

   $$|<y,x \mid F(x)>| = D \times |<x \mid F(x)>|.$$

3. **Conjunction**
   Consider the query $<x \mid F_1(x_1) \wedge F_2(x_2)>$. By adding attributes if needed, it can always be supposed that $x = x_1 \cup x_2$. The algebraic correspondent is then the relation with attributes x whose extension is

   $$|<x \mid F_1(x_1) \wedge F_2(x_2)>| = |<x_1 \mid F_1(x_1)>| \bowtie |<x_2 \mid F_2(x_2)>|.$$

4. **Disjunction**

   $$|<x \mid F_1(x_1) \vee F_2(x_2)>| = |<x \mid F_1(x_1)>| \cup |<x \mid F_2(x_2)>|.$$

5. **Universal quantifier**

   $$|<x \mid \forall y\ F(x,y)>| = |<x,y \mid F(x,y)>| \div D.$$

   As seen in the preceding subsection, acceptable queries are rather of the form $<x \mid \forall y(T(y) \supset F(x,y))>$. Then

   $$|<x \mid \forall y(T(y) \supset F(x,y))>| = |<x,y \mid F(x,y)>| \div |<y \mid T(y)>|.$$

6. **Existential quantifier**

   $$|<x \mid \exists y\ F(x,y)>| = \pi_x(|<x,y \mid F(x,y)>|).$$

7. **Negation**
   If n is the length of x, then

   $$|<x \mid \neg F(x)>| = D^n \setminus |<x \mid F(x)>|.$$

   As seen in the preceding subsection, acceptable queries are rather of the form $<x \mid F_1(x) \wedge \neg F_2(x)>$. Then

   $$|<x \mid F_1(x) \wedge \neg F_2(x)>| = |<x \mid F_1(x)>| \setminus |<x \mid F_2(x)>|.$$

8. **Equivalence and implication**
   Replace $\alpha \equiv \beta$ by $\alpha \supset \beta \wedge \beta \supset \alpha$, and $\alpha \supset \beta$ by $\neg \alpha \vee \beta$.

9. **Equality**
   - $|<x \mid x = x>| = D$
   - $|<x \mid x = a>| = \{a\}$
   - $|<x, y \mid x = y>| = \{(a, a) \mid a \in D\}$

10. **Yes–no queries**
    $|< \mid F(a)>| = \{()\}$ if $a \in |<x \mid F(x)>|$
    $\phantom{|< \mid F(a)>|} = \{\}$ otherwise.

6.3 Logical approach to databases

6.3.1 Definition

Subsection 6.2.4 showed how to associate a relational language $\mathcal{L}$ with a database schema. This section shows how to associate with a database a theory that will be the basis for the logical approach to databases.

There is a very direct correspondence between a tuple a in the extension of a relation r and the ground instance $r(a)$ of the corresponding relation predicate in the language $\mathcal{L}$ associated with the same database.

Definition. The *logical approach* of a database is a first-order logical theory directly describing the relations. The axioms of the theory, written in the relational language associated with the schema of the database, comprise, for each relation r and each n-tuple $(c_1, \ldots, c_n)$ in r, a formula denoted

$$r(c_1, \ldots, c_n).$$

In that axiom, r denotes the relation predicate associated with the relation. In the sequel, there will be no notational distinction between the name of a relation and that of the associated relation predicate.

A database thus admits two equivalent definitions, that directly correspond to each other.

- According to the definition that was qualified as classical in Subsection 6.2.2, because it is well established in the database field independently of any logical modelling, a database comprises domains, relation schemas, relation values or extensions, and integrity constraints.

- According to the logical definition, a database is a set of first-order formulas (a theory) written in the relational language and representing relations. The concepts of schema and constraint are borrowed without modification from the classical definition.

When needed and when context will not suffice, we will indicate explicitly which definition is understood.

6.3.2 Model-theoretic approach

The model-theoretic approach concentrates on the models of the database, defined as a logical theory.

A relational database has numerous models, but only one of them is really interesting, as it contains exactly the information that one wishes the database actually to describe. Every Herbrand model of the database enjoys the following property: the set of tuples that make a relation predicate true contains the set of tuples of the associated relation in the database extension.

The database thus possesses a least Herbrand model, which is the intersection of all the Herbrand models. In a sense, this model identifies with the database itself.

From a logical point of view, the selection, among all existing models, of that least Herbrand model to represent the database expresses in the following manner the CWA, UNA, and DCA assumptions introduced in Subsection 6.2.2.

- The CWA is taken into account by choosing the least Herbrand model: it minimizes the positive information and considers as false each ground instance of a predicate that is not present in the database.

- The UNA is taken into account by the fact that, in a Herbrand interpretation, the constants of the language coincide with those of the domain of the interpretation, so that equality is interpreted as identity.

- The DCA is taken into account by the restriction to a Herbrand model: there are no other constants in the domain of the interpretation than the constants of the language.

As the integrity constraints must be true in the database, the least Herbrand model of the database is also a model of the constraints.

In the following, B will denote a database and $\mathcal{M}_B$ its least Herbrand model.

A query $<\boldsymbol{x} \mid Q(\boldsymbol{x})>$ characterizes the set of tuples $\boldsymbol{c}$ of constants for which $Q(\boldsymbol{c})$ is true in $\mathcal{M}_B$. That set of tuples is denoted $\| <\boldsymbol{x} \mid Q(\boldsymbol{x})> \|$. Formally, this definition is written

$$\| <\boldsymbol{x} \mid Q(\boldsymbol{x})> \| = \{\boldsymbol{c} \mid \ \vDash_{\mathcal{M}_B} Q(\boldsymbol{c})\}.$$

In particular, for yes–no queries:

1. $\|<|Q>\| = \{()\}$, where $()$ denotes the empty tuple, means that Q is true in $\mathcal{M}_B$; the answer to $<|Q>$ is yes in that case,

2. $\|<|Q>\| = \{\}$ means that Q is not true in $\mathcal{M}_B$,

3. $\|<|\neg Q>\| = \{()\}$ means that Q is false in $\mathcal{M}_B$.

Note that (2) and (3) are equivalent and both correspond to a no answer to the query $<|Q>$.

6.3.3 Proof-theoretic approach

Whereas the model-theoretic approach selects, among the models of a database, that (or, in a more general context, those) which best represents the information content, the proof-theoretic approach adds a set of axioms to the database with the same goal of better circumscribing the information represented and its use for query evaluation.

A theory is thus constructed by adding axioms to the (basic theory defining the) database. These axioms are written in the relational language as follows:

- The axioms of the theory defining the database B are as in Subsection 6.3.1: for each relation r and each tuple $c \in r$, the theory contains an axiom $r(c)$.

- The axioms for the closed-world assumption ($CWA(B)$) or the *completion axioms* ($CA(B)$).

  The theory must contain axioms allowing, for example, the proof of a formula $\neg r(a)$ in the case where a does not belong to r. The CWA can be axiomatized in two equivalent ways, either by explicitly including all the negative information in the theory, or by formulating completion axioms.

    – Explicit negative information.
      This formulation of the CWA consists in explicitly asserting all the negative information at the level of ground instances. If $\mathcal{B}_B$ is the Herbrand base of the relation predicates of the database B, the negative information comprises the negation of literals that are in $\mathcal{B}_B$ and not in B. The negative information, noted $CWA(B)$, is thus formally defined as follows:

$$CWA(B) = \{\neg P \mid P \in \mathcal{B}_B \text{ and } P \notin B\}. \tag{6.1}$$

    – Completion axioms.
      They express that the positive information is exactly that contained in the database B. For each relation r, if $c_1 = (c_1^1, \ldots, c_1^n), \ldots, c_p =$

6.3. LOGICAL APPROACH TO DATABASES

$(c_p^1, \ldots, c_p^n)$ denote all the n-tuples of r, then the completion axiom for r is written

$$\forall x \ (r(x) \supset x = c_1 \vee \ldots \vee x = c_p), \tag{6.2}$$

where $x = (x_1, \ldots, x_n)$. Combining this with the assertions concerning r in the database, implication can be replaced by equivalence:

$$\forall x \ (r(x) \equiv x = c_1 \vee \ldots \vee x = c_p). \tag{6.3}$$

In the following, the completion axiom of r will mean interchangeably the implication (6.2) or the equivalence (6.3).

In the case where a relation r is empty in the database, then the completion axiom for r is written:

$$\forall x \ \neg r(x).$$

- The unique-name axioms ($UNA(B)$).
  They explicitly say that all objets $c_1, \ldots, c_q$ in the database are distinct:

  $$c_i \neq c_j \quad \text{for } 1 \leq i < j \leq q.$$

- The domain-closure axiom ($DCA(B)$).
  It says that there are no other objects than those explicitly present in the database:

  $$\forall x \ (x = c_1 \vee \ldots \vee x = c_q).$$

  This axiom expresses that the elements of the considered universe are finite in number, that they are all known, and that they have the constants of the database as names.

Thus, there are two theories, according to whether the negative information is modelled by completion axioms or by the explicit negative information of the CWA; they are defined, respectively, as follows:

$$T_B^{AC} = B \cup CA(B) \cup UNA(B) \cup DCA(B) \tag{6.4}$$

$$T_B^{CWA} = B \cup CWA(B) \cup UNA(B) \cup DCA(B). \tag{6.5}$$

Note that, strictly speaking, B is redundant in (6.4) if the completion axioms are equivalences.

Queries

The answer to a query $<\boldsymbol{x} \mid Q(\boldsymbol{x})>$ for $\boldsymbol{x} = (x_1, \ldots, x_n)$ is the set of n-tuples $\boldsymbol{c} = (c_1, \ldots, c_n)$ for which $Q(\boldsymbol{c})$ is a theorem of the theory $\mathcal{T}_B$ (which is either $\mathcal{T}_B^{CWA}$ or $\mathcal{T}_B^{AC}$). Formally, the following equality holds:

$$\| <\boldsymbol{x} \mid Q(\boldsymbol{x})> \| = \{\boldsymbol{c} \mid \mathcal{T}_B \vdash Q(\boldsymbol{c})\}. \tag{6.6}$$

In other words, answering a query amounts to identifying all its ground instances that are theorems of the theory.

In particular, for yes–no queries:

1. $\| < \mid Q > \| = \{()\}$ means that Q is a theorem of $\mathcal{T}_B$; the answer to $< \mid Q >$ is yes in this case.

2. $\| < \mid Q > \| = \{\}$ means that Q is not a theorem.

3. $\| < \mid \neg Q > \| = \{()\}$ means that $\neg Q$ is a theorem of $\mathcal{T}_B$. Then the answer to the query $\| < \mid Q > \|$ is no.

Note that (2) and (3) are equivalent ($\| < \mid Q > \| = \{\}$ if and only if $\| < \mid \neg Q > \| = \{()\}$). This is consistent with the fact that $\mathcal{T}_B$ is complete (§ 6.3.4).

Constraints

A constraint C is satisfied in the database B if $\mathcal{T}_B \vdash C$, that is, if it is a theorem of the theory.

6.3.4 Correspondence between approaches

This subsection shows that both approaches, model- and proof-theoretic, are equivalent. A preliminary lemma characterizes the models of the theories $\mathcal{T}_B^{CWA}$ et $\mathcal{T}_B^{AC}$.

Lemma 6.1. *The presence of the UNA and DCA axioms in the theories $\mathcal{T}_B^{CWA}$ and $\mathcal{T}_B^{AC}$ entails that the theories only admit models that are isomorphic to Herbrand models.*

Proof. Let $\mathcal{M}$ be a model of one of the theories. It specifies a domain of interpretation Δ and an interpretation function I_c for constants and predicates of $\mathcal{L}$. Remember (§ 6.2.4) that, in an interpretation, equality is interpreted as identity. Let D be the set of constants of $\mathcal{L}$ and let $\#E$ denote the cardinality of set E.

- We first prove the inequality $\#\Delta \leq \#D$.

  If it was true that $\#\Delta > \#D$, then there would exist a $\delta \in \Delta$ such

6.3. LOGICAL APPROACH TO DATABASES

that $\delta \notin I_c(D)$. Then $\delta \neq I_c(d)$ for each $d \in D$, which contradicts the domain-closure axiom.

- Then we prove the inequality $\#\Delta \geq \#D$.
  If it was true that $\#\Delta < \#D$, then there would exist $d_1 \in D, d_2 \in D$ and $\delta \in \Delta$ such that $I_c(d_1) = I_c(d_2) = \delta$, which contradicts the unique-name axiom $d_1 \neq d_2$.

Thus, $\#\Delta = \#D$, and $\mathcal{M}$ is isomorphic to the Herbrand model constructed from $\mathcal{M}$ by replacing each constant δ in Δ by the constant $I_c^{-1}(\delta)$ of D. □

Theorem 6.2. *Theories T_B^{CWA} and T_B^{AC} have as unique Herbrand model the least Herbrand model selected by the model-theoretic approach.*
Proof. First, every Herbrand model of T_B must agree with the base assertions in B and must thus contain them. Then, every Herbrand model of T_B cannot specify a set of tuples larger than that in B, otherwise a completion axiom or the explicit negative information of $CWA(B)$, depending on the theory chosen, would not hold in the model. □

It is interesting to summarize the important properties of theories T_B^{CWA} and T_B^{AC} for ordinary relational databases. Some of those properties will no longer hold for the generalization to deductive databases.

- T_B^{AC} and T_B^{CWA} are categorical, that is, they admit (within an isomorphism) a unique model.

- T_B^{AC} and T_B^{CWA} are equivalent, that is, their theorems are the same.

- T_B^{AC} and T_B^{CWA} are consistent, that is, contradictory facts cannot be deduced from them.

- T_B^{AC} and T_B^{CWA} are complete, that is, for each formula F of the relational language, either F or $\neg F$ is a theorem of the theory.

These properties are direct consequences of Lemma 6.1 and Theorem 6.2.

The following theorem shows that the bridge established in Subsection 6.2.6 between relational language and relational algebra agrees with the semantics of relational databases as it is defined in the model- and proof-theoretic approaches.

Theorem 6.3. *If $F(\boldsymbol{x})$ is a formula of $\mathcal{L}$, then*

$$\| <\boldsymbol{x} \mid F(\boldsymbol{x})> \| = | <\boldsymbol{x} \mid F(\boldsymbol{x})> | .$$

Proof. It suffices to establish a proof for each translation rule in Subsection 6.2.6. Such proofs are immediate in the model-theoretic approach, where the least Herbrand model is the extension of the database (§ 6.3.2). □

In other words, the answers to a query computed in the model- and in the proof-theoretic approaches coincide with the extension of the relation obtained by evaluating the algebraic correspondent of the query.

6.3.5 Summary

To summarize, we have associated with a classical relational database a relational language $\mathcal{L}$, which was used to construct a logical view of a database formalizing the traditional definition. Then, the model-theoretic approach selected a particular model for that database. Finally, in the proof-theoretic approach, two equivalent theories, $\mathcal{T}_B^{AC}$ and $\mathcal{T}_B^{CWA}$, were constructed, to completely axiomatize the information described by the database and its utilization for query evaluation.

For the relational databases, which are the subject of this section, the model- and proof-theoretic approaches are clearly equivalent and, if it did not go any further, the distinction would not be very useful. The information that can be deduced from the relational database is essentially, with an adequate axiomatization, the database itself. This will no longer be the case in the sequel.

We will continue to use the name 'database' for logical theories whose axioms directly characterize the relations. For relational databases, as this section has shown, such a theory contains base assertions describing the extension of the relations. For the deductive databases of Sections 6.4 and 6.5, a database will contain, in addition to base assertions, a set of rules defining virtual relations (or predicates). Finally, when incomplete information is introduced, in Section 6.6, a database will contain additional formulas describing the incomplete information available about the extension of the relations.

The model-theoretic approach will focus on the models of databases thus defined and select that (or those) model(s) which best represent the information that the database is intended to model.

The proof-theoretic approach will add axioms to the database to construct theories representing the information content and mechanisms for query evaluation.

6.4 Definite deductive databases

6.4.1 Introduction

In Subsection 6.3.1, a relational database is defined as a logical theory containing ground instances of relation predicates expressing the extension of the relations. A relational database also contains integrity constraints that, as far as their static role is concerned (the only one that is considered here, see § 6.2.2), synthesize in the relational schema a part of the information of the extension of the relations.

Example 6.17. Consider a relational database with relation predicates named *parent*, *father*, and *mother* and with the following integrity constraints:

$$\forall x \forall y \, (father(x,y) \supset parent(x,y)),$$
$$\forall x \forall y \, (mother(x,y) \supset parent(x,y)).$$

The static information expressed by the constraints can be deduced from the extension of the relations: every individual known to be a father or a mother (from the *father* or *mother* relation) is also explicitly a parent (in the *parent* relation). In addition, in general, there can also exist parents who are not explicitly a father or a mother.

Deductive databases aim at avoiding such a redundancy between some constraints and the extension of relations. Part of the information that relational databases can only express as integrity constraints is now expressed as intensional definitions of a new kind of predicate. By adding these definitions to the theory (i.e., to the database) and suppressing the part of the extension implied by them, a new, more compact theory is obtained. No information is lost, since the theorems that can be proved from the new theory can also be proved from the initial theory.

More precisely, some constraints become deduction rules and some relational predicates become virtual predicates, i.e., predicates whose extension is no longer explicit in the database but only implicit through the deduction rules.

Example 6.18. In Example 6.17, both constraints can be treated as deduction rules that define intensionally the virtual predicate *parent*. It becomes possible to remove from the database, without losing any information, each instance *parent(a, b)* such that either *father(a, b)* or *mother(a, b)* also appears in the database. If a query addresses the *parent* predicate, its extension, or part of it, is deduced from the deduction rules and the *father* and *mother* relations.

This section first gives a formal definition for the concepts of deduction rule and deductive database. Then, a simple but interesting subclass of deductive databases, consisting of the definite deductive databases, is characterized. It is then shown how a model and a theory can be associated with a definite deductive database and how queries can be evaluated.

6.4.2 Deductive databases

Definition. A *deduction rule* defining a *virtual* predicate h is a logical formula of the form

$$\forall var(t)\, (h(t) \leftarrow \exists y\, (b_1 \wedge \ldots \wedge b_n)) \qquad (6.7)$$

where

- the $\leftarrow$ connective is the inverse implication: $A \leftarrow B =_{def} B \supset A$,
- n is a positive integer,
- the b_i's are literals,
- t contains the arguments, variables or constants, of h,
- $var(t)$ contains all the variables of t,
- y contains all variables occurring in $b_1 \wedge \ldots \wedge b_n$ and not in $var(t)$.

Note that $var(t) \cup y$ contains all the variables of the rule.

The literal $h(t)$ is the *head* or *consequent* of the rule, and $\exists y\, (b_1 \wedge \ldots \wedge b_n)$ is the *body* or *antecedent* of the rule. Quantifiers will be omitted whenever this does not impair clarity.

Example 6.19. Consider the following deduction rules defining virtual predicates named *parent*, *grandparent*, *a-f* (ancestor-friend), *o-a* (odd-ancestor), and *e-a* (even-ancestor):

$$r_1 : parent(x,y) \leftarrow father(x,y)$$
$$r_2 : parent(x,y) \leftarrow mother(x,y)$$
$$r_3 : grandparent(x,y) \leftarrow parent(x,z) \wedge parent(z,y)$$
$$r_4 : a\text{-}f(x,y) \leftarrow friend(x,y)$$
$$r_5 : a\text{-}f(x,y) \leftarrow parent(x,z) \wedge a\text{-}f(z,y)$$
$$r_6 : o\text{-}a(x,y) \leftarrow parent(x,y)$$
$$r_7 : o\text{-}a(x,y) \leftarrow parent(x,z) \wedge e\text{-}a(z,y)$$
$$r_8 : e\text{-}a(x,y) \leftarrow parent(x,z) \wedge o\text{-}a(z,y)$$

6.4. DEFINITE DEDUCTIVE DATABASES

The meaning of the last three predicates is as follows: $a\text{-}f(x,y)$ means that x is an ancestor of a friend of y; $o\text{-}a(x,y)$, that x is an ancestor of y with an odd number of generations separating x and y in the genealogy (i.e., x is a parent of y, a great-grandparent of y, etc.); $e\text{-}a(x,y)$, that x is an ancestor of y with an even number of generations separating them.

Rules define dependency relations among predicates (*grandparent* is defined in terms of *parent*) that can be cyclic (*a-f* is defined in terms of itself). The following definitions formalize these concepts.

Definitions. A predicate p *directly depends* on a predicate q if q occurs in the body of a rule whose head is an instance of p. For example, *parent* directly depends on *mother*.

A predicate p *depends* on a predicate q either if p directly depends on q, or if p depends on a predicate r directly depending on q. For example, *a-f* depends on *father*.

A predicate is *recursive* if it depends on itself. For example, *a-f*, *o-a*, and *e-a* are recursive.

A predicate p *strictly depends* on a predicate q if p depends on q and q does not depend on p. For example, *e-a* strictly depends on *parent*.

Two predicates are *mutually recursive* if p depends on q and q depends on p. For example, *o-a* and *e-a* are mutually recursive.

A deduction rule is *recursive* if its antecedent contains a predicate that depends on the consequent of the rule. For example, rules r_5, r_7 and r_8 are recursive.

A deduction rule is *linear* if at most one of the predicates in its body depends on the head of the rule. All rules of Example 6.19 are linear.

Deductive databases are now defined formally.

Definition. A *deductive database* B is a theory composed of an *extensional database* (sometimes denoted B_E) and an *intensional database* (denoted B_I).

The extensional database is a set of ground instances of atoms defining the extensions of the *base predicates* that is, a relational database in the sense of Section 6.2. The intensional database is a set of deduction rules defining the *virtual predicates*

The definition of the relational language of Subsection 6.2.4 still holds for a deductive database, with the virtual predicates included in the language.

A predicate is *mixed* if its definition is partly extensional and partly intensional.

We can suppose, without loss of generality, that there are no mixed predicates. Indeed, given a mixed predicate p defined by a set of ground instances $p(a_1), \ldots, p(a_n)$ and by a set of deduction rules, it is possible to construct a deductive database without mixed predicates that is equivalent with the initial database as follows. The rule

$$p(x) \leftarrow p'(x)$$

is added to the intensional database, and each ground instance $p(a_i)$ is replaced by $p'(a_i)$ in the extensional database. The predicate p is now virtual, while the new predicate p' is a base predicate.

Deductive databases are studied in their general form in Section 6.5. A subclass with some remarkable properties, namely, the subclass containing the *definite deductive databases*, is now studied in more detail.

Definition. A deduction rule is *definite* if the literals in its antecedent are all positive.

In spite of the simplicity of these rules, many interesting applications can be represented with definite deduction rules. Note that a definite deduction rule

$$h(t) \leftarrow b_1 \wedge \ldots \wedge b_n$$

is equivalent to a Horn clause

$$h(t) \vee \neg b_1 \vee \ldots \vee \neg b_n.$$

Definition. A deductive database is *definite* if all its deduction rules are definite.

In the following, we usually speak of definite databases rather than of definite deductive databases. Note that a definite database can be seen as a set of Horn clauses, as both the elements of the extensional database and the elements of the intensional database can be expressed as Horn clauses. Example 6.19 defines the rules of a definite database.

The language used to define definite deduction rules is often called Datalog [Ullman 86]. Datalog is a purely declarative version of Prolog without functions, i.e., a language of Horn clauses without functions. Hence, a set of expressions in Datalog represents a definite database.

Deductive databases that are not definite are sometimes called 'non-Horn' or 'indefinite'. To avoid a negative terminology, we will most often prefer the generic term, 'deductive database', for the general case.

The class of definite deductive databases has an interesting subclass, consisting of the hierarchical databases.

6.4. DEFINITE DEDUCTIVE DATABASES

Definition. A definite deductive database is *hierarchical*[4] if none of its virtual predicates is recursive.

Rules r_1, r_2 and r_3 of Example 6.19 define a hierarchical database. Hierarchical databases are essentially classical relational databases with *view* definitions. A view is an expression, for example of the relational algebra, that defines a new relation in terms of other relations [Date 81], [Ullman 82].

This section studies definite databases and their subclass of hierarchical databases. It is shown that the same semantics can be associated with a definite database in both the model- and the proof-theoretic approaches. The proof-theoretic approach is shown to be particularly well suited to hierarchical databases and to lead to a simple query evaluation algorithm. On the contrary, if rules are recursive, the model-theoretic approach is more fruitful.

6.4.3 Model-theoretic approach

As for relational databases (§ 6.3.2), the model-theoretic approach selects, among the models of a definite database, one model that best represents its intended semantics. In particular, this model must represent the definite database under the three assumptions stated in Subsection 6.2.2. The domain-closure assumption and the unique-name assumption restrict the interesting models to the Herbrand models of the definite database. The closed-world assumption minimizes the number of elements of the Herbrand base that are true in the models of the database. We will show that, like a relational database, a definite database has a single minimal model.

We first state an important property of the Herbrand models of a definite database, the *model intersection* property [Van Emden and Kowalski 76].

Theorem 6.4. *The intersection of all Herbrand models of a definite database B is also a Herbrand model.*

Proof. It is sufficient to prove that the intersection of two Herbrand models of B is still a Herbrand model. We prove this by contradiction. Let $\mathcal{M}_1$ and $\mathcal{M}_2$ be two Herbrand models of B and suppose that their intersection is not a Herbrand model. Let $p(a) \vee \neg p_1(a_1) \vee \ldots \vee \neg p_n(a_n)$ be a ground instance of a deduction rule that is true in both $\mathcal{M}_1$ and $\mathcal{M}_2$ and false in their intersection $\mathcal{M}_1 \cap \mathcal{M}_2$.

Clearly, this clause is false in $\mathcal{M}_1 \cap \mathcal{M}_2$ if and only if $p(a)$ is false and all $p_i(a_i)$ are true in $\mathcal{M}_1 \cap \mathcal{M}_2$. Hence, $p(a)$ is false in at least one of $\mathcal{M}_1$ or

---

[4]This notion of hierarchical *database* is not directly related to the hierarchical *data model* of traditional databases [Date 81], [Ullman 82].

$\mathcal{M}_2$, while all $p_i(a_i)$ are true in both $\mathcal{M}_1$ and $\mathcal{M}_2$. But this implies that $p(a) \vee \neg p_1(a_1) \vee \ldots \vee \neg p_n(a_n)$ is false in at least one of $\mathcal{M}_1$ or $\mathcal{M}_2$. This is in contradiction with the hypotheses. □

The intersection of all Herbrand models of B is noted $\mathcal{M}_B$ and is called the *least Herbrand model* of B. $\mathcal{M}_B$ can also be characterized as the set of ground instances of atoms that are logical consequences of B [Van Emden and Kowalski 76]. This property is expressed by the following theorem.

Theorem 6.5. $\mathcal{M}_B = \{A \in \mathcal{B}_B \mid B \vDash A\}$.
Proof.
$B \vDash A$
 $\Leftrightarrow B \cup \{\neg A\}$ is inconsistent
 $\Leftrightarrow B \cup \{\neg A\}$ has no Herbrand model
 $\Leftrightarrow \neg A$ is false in all Herbrand models of B
 $\Leftrightarrow A$ is true in all Herbrand models of B
 $\Leftrightarrow A \in \mathcal{M}_B$. □

The model-theoretic approach, based on these results, selects, for a definite database B, its least Herbrand model $\mathcal{M}_B$ to characterize precisely the information that the database is intended to represent. The value of a query consists of its ground instances that are true in $\mathcal{M}_B$. A method for the systematic construction of $\mathcal{M}_B$ is given in Subsection 6.4.6.

6.4.4 Proof-theoretic approach with $\mathcal{T}_B^{CWA}$

The proof-theoretic approach gives the semantics of a database as a theory that consists of the database itself and of axioms expressing the closed-world assumption CWA, the unique-name assumption UNA, and the domain-closure assumption DCA. As for relational databases (§ 6.3.3), the possible formulations of the CWA lead to two theories, $\mathcal{T}_B^{CA}$ and $\mathcal{T}_B^{CWA}$. This subsection describes the construction of $\mathcal{T}_B^{CWA}$ for definite databases. The construction of $\mathcal{T}_B^{CA}$ will be described separately for hierarchical databases (§ 6.4.5) and for the general definite databases with recursive rules (§ 6.4.6).

Like relational databases, definite databases contain positive information only, namely the elements of the extensional database and the atoms that can be derived with the deduction rules of the intensional database. Indeed, the Herbrand base $\mathcal{B}_B$ of B is a Herbrand model of B. Hence, no negative literal can be a logical consequence of B. The closed-world assumption provides a way to deduce negative information. For a relational database, all the elements of the Herbrand base that are not part of the database are supposed to be false. For definite databases, this definition is generalized: every element of

6.4. DEFINITE DEDUCTIVE DATABASES

the Herbrand base that cannot be derived from the database is supposed to be false.

Definition. The definition (6.1) of Subsection 6.3.3 generalizes for a definite database B with Herbrand base $\mathcal{B}_B$ as follows:

$$CWA(B) = \{\neg P \mid P \in \mathcal{B}_B \text{ and } B \not\vdash P\}. \tag{6.8}$$

The theory T_B^{CWA} is defined as in (6.5), i.e.,

$$T_B^{CWA} = B \cup CWA(B) \cup DCA(B) \cup UNA(B),$$

where $DCA(B)$ is the domain-closure axiom of B, and $UNA(B)$ its unique-name axioms.

The theory T_B^{CWA} has some interesting properties. The following theorem can immediately be derived from Theorem 6.4 above and from Theorem 6.15 given in Section 6.5.

Theorem 6.6. *The theory T_B^{CWA} is consistent.*

The following theorem states that the ground instances of atoms that can be derived from the theory T_B^{CWA} are exactly those that can be derived from the definite database B.

Theorem 6.7. $T_B^{CWA} \vdash p(a)$ *if and only if* $B \vdash p(a)$.
Proof. The 'only if' part is immediate because $B \subset T_B^{CWA}$ and first-order logic is monotonic. A contradiction can be derived if the 'if' part is supposed false. Suppose that $B \not\vdash p(a)$ and $T_B^{CWA} \vdash p(a)$. If $B \not\vdash p(a)$ then $\neg p(a) \in CWA(B) \subset T_B^{CWA}$. Hence $T_B^{CWA} \vdash p(a)$ and $T_B^{CWA} \vdash \neg p(a)$, and hence T_B^{CWA} is inconsistent. This is in contradiction with Theorem 6.6. □

These results lead to the equivalence of the model- and proof-theoretic approaches (with the theory T_B^{CWA}).

Theorem 6.8. *The theory T_B^{CWA} is categorical and its only Herbrand model is $\mathcal{M}_B$, the least model of B.*
Proof. According to Lemma 6.1, the theory T_B^{CWA} admits Herbrand models only. A Herbrand model of T_B^{CWA} contains at least all elements of $\mathcal{B}_B$ that are logical consequences of T_B^{CWA} and hence of B. On the other hand, the elements of $\mathcal{B}_B$ for which $B \not\vdash p(a)$ cannot be in any Herbrand model of T_B^{CWA}. Indeed, if $B \not\vdash p(a)$ then $\neg p(a) \in CWA(B)$ and hence $T_B^{CWA} \vdash \neg p(a)$. The theory T_B^{CWA} has therefore a single model; this model contains exactly the elements

of $\mathcal{B}_B$ that are logical consequences of B. Hence, according to Theorem 6.5, the model of T_B^{CWA} is $\mathcal{M}_B$. □

The following corollary immediately follows from Theorem 6.8.

Corollary 6.9. *The theory T_B^{CWA} is complete.*

Note that the construction of the theory T_B^{CWA} for definite databases is not as easy as for relational databases. Indeed, all ground instances of atoms that cannot be derived from the database must be constructed explicitly and their negations must be included in T_B^{CWA}. The utility of this construction is therefore more theoretical than practical. Practical query evaluation algorithms are presented in the following subsections.

6.4.5 Hierarchical databases

Hierarchical databases (i.e., definite databases without recursion) are simple generalizations of relational databases. Results obtained for the latter extend without major problems to hierarchical databases.

The definition of T_B^{CWA} of the preceding subsection is valid for any definite database, and thus, in particular, for a hierarchical database. This subsection describes the construction of a theory T_B^{CA} for a hierarchical database and shows that it is equivalent to T_B^{CWA}. It also shows that T_B^{CA} leads to a practical query evaluation algorithm.

COMPLETION

For a relational database B, a completed theory T_B^{CA} was constructed with completion axioms; it is equivalent to T_B^{CWA}, which directly expresses the closed-world assumption (§ 6.3.3). A generalization of this construction of T_B^{CA} for a hierarchical database B is now given. For the extensional database of B, the theory is constructed as for relational databases. We therefore concentrate on the intensional database.

The deduction rules of the intensional database B_I are implications. Adding the inverse implications ensures the completeness of the theory T_B^{CA}.

In effect, consider a deduction rule

$$\forall var(t)\, (p(t) \leftarrow \exists y\, (b_1 \wedge \ldots \wedge b_n))$$

where t is, in general, composed of variables and constants. It can be assumed, without loss of generality, that when the ith component of t is a variable, that variable is x_i. This rule can be rewritten in a *normal form* in which all the

6.4. DEFINITE DEDUCTIVE DATABASES

arguments of the consequent are variables:

$$\forall \boldsymbol{x} \, (p(\boldsymbol{x}) \leftarrow \exists \boldsymbol{y} \, (x_{i_1} = t_{i_1} \wedge \ldots \wedge x_{i_m} = t_{i_m} \wedge b_1 \ldots \wedge b_n))$$

where $t_{i_1}, \ldots, t_{i_m}$ are all the constants of $\boldsymbol{t}$.

If a predicate p occurs in the consequent of k rules in normal form:

$$\forall \boldsymbol{x} \, (p(\boldsymbol{x}) \leftarrow E_j) \text{ for } 1 \leq j \leq k,$$

then the equivalence[5]

$$\forall \boldsymbol{x} \, (p(\boldsymbol{x}) \leftrightarrow E_1 \vee \ldots \vee E_k)$$

is included in the theory T_B^{CA}. This equivalence is called the *completion of the predicate p*[6]. In the sequel, universal quantifiers in front of completions are omitted.

Definition. The theory T_B^{CA} is defined as

$$T_B^{CA} = B \cup CA(B) \cup DCA(B) \cup UNA(B),$$

where $CA(B)$ denotes the set of completions of B.

The theory T_B^{CA} is obtained by replacing in T_B^{CWA} the explicit negative information of $CWA(B)$ by the completions of the predicates of B.

Example 6.20. Let B be the hierarchical part of the database of Example 6.19. To construct the theory T_B^{CA}, the following equivalences are added to B:

$$grandparent(x, y) \leftrightarrow \exists z \, (parent(x, z) \wedge parent(z, y)) \qquad (6.9)$$
$$parent(x, y) \leftrightarrow mother(x, y) \vee father(x, y) \qquad (6.10)$$

The first equivalence states that each *parent* of a *parent* is a *grandparent* (this is expressed by the deduction rule) and that there are no other *grandparents* (this is added by the completion). The second equivalence states that every *father* or *mother* is a *parent* and that there are no other *parents*.

In the remainder of this subsection, the theories T_B^{CWA} and T_B^{CA} are shown to be equivalent. First, we show that T_B^{CA} has a single model.

Theorem 6.10. The theory T_B^{CA} is categorical.
Proof. According to Lemma 6.1, only Herbrand models should be considered. It is sufficient to show that, for any element $p(a)$ of the Herbrand base,

---
[5] $\forall \boldsymbol{x} \, \neg p(\boldsymbol{x})$ cannot be the completion of a virtual predicate.
[6] See also 5.4

either $T_B^{CA} \vdash p(a)$, or $T_B^{CA} \nvdash p(a)$. Indeed, this partitions the Herbrand base into two subsets: a set of elements $p(a)$ that are true in all models of T_B^{CA} and a set of elements $p(a)$ that are false in all models of T_B^{CA}. In that case, T_B^{CA} must have exactly one Herbrand model.

We prove it by induction. For base predicates, the property derives from the corresponding property of relational databases (Theorem 6.2).

Suppose that the property holds for all predicates on which p strictly depends. Let $\forall \boldsymbol{x}\, (p(\boldsymbol{x}) \leftrightarrow E_1 \vee \ldots \vee E_k)$ be the completion of p. Then, for any ground instance $p(a)$, we have the equivalence

$$p(a) \leftrightarrow (E_1 \vee \ldots \vee E_k)[\boldsymbol{x}/a]. \tag{6.11}$$

Moreover, $DCA(B)$ implies, for any formula F, the equivalence

$$\exists y\, (F(y)) \equiv F(a_1) \vee \ldots \vee F(a_m),$$

where $a_1, \ldots, a_m$ are all the tuples of constants with the same length as y. Hence, each $E_j[\boldsymbol{x}/a]$, for $1 \leq j \leq k$, in (6.11), can be replaced by a ground formula containing only predicates on which p strictly depends. By the induction hypothesis, either $T_B^{CA} \vdash (E_1 \vee \ldots \vee E_k)[\boldsymbol{x}/a]$, or $T_B^{CA} \vdash \neg(E_1 \vee \ldots \vee E_k)[\boldsymbol{x}/a]$, and hence, either $T_B^{CA} \vdash p(a)$, or $T_B^{CA} \vdash \neg p(a)$. □

The following theorem states that every element of $\mathcal{B}_B$ that can be derived from T_B^{CA} can also be derived from B and conversely.

Theorem 6.11. $T_B^{CA} \vdash p(a)$ if and only if $B \vdash p(a)$.

Proof. The implication $B \vdash p(a) \Rightarrow T_B^{CA} \vdash p(a)$ holds by the monotonicity of first-order logic. Only the inverse implication $T_B^{CA} \vdash p(a) \Rightarrow B \vdash p(a)$ remains to be proved.

This is proved by induction. If p is a base predicate, the property derives from the corresponding property of relational databases (§ 6.3.4). Suppose that the property holds for every predicate on which p strictly depends. Let $p(\boldsymbol{x}) \leftrightarrow E_1 \vee \ldots \vee E_k$ be the completion of p, where

$$E_i \equiv \exists y\, (x_{i1} = t_{i1} \wedge \ldots \wedge x_{im_i} = t_{im_i} \wedge b_{i1} \wedge \ldots \wedge b_{in_i})$$

for $i = 1, \ldots, k$, and let

$$p(t_i) \leftarrow \exists y\, (b_{i1} \wedge \ldots \wedge b_{in_i}) \tag{6.12}$$

be the deduction rule of B associated with E_i. Then, the hypothesis $T_B^{CA} \vdash p(a)$ and the completion of p imply $T_B^{CA} \vdash (E_1 \vee \ldots \vee E_k)[\boldsymbol{x}/a]$. Since T_B^{CA} is categorical, $T_B^{CA} \vdash E_i[\boldsymbol{x}/a]$ for some $i \in \{1, \ldots, k\}$. Then, by definition of

6.4. DEFINITE DEDUCTIVE DATABASES 311

E_i, $T_B^{CA} \vdash b_{il}[x/a, y/b]$ for some b and for each $l \in \{1, \ldots, n_i\}$, and also $T_B^{CA} \vdash a_{ij} = t_{ij}$ for each $j \in \{1, \ldots, m_i\}$. By the properties of equality, $T_B^{CA} \vdash b_{il}[t_i/a, y/b]$, which implies $B \vdash b_{il}[t_i/a, y/b]$ by the induction hypotheses. The deduction rule (6.12) leads to $B \vdash p(a)$. □

The equivalence of the theories T_B^{CA} and T_B^{CWA} follows from Theorem 6.10 and Theorem 6.11.

Theorem 6.12. $T_B^{CA} \vdash F$ if and only if $T_B^{CWA} \vdash F$, for any formula F.
Proof. Indeed, both T_B^{CA} and T_B^{CWA} are categorical. Theorem 6.11 above and Theorem 6.7 about T_B^{CWA} imply that the models of both theories are necessarily the same. Hence, T_B^{CA} and T_B^{CWA} are equivalent. □

The complete theory T_B^{CA} defines the semantics of a hierarchical database under the closed-world assumption with the advantage over T_B^{CWA} that it is easily constructed. As shown below, a practical query evaluation algorithm for hierarchical databases can be derived from this theory.

QUERY EVALUATION

It follows immediately from Theorems 6.8 and 6.12 that query evaluation is equivalent in the model- and proof-theoretic approaches. Below, we show how a practical query evaluation method can be based on T_B^{CA}.

The answer to a query $< x \mid Q(x) >$ is defined as for a relational database, that is:

$$\| < x \mid Q(x) > \| = \{c \mid T_B^{CA} \vdash Q(c)\}.$$

With this definition, it is in principle possible to find the answer to a query with a theorem prover, although, as usual, this is in general very expensive.

A more efficient method consists in replacing each virtual predicate of the query by the right-hand side of its completion in T_B^{CA}. This process is repeated until the query contains base predicates only. The query can then be evaluated as in a relational database.

Example 6.21. Consider again the completions of the predicates *parent* and *grandparent* of Example 6.20 and the query

$$< x \mid grandparent(x, Elros) > \qquad (6.13)$$

(find all *grandparents* of *Elros*). Replacing in the query (6.13) the predicate *grandparent* by the right-hand side of the equivalence (6.9) leads to the equivalent query

$$< x \mid \exists z \, (parent(x, z) \wedge parent(z, Elros)) > . \qquad (6.14)$$

The virtual predicate *parent* is defined in the theory by the equivalence (6.10) and the query (6.14) can be transformed into

$$< x \mid \exists z \left((father(x,z) \vee mother(x,z)) \right.$$
$$\left. \wedge \, (father(z, Elros) \vee mother(z, Elros)) \right) >.$$

This query consists only of base predicates and, hence, it can be evaluated in the same way as a query in a relational database, for example by translating it into an expression of the relational algebra.

6.4.6 Recursive rules in definite databases

Definite databases generalize hierarchical databases by allowing recursion in the deduction rules. This subsection addresses the new problems raised by recursion. It is no longer possible to construct a simple completed theory as for hierarchical databases. Hence, the query evaluation algorithm presented in Subsection 6.4.5 is no longer valid for recursive queries. Therefore, query evaluation is tackled in the model-theoretic approach. We present a method for constructing the least Herbrand model and a basic query evaluation algorithm.

PROBLEMS WITH THE COMPLETION OF RECURSIVE RULES

The presence of recursive rules makes the construction of a theory T_B^{CA} from completion axioms considerably more complex than for hierarchical databases, as shown by the following example.

Example 6.22. With a base predicate $husband(x,y)$ ('x is the husband of y'), the virtual predicate $married(x,y)$ ('x is married to y') can be defined by the following deduction rules:

$$married(x,y) \leftarrow husband(x,y)$$
$$married(x,y) \leftarrow married(y,x).$$

This definite database B is not hierarchical. When completing the rules as in hierarchical databases (§ 6.4.5), the following equivalence is constructed:

$$married(x,y) \leftrightarrow husband(x,y) \vee married(y,x)$$

and it is included in T_B^{CA}. This theory is not complete and, hence, not categorical. Indeed, if, for two constants a and b, neither $husband(a,b)$ nor $husband(b,a)$ belong to the extensional part B_E of B, then it becomes impossible to derive either $married(a,b)$ or its negation $\neg married(a,b)$. In some models of the theory, $married(a,b)$ holds, while in others, it does not.

6.4. DEFINITE DEDUCTIVE DATABASES

On the other hand, the theory T_B^{CWA} is always complete. In the example, it is easy to see that if $husband(a,b) \notin B_E$ and $husband(b,a) \notin B_E$, then $T_B^{CWA} \vdash \neg married(a,b)$.

This example shows that, if a definite database is completed like a hierarchical database, the equivalence between T_B^{CA} and T_B^{CWA} is lost in general. Hence, the completion method adopted for hierarchical databases is no longer valid for general definite databases.

The query evaluation algorithm based on T_B^{CA} for hierarchical databases is no longer valid in the presence of recursive predicates either. The replacement of virtual predicates in queries by their definitions given by their completions leads to an endless loop when faced with a recursive predicate. Hence, another query evaluation algorithm is necessary. To obtain it, the model-theoretic approach is reconsidered and a method to construct the least Herbrand model is given.

CONSTRUCTION OF THE LEAST HERBRAND MODEL

The results presented below are primarily due to Van Emden and Kowalski in a more general context than the one considered here [Van Emden and Kowalski 76]. We give simplified proofs adapted to databases.

In order to construct the least Herbrand model, a mapping from the set of Herbrand interpretations to itself is defined as follows.

Definition. Let B be a definite database and let $\mathcal{B}_B$ be its Herbrand base. The *immediate consequence mapping* $T_B : 2^{\mathcal{B}_B} \to 2^{\mathcal{B}_B}$ is defined as follows. For a Herbrand interpretation $I \subset \mathcal{B}_B$, the value $T_B(I)$ is given by

$$T_B(I) = I \cup \{A \in \mathcal{B}_B \mid A \leftarrow A_1 \wedge \ldots \wedge A_n \text{ is a ground instance of a rule of } B \text{ and } \{A_1, \ldots, A_n\} \subset I\}.$$

Herbrand models can be characterized with the mapping T_B. The following theorem shows that Herbrand models of B are Herbrand interpretations that contain the extensional part B_E of B and that are *fixpoints* of T_B [Lloyd 87].

Theorem 6.13. *Given a definite database B and a Herbrand interpretation I of B, then I is a model of B if and only if B_E is as subset of I and I is a fixpoint of T_B, i.e., I satisfies the equation $T_B(I) = I$.*
Proof. If $B_E \subset I$, then I is a model of B_E and conversely. It remains to be proved that I is a model of B_I if and only if $T_B(I) = I$. The interpretation I is a model of B_I if and only if, for each ground instance $A \leftarrow A_1 \wedge \ldots \wedge A_n$ of a rule of the intensional part B_I of B, the hypothesis $\{A_1, \ldots, A_n\} \subset I$

implies $A \in I$, that is, if and only if $T_B(I) \subset I$. Since $I \subset T_B(I)$ is true by construction, I is a model of B_I if and only if $T_B(I) = I$. □

From a mapping T, mappings T^α are defined as follows:

$$T^0(I) = I,$$
$$T^n(I) = T(T^{n-1}(I)) \text{ for } n = 1, 2, \ldots,$$
$$T^\omega(I) = \bigcup_{n=0}^\infty T^n(I).$$

Note that the mapping T_B is monotonic, i.e., $I_1 \subset I_2$ implies $T_B(I_1) \subset T_B(I_2)$. Hence, the property $I \subset T_B(I)$ implies $T_B^n(I) \subset T_B^{n+1}(I)$ for any n. Since the set $\mathcal{B}_B$ is finite, there exists a positive integer n_0 such that $T_B^{n_0}(B_E) = T_B^{n_0+1}(B_E)$ and, hence,

$$T_B^{n_0}(B_E) = T_B^\omega(B_E).$$

The following theorem characterizes the least Herbrand model $\mathcal{M}$ of B in terms of T_B.

Theorem 6.14. $\mathcal{M}_B = T_B^\omega(B_E)$.
Proof. Let n_0 be a positive integer with the property $T_B^{n_0}(B_E) = T_B^\omega(B_E)$ mentioned above. From the inclusion $B_E \subset T_B^{n_0}(B_E)$ and the equalities $T_B^{n_0}(B_E) = T_B^{n_0+1}(B_E) = T_B(T_B^{n_0}(B_E))$, it follows, in view of Theorem 6.13, that $T_B^{n_0}(B_E)$ is a model of B. By definition of $\mathcal{M}_B$, this implies $\mathcal{M}_B \subset T_B^{n_0}(B_E)$.

To prove the theorem, the inverse inclusion $T_B^{n_0}(B_E) \subset \mathcal{M}_B$ remains to be shown. The proof is given by induction. First, $T_B^0(B_E) = B_E \subset \mathcal{M}_B$. Further, when supposing $T_B^{n-1}(B_E) \subset \mathcal{M}_B$, then, $T_B^n(B_E) = T_B(T_B^{n-1}(B_E)) \subset T_B(\mathcal{M}_B)$ follows since T_B is monotonic. But, since $\mathcal{M}_B$ is a model, $T_B(\mathcal{M}_B) = \mathcal{M}_B$ (Theorem 6.13). And hence, $T_B^n(B_E) \subset \mathcal{M}_B$ for any n, and, in particular, $T_B^{n_0}(B_E) \subset \mathcal{M}_B$. □

QUERY EVALUATION IN A DEFINITE DATABASE

Subsection 6.2.6 has shown how to transform a query expressed in the relational language into an expression of the relational algebra. A new operator, the fixpoint operator, noted FP, is added to the relational algebra to express queries containing virtual predicates.

Definition. For a virtual predicate p of a definite database B, the *fixpoint* $FP(p)$ is defined as follows:

$$FP(p) = \{a \mid p(a) \in T_B^\omega(B_E)\}.$$

6.4. DEFINITE DEDUCTIVE DATABASES

The algebraic correspondent of a query on a virtual predicate p is defined as follows:
$$|< \boldsymbol{x} \mid p(\boldsymbol{x},a) >| = \pi_{\boldsymbol{x}}\sigma_{y=a} FP(p(\boldsymbol{x},y)).$$

Example 6.23. With the virtual predicate *a-f* of Example 6.19, the following correspondence is obtained for the query 'Who are the friends of descendants of *Eärendil* ?':
$$|< y \mid \textit{a-f}(\textit{Eärendil}, y) >| = \pi_y \sigma_{x=\textit{Eärendil}} FP(\textit{a-f}(x,y)).$$

We shall see in the sequel how the fixpoint $FP(p)$ of a virtual predicate can be evaluated. The evaluation of a fixpoint, possibly preceded by a selection and/or a projection, gives an answer that is exactly the same as the one obtained by the model-theoretic approach. Thus, in our example, we have
$$\| < y \mid \textit{a-f}(\textit{Eärendil}, y) > \| = \pi_y \sigma_{x=\textit{Eärendil}} FP(\textit{a-f}(x,y)).$$

The answer to an atomic query can in principle be obtained by selecting in the least Herbrand model those elements that satisfy the query. Such a method needs the explicit and complete construction of the least Herbrand model, a very expensive task in the case of a large database. To avoid this problem, much research effort was put into the development of evaluation algorithms that do not need the complete construction of the least Herbrand model. This research resulted in the development of a large number of algorithms.

A simple but in general not very efficient method to evaluate an atomic query is the so-called *naïve* method. The idea is to construct a subset of the least Herbrand model that is in some way linked with the query.

Example 6.24. Consider the following query on the definite database of Example 6.19:
$$< y \mid \textit{a-f}(\textit{Eärendil}, y) >.$$
A mapping $T_{\textit{a-f}}$ can be defined for the rules defining the predicate *a-f*, namely,
$$\textit{a-f}(x,y) \leftarrow \textit{friend}(x,y),$$
$$\textit{a-f}(x,y) \leftarrow \textit{parent}(x,z) \wedge \textit{a-f}(z,y).$$
The naïve method computes $FP(\textit{a-f}) = T^{\omega}_{\textit{a-f}}(B_E)$. Since the set of rules defining *a-f* is by itself a definite database, the theorems given in Subsection 6.4.6 can be applied to this set and the associated mapping $T_{\textit{a-f}}$. Hence, there exists a positive integer n_0 such that
$$T^{\omega}_{\textit{a-f}}(B_E) = T^{n_0}_{\textit{a-f}}(B_E).$$

Thus, to find the set $T_{a\text{-}f}^{\omega}(B_E)$, it is sufficient to apply $T_{a\text{-}f}$ iteratively until no new ground instance of $a\text{-}f(x, y)$ is generated.

Suppose that B_E contains the following elements:

$$friend(EUadan, Aragorn) \quad parent(Elrond, EUadan)$$
$$friend(E\ddot{a}rendil, Finrod) \quad parent(E\ddot{a}rendil, Elrond)$$
$$friend(Elwing, Galadriel) \quad parent(Nimloth, Elwing).$$

The set $T_{a\text{-}f}^{\omega}(B_E)$ is obtained through the following computation:

$$T_{a\text{-}f}^{0}(B_E) = B_E$$
$$T_{a\text{-}f}^{1}(B_E) = B_E \cup \{ a\text{-}f(EUadan, Aragorn), a\text{-}f(E\ddot{a}rendil, Finrod),$$
$$\qquad a\text{-}f(Elwing, Galadriel)\}$$
$$T_{a\text{-}f}^{2}(B_E) = B_E \cup \{ a\text{-}f(EUadan, Aragorn), a\text{-}f(E\ddot{a}rendil, Finrod),$$
$$\qquad a\text{-}f(Elwing, Galadriel), a\text{-}f(Elrond, Aragorn),$$
$$\qquad a\text{-}f(Nimloth, Galadriel)\}$$
$$T_{a\text{-}f}^{3}(B_E) = B_E \cup \{ a\text{-}f(EUadan, Aragorn), a\text{-}f(E\ddot{a}rendil, Finrod),$$
$$\qquad a\text{-}f(Elwing, Galadriel), a\text{-}f(Elrond, Aragorn),$$
$$\qquad a\text{-}f(Nimloth, Galadriel), a\text{-}f(E\ddot{a}rendil, Aragorn)\}$$
$$T_{a\text{-}f}^{4}(B_E) = T_{a\text{-}f}^{3}(B_E).$$

The answer is extracted from $T_{a\text{-}f}^{3}$ by an appropriate selection and projection:

$$|< y \mid a\text{-}f(E\ddot{a}rendil, y) >| = \{Aragorn, Finrod\}.$$

A disadvantage of the naïve method is that it generates at each step all instances already generated in the previous steps. A straightforward improvement consists in repeating the evaluation of recursive rules only. In the example, this improvement avoids the generation at each step of the instances $a\text{-}f(EUadan, Aragorn)$, $a\text{-}f(E\ddot{a}rendil, Finrod)$ and $a\text{-}f(Elwing, Galadriel)$. On the other hand, instances generated by a recursive rule in some step are still generated in later steps. This can be avoided by using the *differential* method [Bancilhon 85]. Essentially, that method only uses at each step, as far as possible, instances generated at the preceding step. In the example, this means that the mapping is, in each step, only applied to the new instances generated in the preceding step. In successive steps, the following ground instances are generated:

$$\{a\text{-}f(EUadan, Aragorn), a\text{-}f(E\ddot{a}rendil, Finrod), a\text{-}f(Elwing, Galadriel)\}$$
$$\{a\text{-}f(Elrond, Aragorn), a\text{-}f(Nimloth, Galadriel)\}$$
$$\{a\text{-}f(E\ddot{a}rendil, Aragorn)\}.$$

The answer is then selected in the set of all instances generated during the process.

Another disadvantage of the naïve method, which is not remedied by the differential method, is the generation of a lot of superfluous instances. In the example, as seen above,

$$|< y \mid a\text{-}f(E\ddot{a}rendil, y) >| = \pi_y \sigma_{x=E\ddot{a}rendil} FP(a\text{-}f(x, y)).$$

So, $FP(a\text{-}f(x,y))$ is evaluated before applying the selection. Hence, the complete extension of $a\text{-}f$ is computed, even if only a subset of this extension is asked for. In general, it is not possible to commute the selection operator and the fixpoint operator, or, in other words, to instantiate the deduction rules and use these instantiated forms during evaluation. For example, the rules

$$a\text{-}f(E\ddot{a}rendil, y) \leftarrow friend(E\ddot{a}rendil, y)$$
$$a\text{-}f(E\ddot{a}rendil, y) \leftarrow (parent(E\ddot{a}rendil, z) \wedge a\text{-}f(z, y))$$

only lead to generating

$$a\text{-}f(E\ddot{a}rendil, Finrod)$$

and the answer thus obtained is not complete.

For some years, much research effort has been put into the development of algorithms that, essentially, combine the evaluation of a selection and a fixpoint operator. Some of these algorithms only apply to rules of some limited form; in general, they are limited to linear rules [Chang 80], [Henschen and Naqvi 84]. Others do not halt in presence of cycles[7][Chang 80], [Henschen and Naqvi 84], [Saccà and Zaniolo 86]. Finally, there exist algorithms that always obtain complete answers and that generate only relevant instances [Vieille 86], [Rohmer et al. 86], [Beeri and Ramakrishnan 87], [Hulin 88a]. These algorithms are quite complex and are therefore not described here. An overview is presented in [Roelants 87] and [Demolombe and Royer 87].

6.5 Negation in deductive databases

6.5.1 Introduction

Rules in definite deductive databases are restricted to be Horn clauses. As shown in the preceding section, those databases enjoy nice properties such as the existence of a natural semantics and of realistic, sound, and complete evaluation mechanisms for answering queries. However, the expressive power of definite deductive databases is limited. The expressive power is significantly enhanced by introducing negation in rule bodies, as shown by the following example.

---

[7]An example of a cycle for the relation *friend* in the extensional database is the simultaneous presence of *friend(Eärendil, Finrod)* and *friend(Finrod, Eärendil)*.

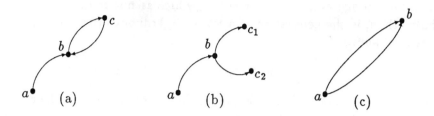

Figure 6.2 : (a) cycle, (b) branching, (c) confluence.

Example 6.25. Consider a coloured graph defined with the base predicates $edge(x, y)$ ('y an immediate successor of x') and $coloured(n, v)$ ('node n has colour v'). An edge is said to be *acceptable* if it connects two differently coloured nodes. With general (non-Horn) clauses, an intuitive perception of negation suggests writing

$$acceptable\text{-}edge(x, y) \leftarrow edge(x, y) \land coloured(x, z) \land \neg coloured(y, z)$$

but, as will be seen later, correctness depends on an adequate definition of the semantics of negation.

Two nodes x and y are connected by a *path* in the coloured graph if there is a succession of acceptable edges such that x is the origin of the first edge, y is the end of the last edge, and the end of an edge is the origin of the following edge. This can be expressed with the Horn clauses

$$path(x, y) \leftarrow acceptable\text{-}edge(x, y),$$
$$path(x, y) \leftarrow acceptable\text{-}edge(x, z) \land path(z, y).$$

The information that two nodes in the graph are connected by a path but do not belong to the same cycle, can be expressed by the (non-Horn) rule

$$acyclic\text{-}path(x, y) \leftarrow path(x, y) \land \neg path(y, x).$$

A node a is the origin of a *star* in the graph if all the paths of origin a go away from a in star-like fashion, i.e., without cycling on themselves, and without branching or confluence. The three situations preventing a from being the origin of a star are schematized, respectively, by (a), (b) and (c) in Figure 6.2. A directed arc (a, b) represents a path from a to b in the graph.

This can be expressed by the following rules:

$$star\text{-}failure(x) \leftarrow path(x, z) \land path(z, y) \land path(y, z),$$
$$star\text{-}failure(x) \leftarrow path(x, z) \land acceptable\text{-}edge(z, z_1)$$
$$\land\ acceptable\text{-}edge(z, z_2) \land z_1 \neq z_2,$$
$$star\text{-}failure(x) \leftarrow path(x, z_1) \land path(x, z_2)$$

6.5. NEGATION IN DEDUCTIVE DATABASES

$$\land \ acceptable\text{-}edge(z_1, z) \land acceptable\text{-}edge(z_2, z)$$
$$\land \ z_1 \neq z_2,$$
$$star(x) \leftarrow \neg star\text{-}failure(x).$$

The first three rules express the three conditions of failure, the fourth one expresses that x is the origin of a star in the graph if no failure condition is satisfied. The expressive power is actually enhanced, as $star(x)$ is defined by conditions that must be verified on all paths of origin x. Such a definition is not expressible with Horn clauses.

The aim of this section is to describe a semantics that correctly conveys intuition for the deductive databases illustrated by the examples above.

Still, negations in the body of rules sometimes lead to confusing definitions, as shown in the following example.

Example 6.26. Let $base\text{-}prop$ be a base predicate expressing some given property of the nodes of a graph and let $coloured$ be a predicate defined by the following rules:

$$coloured(y) \leftarrow base\text{-}prop(y),$$
$$coloured(y) \leftarrow edge(x, y) \land \neg coloured(x).$$

The definition of $coloured$ seems to indicate a kind of alternation in the colouring of the nodes in the graph: the end of an edge whose origin is not coloured must itself be coloured. That rule is not intuitively clear: indeed, how can it be interpreted if the database contains the facts $edge(a, b)$ and $edge(b, a)$ but contains neither $base\text{-}prop(a)$ nor $base\text{-}prop(b)$? Then either a or b or both a and b can be coloured.

The problem raised by Example 6.26 is whether a satisfactory semantics can be given to the rules. Moreover, how would such a semantics interpret the vague idea of alternation that the rules suggest ?

6.5.2 The closed-world assumption and deductive databases

When negations are present in the body of rules, the closed-world assumption generally leads to an inconsistent theory.

Consider a deductive database B whose extensional component is the set $\{edge(a, b), coloured(a, v)\}$ and whose deduction rules are the rules of Example 6.25. No new ground instances of predicates of B can be derived with the rules. If $CWA(B)$ is defined as in (6.8) for definite deductive databases, that is,

$$CWA(B) = \{\neg P \mid P \in \mathcal{B}_B \text{ and } B \nvdash P\},$$

then $CWA(B)$ contains, among others, the negative literals $\neg coloured(b,v)$ and $\neg acceptable\text{-}edge(a,b)$. The theory $B \cup CWA(B)$ is then inconsistent. Indeed, the ground instance $acceptable\text{-}edge(a,b)$ can be derived from it with the literals $edge(a,b)$, $coloured(a,v)$, and $\neg coloured(b,v)$. Inconsistency similarly arises with the database of Example 6.26.

The presence of negative literals in rule bodies explains the inconsistency of the completion with the CWA. Indeed, negative literals added to the theory by the CWA can lead to the generation of new positive literals that contradict $CWA(B)$.

The conditions of failure are specified by the following theorem. Its proof is given in a slightly more general framework where theories are finite sets of function-free clauses[8].

Theorem 6.15. Let B be a finite set of function-free clauses. Then the following three propositions are equivalent:
(1) the theory $B \cup CWA(B)$ is inconsistent;
(2) there exist ground atoms $A_1, \ldots, A_n$ such that $B \vdash A_1 \vee \ldots \vee A_n$ and $B \nvdash A_i$ for $i = 1, \ldots, n$;
(3) the intersection of all the Herbrand models of B is not a Herbrand model of B (the model intersection property of Theorem 6.4 no longer holds).

Proof. First, let us show that $(1) \Rightarrow (3)$, or, equivalently, that when the intersection $\mathcal{M}$ of all the Herbrand models of B is itself a model of B, then $B \cup CWA(B)$ is consistent. By the same argument as in the proof of Theorem 6.5, it is easy to show that $\mathcal{M}$ is the set of ground atoms that are logical consequences of B. It follows that $\mathcal{M}$ is a model of $CWA(B)$. Indeed, no ground atom whose negation belongs to $CWA(B)$ can belong to $\mathcal{M}$. If, according to the hypothesis, $\mathcal{M}$ is a model of B, then it is also a model of the theory $B \cup CWA(B)$ which is thus consistent.

Next, let us show that $(3) \Rightarrow (2)$. Let $\mathcal{M}_1, \ldots, \mathcal{M}_m$ be the Herbrand models of B. Let $\mathcal{M}$ denote their intersection: $\mathcal{M} = \bigcap_{j=1}^{m} \mathcal{M}_j$, and suppose $\mathcal{M}$ is not a model of B. Let C be the clause consisting of the disjunction of all the atoms of $\bigcup_{j=1}^{m} \mathcal{M}_j \setminus \mathcal{M}$. The set $B \cup \{\neg C\}$ is inconsistent. Indeed, as $\mathcal{M}$ is not a model of B, the sets $\mathcal{M}_j \setminus \mathcal{M}$ are non-empty and, consequently, C contains at least one element out of each Herbrand model of B. One can therefore conclude that $B \cup \{\neg C\}$ is inconsistent because a clausal form is inconsistent if and only if it is false in all its Herbrand interpretations (see § 1.2.9 in [Thayse 88]). Then, $B \vdash C$. Moreover, if c is an atom of C, then $B \nvdash c$. Indeed, c cannot belong to each $\mathcal{M}_j$, otherwise it would belong to $\mathcal{M}$ and thus would not belong to C.

Last, let us show that $(2) \Rightarrow (1)$. If $B \nvdash A_i$, then $\neg A_i \in CWA(B)$ for $i =$

---

[8]Unlike deductive databases, such theories can contain clauses where all literals are negative.

6.5. NEGATION IN DEDUCTIVE DATABASES

$1, \ldots, n$. It follows that the formula $\neg(A_1 \lor \ldots \lor A_n)$ is a logical consequence of $CWA(B)$. Compatibility with the hypothesis $B \vdash A_1 \lor \ldots \lor A_n$ requires $B \cup CWA(B)$ to be inconsistent. □

Note that the second condition of the theorem can be rephrased equivalently as: there exists a ground clause, not reduced to a single literal, that can be derived from B and such that none of its subclauses is itself derivable from B.

That kind of *irreducible* disjunctive information can be present in databases where negations appear in the body of the rules. The rule defining *acyclic-path* in Example 6.25 can be rewritten equivalently:

$$path(y, x) \lor acyclic\text{-}path(x, y) \leftarrow path(x, y).$$

Instances of *path* usually generate with this rule disjunctive information that makes the theory $B \cup CWA(B)$ inconsistent.

The classical model-theoretic approach is also doomed by Theorem 6.15: in most cases, the concept of least Herbrand model is meaningless in indefinite deductive databases. The rule of Example 6.26 can be rewritten equivalently as follows:

$$coloured(x) \lor coloured(y) \leftarrow edge(x, y),$$

and it must be interpreted as requiring that the origin or the end of each edge be coloured. If the extensional database is $B_E = \{edge(a, b)\}$, then both sets $\{edge(a, b), coloured(a)\}$ and $\{edge(a, b), coloured(b)\}$ are Herbrand models of B. Yet, their intersection is not a model of B. There is thus no least Herbrand model for that database. However, there always exist some *minimal* models, i.e., some Herbrand models such that none of their proper subsets is itself a Herbrand model. In the example above, the two models described are the only minimal Herbrand models of the database.

Two approaches can be followed to associate a semantics with deductive databases. On one hand, the CWA can be weakened by excluding the negative literals leading to inconsistencies. That approach results in the construction of incomplete theories: there exist elements of the Herbrand base such that neither they nor their negations are derivable from the theory. The *generalized closed-world assumption* (GCWA) belongs to that approach [Minker 82], [Shepherdson 88b]. It is studied extensively in Subsection 6.5.3. As will be seen, the GCWA is applicable to any deductive database.

On the other hand, one can favour a minimal model that is considered more 'natural' than the others. In that approach, theories remain complete. However, the approach in question is only applicable for restricted classes of databases. One of them, the class of *stratified* deductive databases [Apt et al. 88], [Przymusinski 88a], will be studied in Subsection 6.5.4.

6.5.3 The generalized closed-world assumption

As shown in Theorem 6.8, the closed-world assumption applied to definite deductive databases leads to evaluating queries in the least Herbrand model of the databases.

In general, as already observed, a deductive database has no least Herbrand model but has several minimal Herbrand models. That multiplicity is a consequence of incompleteness in the knowledge of the world modelled by the database. Because of the irreducible disjunctive information present in the database, there exist several possibilities for minimizing the positive information, instead of just one as for definite deductive databases. Each minimal Herbrand model can be seen as a possible representation of the world modelled by the database where the positive information has been minimized. Under the *generalized closed-world assumption* (GCWA), the formulas of the relational language are evaluated with respect to the set of minimal Herbrand models of the database. More precisely, a formula is considered
- true if it is true in all the minimal Herbrand models,
- false if it is false in all the minimal Herbrand models,
- indefinite if it is true in some minimal Herbrand models and false in the others.

That generalization of CWA belongs to the model-theoretic approach to databases which was introduced in Subsection 6.3.2.

SEMANTIC AND SYNTACTIC DEFINITIONS

We follow the presentations of [Minker 82] and [Shepherdson 88b]; by so doing, we adopt a definition of the GCWA which relies on the proof-theoretic approach and which associates, with any deductive database B, a theory T_B^{GCWA}. It will be seen later that the model- and proof-theoretic approaches to GCWA are *not* equivalent.

If B is a deductive database, then $GCWA(B)$ is defined as follows:

$$GCWA(B) = \{\neg A \mid A \in \mathcal{B}_B \text{ and } A \text{ does not belong to any minimal Herbrand model of } B\}.$$

This definition of $GCWA(B)$ is called the *semantic* definition, because it is based on the models of the database.

Note that $GCWA(B)$ is a generalization of $CWA(B)$; indeed, when B_I is a set of Horn clauses, there is a single minimal model: the least Herbrand model.

Under the GCWA, queries to the database are answered with respect to the theory

$$T_B^{GCWA} = B \cup GCWA(B) \cup DCA(B) \cup UNA(B).$$

6.5. NEGATION IN DEDUCTIVE DATABASES

In Example 6.26, with $B_E = \{edge(a,b), edge(b,c), coloured(b)\}$, there is a single minimal Herbrand model, namely B_E itself. In that situation, $GCWA(B)$ is comprised of the negations of the elements of $\mathcal{B}_B \setminus B_E$ and coincides with $CWA(B)$. Theorem 6.15 guarantees the consistency of $B \cup CWA(B)$. If the fact $edge(d,a)$ is added to B_E, then there are two minimal Herbrand models: $B_E \cup \{coloured(a)\}$ and $B_E \cup \{coloured(d)\}$. The theory $\mathcal{T}_B^{GCWA}$ is no longer complete because it does not permit us to determine the truth value of $coloured(a)$ and $coloured(d)$. Note, however, that $\mathcal{T}_B^{GCWA} \vdash \neg coloured(c)$. Thus $\mathcal{T}_B^{GCWA}$ is richer than B.

If $B_E = \{edge(a,b)\}$, then there now exist two minimal Herbrand models: $\{edge(a,b), coloured(a)\}$ and $\{edge(a,b), coloured(b)\}$. It follows that the set $GCWA(B)$ contains neither $coloured(a)$ nor $coloured(b)$, which are thus indefinite. No new information is thus added by the GCWA concerning predicate $coloured$.

The examples above show the low practical interest of the semantic definition of $GCWA(B)$: determining all the minimal Herbrand models may be computationally very expensive. A more syntactic characterization of the GCWA is needed. The following theorem, for which an elegant proof can be found in [Shepherdson 88b], establishes a basis for such a characterization.

Theorem 6.16. *Let B be a deductive database. Then, a ground atom A belongs to at least one minimal Herbrand model of B if and only if there exist some ground atoms $A_1, \ldots, A_n$ (with $n \geq 0$) such that*

$$B \vdash (A \vee A_1 \vee \ldots \vee A_n) \text{ and } B \not\vdash (A_1 \vee \ldots \vee A_n).$$

Note that when $n = 0$, the condition reduces to $B \vdash A$ and A then belongs to all the Herbrand models. Let $\mathcal{S}_B$ denote the set of ground atoms satisfying the condition of Theorem 6.16. Then $\mathcal{S}_B$ is the union of all the minimal Herbrand models of B, and we are therefore allowed to write

$$GCWA(B) = \{\neg A \mid A \in \mathcal{B}_B \setminus \mathcal{S}_B\}.$$

That new definition is called the *syntactic* definition of $GCWA(B)$. Comparing with Theorem 6.15, it becomes clear that the GCWA avoids adding to B negative ground literals that lead to inconsistencies. Indeed, if $B \vdash A_1 \vee \ldots \vee A_n$ but $B \not\vdash A_i$ for $i = 1, \ldots, n$, then no literal $\neg A_i$ belongs to $GCWA(B)$.

Example 6.27. Consider again Example 6.26 with $B_E = \{edge(a,b), edge(b,c), edge(d,a), coloured(b)\}$. The recursive rule defining the predicate *coloured* can be rewritten equivalently as follows:

$$coloured(x) \lor coloured(y) \leftarrow edge(x,y).$$

The disjunction $coloured(d) \lor coloured(a)$ can then be derived from the fact $edge(d,a)$. As neither $coloured(d)$ nor $coloured(a)$ can be derived from B, they both belong to $\mathcal{S}_B$. More precisely, it can easily be checked that $\mathcal{S}_B = \mathcal{B}_E \cup \{coloured(d), coloured(a)\}$.

PROPERTIES OF THE GENERALIZED CLOSED-WORLD ASSUMPTION

The following theorems characterize the generalized closed-world assumption. They are mainly due to Minker [Minker 82]. Here we give suitable versions of them, with simplified proofs, adjusted to deductive databases.

Theorem 6.17. *The set of minimal Herbrand models of B coincides with the set of minimal Herbrand models of T_B^{GCWA}.*

Proof. Given the semantic definition, no minimal Herbrand model of B contains any atom whose negation belongs to $GCWA(B)$. Each minimal Herbrand model of B is thus a model of T_B^{GCWA}. These models are clearly minimal models of T_B^{GCWA}. Conversely, a minimal model $\mathcal{M}$ of T_B^{GCWA} is a minimal model of B. Otherwise, there would exist a minimal model $\mathcal{M}'$ of B that would be a proper subset of $\mathcal{M}$. Then $\mathcal{M}'$ would also be a model of T_B^{GCWA}, which contradicts the minimality of $\mathcal{M}$. □

The next theorem shows that the GCWA eliminates the risk of inconsistency that was present with the CWA.

Theorem 6.18. *If B is a deductive database, then the theory T_B^{GCWA} is consistent.*

Proof. The Herbrand base of B is also a Herbrand model of B. As it is finite, it necessarily admits a subset which is a minimal Herbrand model of B. That model is also a minimal Herbrand model of T_B^{GCWA}. □

As a consequence of the monotonicity of predicate calculus, every formula which is true in B remains true in T_B^{GCWA}. That statement can be inverted for formulas without negations as the next theorem shows.

Theorem 6.19. *Let B be a deductive database and let F be a closed formula where the only connectives are $\land$ and $\lor$. Then, the hypothesis $T_B^{GCWA} \vdash F$ implies $B \vdash F$.*

Proof. Assume that $T_B^{GCWA} \vdash F$. By virtue of the domain-closure axiom, universal and existential quantifications present in F can be replaced by conjunctions and disjunctions on the set of elements of the domain. Formula F

6.5. NEGATION IN DEDUCTIVE DATABASES

is thus equivalent to a formula composed of disjunctions and conjunctions of elements of the Herbrand domain of B. Suppose, by contradiction, that $B \not\vdash F$. Then $B \cup \{\neg F\}$ is consistent and hence admits a minimal Herbrand model $\mathcal{M}$. This model is then also a minimal Herbrand model of B. Indeed, if $\mathcal{M}'$ is a model of B and a proper subset of $\mathcal{M}$, then $\mathcal{M}'$ falsifies more atoms than $\mathcal{M}$ and thus preserves the truth of $\neg F$. The set $\mathcal{M}'$ is thus also a model of $B \cup \{\neg F\}$, which contradicts the hypothesis that $\mathcal{M}$ is minimal.

Theorem 6.17 allows us to assert that $\mathcal{M}$ is a minimal model of T_B^{GCWA}, which contradicts the hypothesis since $\mathcal{M}$ falsifies F. □

Theorem 6.19 shows that the GCWA does not add new positive information to the database. To contrast with the CWA, remember that negative literals of $CWA(B)$ can often be used in the body of deduction rules to infer new ground instances for the heads of these rules. Such a new positive information always leads to inconsistency.

The semantic and axiomatic approaches to the GCWA are not equivalent. It follows immediately from Theorem 6.17 that any formula derivable from T_B^{GCWA} is true in all the minimal Herbrand models of B. Unfortunately, the converse statement is not valid. Consider a database B reduced to the clause $p \vee q$. The formula $(p \wedge \neg q) \vee (\neg p \wedge q)$ is true in all the minimal Herbrand models of B but is not derivable from T_B^{GCWA} since $GCWA(B) = \{\}$. That discrepancy is due to the fact that T_B^{GCWA} may have other models than the minimal Herbrand models of B. This reveals an important difference between the model- and proof-theoretic approaches to the GCWA: a complete evaluation mechanism for formulas in T_B^{GCWA} (designing one is by itself a hard problem [Bossu and Siegel 85], [Yahya and Henschen 85], [Henschen and Park 88]) would not enable us to find all the formulas that are true in all the minimal Herbrand models of B.

Moreover, the semantics conveyed by the GCWA is not always satisfactory to express the intuition that guides the use of negation in the body of deduction rules.

Example 6.28. Consider again Example 6.25 with

$$B_E = \{edge(a,b), edge(b,c), coloured(a,r), coloured(b,v), coloured(c,j)\}.$$

The rule defining *acceptable-edge* can be rewritten as

$$acceptable\text{-}edge(x,y) \vee coloured(y,z) \leftarrow edge(x,y) \wedge coloured(x,z).$$

In the following two formulas derivable from B:

$$acceptable\text{-}edge(a,b) \vee coloured(b,r),$$
$$acceptable\text{-}edge(b,c) \vee coloured(c,v),$$

no atom can be derived from B. By virtue of the syntactic definition of the GCWA, the truth value of these atoms is thus indefinite in $\mathcal{T}_B^{GCWA}$. This conflicts with the intuition that led to the definition of *acceptable-edge*. Since $coloured(b, r)$ and $coloured(c, v)$ are absent from B_E, it seems reasonable to suppose them false and then to infer that $acceptable\text{-}edge(a, b)$ and $acceptable\text{-}edge(b, c)$ are true.

Such a conflict originates from the fact that someone writing a rule generally interprets it in a specific minimal Herbrand model and not in the set of minimal Herbrand models. In the next subsection, we will characterize a suitable 'specific model' for an interesting subclass of deductive databases.

6.5.4 Stratified databases

STRATIFICATION OF A DATABASE

Note that $GCWA(B)$ only depends upon the logical content of B. The clause $p \vee q$ can indeed be written either as $p \leftarrow \neg q$ or as $q \leftarrow \neg p$, without influencing $GCWA(B)$.

Yet, in certain circumstances, the specific form given to a rule denotes an extra-logical intention which is not always taken into account by a purely logical interpretation.

Example 6.29. Let B be a deductive database whose intensional component consists of the first few rules of Example 6.25, namely:

$acceptable\text{-}edge(x, y) \leftarrow edge(x, y) \wedge coloured(x, z) \wedge \neg coloured(y, z),$
$path(x, y) \leftarrow acceptable\text{-}edge(x, y),$
$path(x, y) \leftarrow acceptable\text{-}edge(x, z) \wedge path(z, y),$
$acyclic\text{-}path(x, y) \leftarrow path(x, y) \wedge \neg path(y, x).$

We now show how the specific form of the rules can constrain the semantics of B. As the predicate *coloured* occurs negatively in the body of the rule defining *acceptable-edge*, it seems natural first to enrich B by adding $CWA(B_E)$ before evaluating *acceptable-edge* and *path*. If

$B_E = \{edge(a, b), edge(b, c), coloured(a, r), coloured(b, v), coloured(c, j)\},$

then the literals $\neg coloured(b, r)$ and $\neg coloured(c, v)$ belong to $CWA(B_E)$. Following the strategy explained above, one can derive the literals $acceptable\text{-}edge(a, b)$, $acceptable\text{-}edge(b, c)$, as well as $path(a, b)$, $path(b, c)$ and $path(a, c)$.

As *path* appears negatively in the body of the rule defining *acyclic-path*, then, before evaluating that rule, the strategy adds to the database the negations of the ground instances of *path* that have not yet been derived, that is, among others, $\neg path(b, a)$, $\neg path(c, b)$ and $\neg path(c, a)$. Then, in accordance

6.5. NEGATION IN DEDUCTIVE DATABASES

with intuition, one can derive the atoms *acyclic-path*(a,b), *acyclic-path*(b,c) and *acyclic-path*(a,c).

So, the predicates *edge*, *coloured*, *acceptable-edge*, *path*, and *acyclic-path* have been partitioned into three successive *strata* {*edge, coloured*}, {*acceptable-edge, path*} and {*path-admissible*}, so that any negative literal occurring in the definition of a predicate of a given stratum is always an instance of a predicate of a preceding stratum.

A deductive database is said to be *stratified* if the set V of its predicates can be decomposed into disjoint subsets $S_1, \ldots, S_r$ with the properties described below. One writes $\text{rank}(l) = j$ to indicate that l is a (positive or negative) instance of a predicate of S_j. For every rule $p \leftarrow l_1 \wedge \ldots \wedge l_q$, the following conditions must be satisfied:

- for every positive literal l_i ($1 \leq i \leq q$) that is an instance of a virtual predicate, $\text{rank}(l_i) \leq \text{rank}(p)$,

- for every negative literal l_i ($1 \leq i \leq q$) that is an instance of a virtual predicate, $\text{rank}(l_i) < \text{rank}(p)$.

The r-tuple $(S_1, \ldots, S_r)$ is called a *stratification* of B and S_i ($1 \leq i \leq r$) are called *strata* of B.

The subsets $S_1 = \{edge, coloured\}$, $S_2 = \{acceptable\text{-}edge, path\}$, $S_3 = \{star\text{-}failure, acyclic\text{-}path\}$, $S_4 = \{star\}$ constitute a stratification for Example 6.25. The subsets $S_1 = \{edge\}$, $S_2 = \{coloured\}$, $S_3 = \{acceptable\text{-}edge, path, star\text{-}failure\}$, $S_4 = \{star\}$, $S_5 = \{acyclic\text{-}path\}$ constitute another stratification for the same database.

The definition above is constructive: it requires that an explicit stratification be exhibited. A more syntactic definition relies on the concept of *mutual recursion* defined in Subsection 6.4.2. Observe that mutual recursion induces an equivalence relation on the set V of predicates of B. A syntactic characterization is supplied by the following theorem.

Theorem 6.20. *A deductive database B is stratified if and only if for every class $M = \{p_1, \ldots, p_n\}$ induced by the relation of mutual recursion, no rule whose head is an instance of p_i contains in its body a negative instance of p_j ($1 \leq i, j \leq n$).*

Proof. If the database is stratified, then, by definition, if p_i and p_j are mutually recursive predicates, they belong to the same stratum. It follows that no rule whose head is an instance of p_i contains in its body a negative instance of p_j. This proves the 'only if' part of the theorem. To establish the 'if' part, a dependence relation $\ll$ is defined on the equivalence classes, in such a way that $M_1 \ll M_2$ if M_2 contains predicates depending on predicates of M_1. That relation is clearly a partial order. Let S_1 be the union of the minimal classes for the dependence relation and let S_{i+1} be the union of the classes depending on classes of S_i ($1 \leq i < r$). It is easy to see that $(S_1, \ldots, S_r)$ is a stratification of B. □

It is sometimes said that a database is stratified when 'negation does not interfere with recursion', that is, when no predicate p_1 is defined in terms of a literal $\neg p_2(t)$, such that the predicate p_2 is itself defined in terms of p_1. Note that a definite deductive database is always stratified and that the database of Example 6.26 is not stratified; indeed, the definition of *coloured* contains an occurrence of $\neg coloured$.

SEMANTICS OF STRATIFIED DATABASES

If $S = (S_1, \ldots, S_r)$ is a stratification of B, if B_E^i is the subset of B_E comprised of the ground atoms of predicates of S_i, and if R_i is the set of rules defining the virtual predicates of S_i, then the sets

$$B_i = \bigcup_{j=1}^{i} (B_E^j \cup R_j)$$

with $i = 1, \ldots, r$ are called the *partial databases* of B. A *partial theory* T_{B_i} is associated with each partial database B_i as follows:

$$T_{B_1}^S = B_1 \cup CWA(B_1) \cup DCA(B) \cup UNA(B),$$

$$T_{B_i}^S = B_E^i \cup R_i \cup T_{B_{i-1}}^S \cup CWA(B_E^i \cup R_i \cup T_{B_{i-1}}^S) \qquad \text{for } i = 2, \ldots, r.$$

The theory T_B^S is defined as $T_B^S = T_{B_r}^S$. Queries to the database B are evaluated in that theory T_B^S.

Note that B_1 is a definite deductive database and hence that $T_{B_1}^S = T_{B_1}^{CWA}$. An inference mechanism immediately follows from that definition. It formalizes the mechanism illustrated with an example at the beginning of this subsection: the CWA is successively applied to the predicates of each stratum while taking into account the negative ground literals generated on the previous strata.

The following example emphasizes again the importance of the extra-logical information conveyed by the specific form of deduction rules.

Example 6.30. Consider the database $\{p \vee q\}$. If it is written in the equivalent form $\{p \leftarrow \neg q\}$, then $(\{q\}, \{p\})$ is the only possible stratification and so $T_B^S \vdash p \wedge \neg q$. On the other hand, if the rule is rewritten $\{q \leftarrow \neg p\}$, then $(\{p\}, \{q\})$ is the only possible stratification and, in this case, $T_B^S \vdash \neg p \wedge q$.

The definition given above for the semantics of stratified databases belongs to the proof-theoretic approach. It is also possible to give a definition in the

6.5. NEGATION IN DEDUCTIVE DATABASES

model-theoretic approach. Some preliminary definitions are necessary.

Definition. The immediate consequence mapping defined in Subsection 6.4.6 is generalized in order to compute the immediate consequences of a subset of the deduction rules of a deductive database B.

If R is a set of deduction rules of B, then the *immediate consequence mapping* T_R is a mapping from $2^{\mathcal{B}_B}$ to $2^{\mathcal{B}_B}$ defined as follows:

$T_R(I) =$
$\quad I \cup \{A \in \mathcal{B}_B \mid$ there exists a ground instance of a rule of R,
$\quad$ of the form $A \leftarrow B_1 \wedge \ldots \wedge B_q$, such that:
$\quad$ – the positive literals of $B_1, \ldots, B_q$ belong to I,
$\quad$ – the negative literals of $B_1, \ldots, B_q$ do not belong to $I\}$.

The mappings T_R^n (for every natural number n) and T_R^ω are defined in the usual way (§ 6.4.6).

Some results obtained in Subsection 6.4.6 immediately generalize. They are summarized in the following theorem, given without proof.

Theorem 6.21.
1. An interpretation I is a model of R if and only if $I = T_R(I)$.
2. For every I, there exists a natural number n_0 such that $T_R^{n_0+1}(I) = T_R^{n_0}(I)$, and thus $T_R^\omega(I) = T_R^{n_0}(I)$.
3. Consequently, $T_R^\omega(I)$ is a model of R.

Now let $S = (S_1, \ldots, S_r)$ be a stratification of a database B. In the model-theoretic approach, a specific model, noted $\mathcal{M}_B$, is associated with B; it is defined by induction as follows:

$$\mathcal{M}_1 = T_{R_1}^\omega(B_E^1),$$
$$\mathcal{M}_i = T_{R_i}^\omega(B_E^i \cup \mathcal{M}_{i-1}) \qquad \text{for } i = 2, \ldots, r,$$
$$\mathcal{M}_B = \mathcal{M}_r.$$

The following theorem shows the equivalence between the model- and proof-theoretic approaches for stratified databases.

Theorem 6.22. The set $\mathcal{M}_i$ is the unique Herbrand model of the theory $T_{B_i}^S$, for $i = 1, \ldots, r$.
Proof. The theorem will be proved by induction. It holds for $i = 1$, as can easily be shown from Theorems 6.8 and 6.14 and from the fact that B_1 is a definite deductive database.

Assume that the theorem holds for some i. Let T_i be $B_E^{i+1} \cup R_{i+1} \cup T_{B_i}^S$. Then, $\mathcal{M}_{i+1}$ must be proved to be the unique model of $T_i \cup CWA(T_i)$.

First, the set $\mathcal{M}_{i+1}$ is a model of T_i. As $\mathcal{M}_{i+1} = T^\omega_{R_{i+1}}(B^{i+1}_E \cup \mathcal{M}_i)$, Theorem 6.21 shows that $\mathcal{M}_{i+1}$ is a model of $B^{i+1}_E \cup R_{i+1}$. On the other hand, as R_{i+1} does not define any predicate of $\bigcup_{j \leq i} S_j$, the subset of $\mathcal{M}_{i+1}$ comprised of the ground instances of predicates of $\bigcup_{j \leq i} S_j$ is $\mathcal{M}_i$. By the induction hypothesis, the set $\mathcal{M}_{i+1}$ is also a model of $T^S_{B_i}$. Hence, it is a model of T_i.

Next, every element of $\mathcal{M}_{i+1}$ is derivable from T_i. Indeed, by the induction hypothesis, $\mathcal{M}_i$ is the unique Herbrand model of $T^S_{B_i}$. Therefore, its elements are derivable from $T^S_{B_i}$ and thus from T_i. It is then easy to see that $\mathcal{M}_{i+1} = T^\omega_{R_{i+1}}(B^{i+1}_E \cup \mathcal{M}_i)$ also consists of elements derivable from T_i.

As any Herbrand model of a theory contains the set of ground atoms derivable from that theory, $\mathcal{M}_{i+1}$ is thus the least Herbrand model of T_i. The model intersection property thus holds in theory T_i. That theory can thus be completed by the CWA without losing consistency.

The completed theory $T^S_{B_{i+1}} = T_i \cup CWA(T_i)$ is thus categorical and admits $\mathcal{M}_{i+1}$ as a unique model. $\square$

Corollary 6.23. *If B is a stratified database and if Q is a closed formula, then the theory T^S_B is categorical and $T^S_B \vdash Q$ if and only if $\models_{\mathcal{M}_B} Q$.*

With the definition above, the semantics of a stratified database seems to depend on the choice of its stratification. We have seen that some databases admit several stratifications and that no stratification can be considered more 'natural' than the others. We show in the following that, in fact, the semantics does not depend on the particular stratification chosen. This fundamental result is due to Apt, Blair and Walker [Apt et al. 88]. We give it a new proof, much shorter than the original proof.

Theorem 6.24. *The semantics of a stratified database B is independent of the choice of the stratification.*

Proof. If p is a predicate of B, let P denote the set of predicates on which p depends and let $P_<$ denote the set of predicates on which p strictly depends.

If I is a Herbrand interpretation of B and if E is a set of predicates of B, let $|I|_E$ denote the subset of atoms of I that are ground instances of predicates contained in E.

As a preamble, note that if p belongs to the ith stratum of a stratification, then $|\mathcal{M}_B|_P = |\mathcal{M}_i|_P$. Indeed, $\mathcal{M}_i \subset \mathcal{M}_B$ and, by construction, in $\mathcal{M}_B$, there are no ground instances of predicates of P that do not already belong to $\mathcal{M}_i$.

Let $(S_1, \ldots, S_r)$ and $(S'_1, \ldots, S'_{r'})$ be two stratifications of B, and let $\mathcal{M}_B$ and $\mathcal{M}'_B$ be the respective models associated with B for these stratifications. The property will be established if, for every p, it can be shown that $|\mathcal{M}_B|_P = |\mathcal{M}'_B|_P$. The proof proceeds by induction on the dependence relation.

6.5. NEGATION IN DEDUCTIVE DATABASES

First, if b is a base predicate, then $|\mathcal{M}_B|_{\{b\}} = |\mathcal{M}'_B|_{\{b\}} = |B_E|_{\{b\}}$.

Next, let p be a virtual predicate. It can be shown that the following induction hypothesis:

$$|\mathcal{M}_B|_{P_<} = |\mathcal{M}'_B|_{P_<} \qquad\text{(IH.1)}$$

implies $|\mathcal{M}_B|_P = |\mathcal{M}'_B|_P$.

There exists an index i such that $p \in S_i$. By the preamble of the proof, it will be sufficient to prove the inclusion $|\mathcal{M}_i|_P \subset |\mathcal{M}'_B|_P$. We proceed again by induction. Assume, by convention, that $\mathcal{M}_0 = \{\}$.

First, we have

$$\begin{aligned}
|B_E^i \cup \mathcal{M}_{i-1}|_P &\subset |B_E^i \cup \mathcal{M}_{i-1}|_{P_<} & (P \setminus P_< \subset S_i) \\
&\subset |\mathcal{M}_B|_{P_<} & (B_E^i \cup \mathcal{M}_{i-1} \subset \mathcal{M}_B) \\
&\subset |\mathcal{M}'_B|_{P_<} & \text{(IH.1)} \\
&\subset |\mathcal{M}'_B|_P & (P_< \subset P)
\end{aligned}$$

With the following induction hypothesis:

$$|T_{R_i}^k(B_E^i \cup \mathcal{M}_{i-1})|_P \subset |\mathcal{M}'_B|_P \qquad\text{(IH.2)}$$

it must be shown that $|T_{R_i}^{k+1}(B_E^i \cup \mathcal{M}_{i-1})|_P \subset |\mathcal{M}'_B|_P$.

Let v be a predicate in $P \setminus P_<$. If

$$v(a) \in (|T_{R_i}^{k+1}(B_E^i \cup \mathcal{M}_{i-1})|_P \setminus |T_{R_i}^k(B_E^i \cup \mathcal{M}_{i-1})|_P),$$

then there exists an instance of a rule of R_i of the form

$$v(a) \leftarrow v_1(a_1) \wedge \ldots \wedge v_j(a_j) \wedge \neg v_{j+1}(a_{j+1}) \wedge \ldots \wedge \neg v_l(a_l)$$

such that:

1. $v_m(a_m) \in |T_{R_i}^k(B_E^i \cup \mathcal{M}_{i-1})|_P$ for $m = 1, \ldots, j$,
2. $v_n(a_n) \notin |T_{R_i}^k(B_E^i \cup \mathcal{M}_{i-1})|_P$ for $n = j+1, \ldots, l$.

Then

1. $v_m(a_m) \in |\mathcal{M}'_B|_P$ for $m = 1, \ldots, j$ by IH.2,
2. $v_n(a_n) \notin |\mathcal{M}'_B|_P$ for $n = j+1, \ldots, l$ as, by the stratification, $v_n \in \bigcup_{s=1}^{i-1} S_s$ and thus, with $V = \{v_{j+1}, \ldots, v_l\}$,

$$\begin{aligned}
|T_{R_i}^k(B_E^i \cup \mathcal{M}_{i-1})|_V &= |\mathcal{M}_{i-1}|_V & \text{(Definition of } T_{R_i}) \\
&= |\mathcal{M}_B|_V & (V \subset \bigcup_{s=1}^{i-1} S_s) \\
&= |\mathcal{M}'_B|_V & (V \subset P_< \text{ and IH.1})
\end{aligned}$$

Since $\mathcal{M}'_B$ is a model of B, it follows that $v(a) \in \mathcal{M}'_B$ and hence that

$$|T_{R_i}^{k+1}(B_E^i \cup \mathcal{M}_{i-1})|_P \subset |\mathcal{M}'_B|_P.$$

It follows that $|\mathcal{M}_i|_P = \bigcup_{k=0}^{\infty} |T_{R_i}^k(B_E^i \cup \mathcal{M}_{i-1})|_P \subset |\mathcal{M}'_B|_P$, which proves the theorem. $\square$

QUERY EVALUATION IN A STRATIFIED DATABASES

Query evaluation can be implemented by a mechanism quite similar to the one described for definite deductive databases. The only operator to be modified is the fixpoint operator. Instead of computing the relation corresponding to a virtual predicate v in the least Herbrand model, the operator will now compute the relation corresponding to v in the unique model $\mathcal{M}_B$ associated with the stratified database. This can be done from a stratification of the database reduced to the facts and rules involving v and the predicates which v depends on. The model $\mathcal{M}_B$ associated with the reduced database can be constructed by a sequence of fixpoint computations, applied to the successive strata, and the relation corresponding to v is extracted from that model $\mathcal{M}_B$.

6.6 Incomplete information

6.6.1 Introduction

Subsection 6.3.4 has shown that a relational database theory is complete under the closed-world, domain-closure, and unique-name assumptions: any formula is either true or false. It is sometimes desirable to broaden the relational database concept[9] to allow the modelling of *incomplete information*. By incomplete information, we mean any information that cannot be expressed as a conjunction of ground instances of predicates of the relational language.

Example 6.31. In a relational database with the *father* relation defined as in Example 6.1, the proposition

Galadriel and Finrod have the same father (but he is unknown)

is written, in the relational language of Subsection 6.2.4:

$$\exists p \, (father(p, Galadriel) \wedge father(p, Finrod)).$$

This incomplete information cannot be expressed solely with ground instances of relation predicates.

This section describes the addition to relational databases of some types of incomplete information and studies some consequences on the query evaluation mechanisms.

There exist many types of incomplete information; some of them are illustrated hereafter.

---

[9]The treatment of incomplete information in deductive databases is still embryonic [Hulin 88].

6.6. INCOMPLETE INFORMATION

- Existentially quantified conjunctive formulas.

    Example 6.32. *Nimloth's father is unknown:*
    $$\exists p \, father(p, Nimloth).$$

The values of some components of a tuple of a relation are completely unknown.

Example 6.33. *Galadriel and Finrod have the same father (but he is unknown):*
$$\exists p \, (father(p, Galadriel) \wedge father(p, Finrod)).$$

Some components of several tuples (of some relations) have identical but unknown values.

Example 6.34. *Nimloth's father is unknown but it is known that he is neither Eärendil nor Elrond:*
$$\exists p \, (father(p, Nimloth) \wedge p \neq E\ddot{a}rendil \wedge p \neq Elrond).$$

The values of some components of a tuple are unknown but it is known that they do not belong to a specific set of constants.

Example 6.35. *Both Nimloth's and Galadriel's fathers are unknown but known to be different:*
$$\exists p_1 \, \exists p_2 \, (father(p_1, Nimloth) \wedge father(p_2, Galadriel) \wedge p_1 \neq p_2).$$

Some components of several tuples (of some relations) have different unknown values.

- Disjunctive formulas.

    Example 6.36. *Galadriel's father is either Finarfin or Hùrin:*
    $$\exists p \, (father(p, Galadriel) \wedge (p = Finarfin \vee p = H\grave{u}rin)),$$

or equivalently:
$$father(Finarfin, Galadriel) \vee father(H\grave{u}rin, Galadriel).$$

The values of some components of a tuple of a relation are unknown but they are known to belong to a specific set of constants.

Example 6.37. *Idril's father is either Fingon's brother or Gil-Galad's uncle:*

$$\exists p\, (father(p, Idril) \wedge (brother(p, Fingon) \vee uncle(p, Gil\text{-}Galad))).$$

General disjunctive information.

There exist several extensions of relational databases catering for certain types of incomplete information. The rest of this section is limited to the types of incomplete information illustrated by Examples 6.32 and 6.33. A more general extension allows, in addition, a proper treatment of the types illustrated by Examples 6.34 and 6.35 [Reiter 84b], [Reiter 86]. There exists another extension for the types of Example 6.36 with some restrictions [Imieliński 86]. Example 6.37 is beyond the reach of currently studied extensions.

In the following sections, the introduction of incomplete information in relational databases is studied along three complementary points of view:

1. **Syntactic point of view:** how can incomplete information be represented appropriately ? In particular, how can it be distinguished from known information ?
2. **Semantic point of view:** how can the semantics of a database containing incomplete information be formalized in both the proof- and model-theoretic approaches ?
3. **Algorithmic point of view:** for a given syntax and semantics, how can query evaluation algorithms be generalized from those designed for ordinary relational databases ?

6.6.2 Syntax

MARKED NULL VALUES

The existential quantifiers in existentially quantified conjunctive formulas, as in Examples 6.32 and 6.33, can be removed by Skolemization (see § 1.2.8 in [Thayse 88]): each existential variable is replaced by a constant, noted ω_i (with $i = 1, 2, \ldots$), distinct from all other constants initially present in the database. Those new constants are commonly called *marked null values* [Reiter 84b], [Reiter 86], [Imieliński and Lipski 84a]. After Skolemization, the information can be represented as an ordinary relation, as shown in the following example.

Example 6.38. The incomplete information of Examples 6.32 and 6.33 is

6.6. INCOMPLETE INFORMATION

represented by the following existentially quantified conjunctive formulas:

$$\exists p \; father(p, Nimloth),$$
$$\exists p \; father(p, Galadriel) \wedge father(p, Finrod).$$

After Skolemization, they become:

$$father(\omega_1, Nimloth),$$
$$father(\omega_2, Galadriel) \wedge father(\omega_2, Finrod),$$

represented by the relation:

$$father(\omega_1, Nimloth),$$
$$father(\omega_2, Galadriel),$$
$$father(\omega_2, Finrod).$$

Marked null values differ from ordinary constants. They denote values that are incompletely identified in the database B; these values can be ordinary constants present in B or completely new ones. Two occurrences of the same constant ω_i in B denote the same ordinary constant.

Definition. A *database with marked null values* is a set of ground instances of relation predicates, as in an ordinary relational database (§ 6.3.1), except that the set of constants $\mathcal{D}$ of the database is now:

$$\mathcal{D} = \mathcal{C} \cup \{\omega_1, ..., \omega_n\}, \;\; \text{with} \;\; \mathcal{C} \cap \{\omega_1, ..., \omega_n\} = \{\},$$

where $\mathcal{C}$ denotes the set of *ordinary* constants and where $\{\omega_1, ..., \omega_n\}$ is the set of marked null values.

CODD'S NULL VALUES

A coarser technique than marked null values allows only one symbol ω_o to denote unknown values. The occurrences of ω_o in the database are commonly called *Codd's null values* [Codd 79], [Grant 77], [Grant 79], [Biskup 83], [Keller 86], [Imieliński and Lipski 84a].

Each occurrence of ω_o represents an unknown value (not necessarily the same as that represented by any other occurrence of ω_o). The unknown value can be different from all known constants in $\mathcal{C}$ (as it was the case with marked null values too).

Example 6.39. The incomplete information '*Nimloth's father and Eöl's child are both unknown*' is represented by

$$father(\omega_o, Nimloth),$$
$$father(Eöl, \omega_o).$$

Definition. A *database with Codd's null values* is a set of ground instances of relation predicates such that:

$$\mathcal{D} = \mathcal{C} \cup \{\omega_o\}, \text{ with } \omega_o \notin \mathcal{C},$$

where $\mathcal{C}$ denotes the set of *ordinary* constants and where ω_o is a unique symbol that denotes Codd's null values.

Some consequences of this syntactic weakening of marked null values are described below. For example, with Codd's null values, it is not possible to express that the same unknown value appears in several tuples of a database. It will also be seen that, with certain precautions, Codd's null values can be seen as a special case of marked null values.

Example 6.40. In Example 6.38, the information *'Nimloth's father is unknown, and Galadriel and Finrod have the same unknown father'* is only approximately represented as a relation with Codd's null values by:

$$father(\omega_o, Nimloth),$$
$$father(\omega_o, Finrod),$$
$$father(\omega_o, Galadriel).$$

The information that Finrod's father is the same as Galadriel's is lost.

6.6.3 Model-theoretic approach

The model-theoretic approach associates with any database B an appropriate set of models. In previous sections, except in the context of the generalized closed-world assumption (§ 6.5.3), the model-theoretic approach has always succeeded in choosing a single model to describe all the information in B. This is no longer possible in databases with null values. Each model describes a relational database with complete information (i.e., an ordinary relational database as defined in Section 6.2) compatible with B. Then, a formula is said to be true (or false) in B if it is true (or false) in all models of B.

Unfortunately, the number of models, large in general, complicates the efficient evaluation of queries. In particular, the plurality of models makes incomplete the traditional evaluation techniques.

MARKED NULL VALUES

The models of a database B with the marked null values $\omega_1, ..., \omega_n$ will be defined on the set of constants (i.e., the Herbrand universe):

$$\mathcal{D}^* = \mathcal{C} \cup \{\omega_1^*, ..., \omega_n^*\},$$

6.6. INCOMPLETE INFORMATION

where the w_i^*'s are constants distinct from all other constants in C. The w_i^*'s represent the possible values for the w_i's outside of C. In $\mathcal{D}^*$, the behaviour of the w_i^*'s is identical to that of ordinary constants.

A model $\mathcal{M}_B$ is built from B by giving a value from $\mathcal{D}^*$ to each marked null value:
$$\mathcal{M}_B = B\theta \text{ for a substitution } \theta : \{w_1, ..., w_n\} \to \mathcal{D}^*.$$

More precisely, $B\theta$ is an ordinary relational database having a least Herbrand model which is the database $B\theta$ itself (§ 6.3.2). This defines a model $\mathcal{M}_B$ for each possible substitution θ. The set of models $\mathcal{M}_B$ associated with B is denoted Ξ_B.

Example 6.41. Let B be the database with marked null values of Example 6.38:
$$father(w_1, Nimloth),$$
$$father(w_2, Finrod),$$
$$father(w_2, Galadriel).$$

Here, $C = \{Finrod, Nimloth, Galadriel\}$. The following relational databases are models in Ξ_B:

$father(Finrod, Nimloth)$	$father(w_1^*, Nimloth)$
$father(w_1^*, Finrod)$	$father(w_2^*, Finrod)$
$father(w_1^*, Galadriel)$	$father(w_2^*, Galadriel)$
$\theta = \{(w_1, Finrod), (w_2, w_1^*)\}$	$\theta = \{(w_1, w_1^*), (w_2, w_2^*)\}$

while the following ones are not:

$father(w_1^*, Nimloth)$	$father(w_1^*, Nimloth)$
$father(w_2^*, Finrod)$	$father(w_2^*, Finrod)$
$father(Finrod, Galadriel)$	$father(w_1^*, Galadriel)$

Note that all models in Ξ_B do not necessarily contain an occurrence of each of the new constants w_i^*. Some models may even contain no w_i^* at all.

A query has the same form $<\boldsymbol{x} \mid F(\boldsymbol{x})>$ as for ordinary relational databases (§ 6.2.4), except that it may now contain marked null values.

The answer to the query $<\boldsymbol{x} \mid F(\boldsymbol{x})>$ is defined as follows:

$$\| <\boldsymbol{x} \mid F(\boldsymbol{x})> \| = \{\boldsymbol{a} \in \mathcal{D} \mid \vDash_{B\theta} F(\boldsymbol{a})\theta \text{ for every } B\theta \in \Xi_B\}$$

with $\boldsymbol{a} = (a_1, ..., a_k)$ and $\mathcal{D} = \mathcal{D}^k$, if $\boldsymbol{x} = (x_1, ..., x_k)$.

The answer to the query contains the tuples $\boldsymbol{a}$, arguments of the ground instances $F(\boldsymbol{a})$ of F, for which formula $F(\boldsymbol{a})\theta$ is true in all models $B\theta$ in Ξ_B.

Note that, in general, a formula can be true in some models $B\theta$ and false in some others.

Example 6.42. Consider the following query Q on the database of Example 6.38:
$$Q \; = \; <x,y \; | \; \exists z \, (father(z,x) \wedge father(z,y))>.$$
Its answer is
$$\|Q\| = \{(Finrod, Galadriel), (Galadriel, Finrod)\}.$$

Codd's null values

Given a database B with Codd's null values, a database B° with marked null values is built by numbering the occurrences of ω_o in B, that is, by replacing each occurrence of ω_o by a different marked null value ω_i in the set $\{\omega_1, ..., \omega_n\}$, where n is the number of occurrences of ω_o in B.

The set Ξ_B of models associated with B is defined as the set of the models associated with the database B°, that is, $\Xi_B = \Xi_{B^\circ}$.

A query has the same form $<x \; | \; F(x)>$ as for ordinary relational databases (§ 6.2.4). Its only constants belong to $\mathcal{C}$. The answer to it is defined as follows:
$$\| <x \; | \; F(x)> \| \; = \; \{a\sigma \in \mathcal{D} \; | \; \models_{B^\circ\theta} F(a)\theta \text{ for every } B^\circ\theta \in \Xi_{B^\circ}\}.$$

Here, $\theta : \{\omega_1, ..., \omega_n\} \to \mathcal{D}^*$ is a substitution that transforms marked null values into ordinary or ω_i^* constants and $\sigma = \{(\omega_1, \omega_o), ..., (\omega_n, \omega_o)\}$ is the substitution that transforms marked null values back into Codd's null values.

In queries, ω_o's are not allowed because an occurrence of ω_o in a query could not be identified with certainty with any occurrence of ω_o in B.

Example 6.43. Consider the following database with Codd's null values:
$$B = \{spouses(Finarfin, Eärwen, \omega_o), spouses(Elrond, \omega_o, 3)\}$$
and the query 'Who is Elrond's wife and how many children does she have, and who is Eärwen's husband and how many children does he have?' written
$$Q = <x,y \; | \; spouses(Elrond, x, y) \vee spouses(x, Eärwen, y)>.$$
We have
$$B^\circ = \{spouses(Finarfin, Eärwen, \omega_1), spouses(Elrond, \omega_2, 3)\},$$

6.6. INCOMPLETE INFORMATION

and

$$\begin{aligned} \|Q\| &= \| <x,y \mid spouses(Elrond, x, y) \vee spouses(x, E\ddot{a}rwen, y)> \| \\ &= \{(\omega_2, 3)\{(\omega_1, \omega_o), (\omega_2, \omega_o)\}, (Finarfin, \omega_1)\{(\omega_1, \omega_o), (\omega_2, \omega_o)\}\} \\ &= \{(\omega_o, 3), (Finarfin, \omega_o)\}, \end{aligned}$$

where $\{(\omega_1, \omega_o), (\omega_2, \omega_o)\}$ is the appropriate substitution σ.

6.6.4 Query evaluation in the model-theoretic approach

As the set Ξ_B is very large in general, it is not reasonable in practice to compute the answer to a query by evaluating it in all models in Ξ_B.

Query evaluation algorithms, when null values are present, try to compromise between efficiency on the one hand, and soundness and correctness on the other, while accepting a class of queries as broad as possible. Query evaluation algorithms, described hereafter, are based on operators of the relational algebra as defined in Subsection 6.2.3 with as few modifications as possible.

MARKED NULL VALUES

Only the selection operator has to be modified; it is redefined as follows. If $f(A)$ is the qualification of the selection, then:

$$\sigma_{f(A)}(r) = \{a \in r \mid \forall x_1 ... \forall x_{n(a)} f(a)\{(\omega_1, x_1), ..., (\omega_{n(a)}, x_{n(a)})\}\}$$

where $\omega_1, ..., \omega_{n(a)}$ denote the marked null values of a.

Example 6.44. For the database

$$B = \{r(a, \omega_1, c), r(a, \omega_1, a)\}$$

we have

$$\sigma_{(1=a \wedge 2 \neq b) \vee (2=b \wedge 3 \neq 1)}(r) = \{(a, \omega_1, c)\}.$$

For the first tuple, the condition reduces to $\forall x_1 (x_1 \neq b \vee x_1 = b)$, which is trivially true, while for the second tuple, it reduces to $\forall x_1 (x_1 \neq b)$ which is not true in all generality.

The following results hold for marked null values [Imieliński and Lipski 84a]:

1. The answer to a query Q is sound if it is computed from its algebraic correspondent $|Q|$ as described in Subsection 6.2.6 with the new selection operator, provided that $|Q|$ does not involve set difference. In other words, $|Q| \subset \|Q\|$ holds in this context.

340   CHAPTER 6.   LOGIC AND DATABASES

2. However, this query evaluation technique is complete (i.e., $|Q| \supset \|Q\|$) only for queries whose algebraic correspondent only contains projections, positive selections[10], unions, and joins. Positive existentially quantified conjunctive queries, for which $|Q| = \|Q\|$ holds, are part of the allowed queries. It is shown in [Imieliński and Lipski 81] and [Imieliński and Lipski 84b] that there exist generalizations of the relational algebra for which broader classes of queries can be handled.

These results are illustrated in the following two examples.

Example 6.45.   With the database
$$B = \{r(a, w_1, c), r(a', w_1, c)\}$$
and the query
$$Q = <x, z \mid \exists y \ (r(x, y, c) \wedge r(a, y, z))>,$$
the evaluation algorithm described above leads to

$$
\begin{aligned}
|Q| &= \pi_{15}(\sigma_{3=c}(r) \underset{2=2}{\bowtie} \sigma_{1=a}(r)) \\
&= \pi_{15}\left(\{(a, w_1, c)\} \underset{2=2}{\bowtie} \{(a, w_1, c)\}\right) \\
&= \pi_{15}(\{(a, w_1, c, a, c)\}) \\
&= \{(a, c)\}.
\end{aligned}
$$

This is the same answer as that obtained with the model-theoretic approach as can be seen by building the set Ξ_B and evaluating the query in every model in Ξ_B.

The following example illustrates the case where the algebraic correspondent of a query contains projections and negative selections[11].

Example 6.46.   With the database of Example 6.45, the query
$$Q =<z \mid \exists x \ \exists y \ (r(x, y, z) \wedge ((x = a \wedge y = b) \vee (x = a' \wedge y \neq b)))>$$
evaluates to
$$
\begin{aligned}
|Q| &= \pi_3(\sigma_{(1=a \wedge 2=b) \vee (1=a' \wedge 2 \neq b)}(r)) \\
&= \pi_3(\{\}) \\
&= \{\}.
\end{aligned}
$$

This answer is not complete as $c \in \|Q\|$. Indeed, for all models $B\theta$ of B,

---
[10] Positive selections are built from $=, \wedge$ and $\vee$ only, that is, without the connective $\neg$.
[11] Negative selections contain, among others, the $\neg$ connective.

6.6. INCOMPLETE INFORMATION

- either $w_1^* = b$, then, as $r(a, w_1^*, c) \in B\theta$,
  we have $\models_{B\theta} r(a, w_1^*, c) \land a = a \land w_1^* = b$;
- or $w_1^* \neq b$, then, as $r(a', w_1^*, c) \in B\theta$,
  we have $\models_{B\theta} r(a', w_1^*, c) \land a' = a' \land w_1^* \neq b$.

CODD'S NULL VALUES

As with marked null values, only the selection operator has to be redefined. If a is a tuple of a relation r and if a° is the tuple obtained by replacing the occurrences of w_o in a by different marked null values w_i (with $1 \leq i \leq n(a)$), then the selection in r with qualification $f(A)$ is defined by:

$$\sigma_{f(A)}(r) = \{ a \in r \mid \forall x_1 ... \forall x_{n(a)} f(a^\circ)\{(w_1, x_1), ..., (w_{n(a)}, x_{n(a)})\} \}$$

The following results hold for Codd's null values [Imieliński and Lipski 84a]:

1. The answer to a query Q is sound if it is computed from its algebraic correspondent $|Q|$ as described in Subsection 6.2.6 with the new selection operator, provided that $|Q|$ does not involve set difference. In other words, $|Q| \subset \|Q\|$ holds in this context.
2. However, this query evaluation technique is complete (i.e., $|Q| \supset \|Q\|$) only for queries whose algebraic correspondent only involves projection and (positive and negative) selection.

These results are illustrated in the following two examples.

Example 6.47. Consider the database

$$B = \{r(c, w_o, b), r(w_o, b, c)\}$$

and the query

$$Q = < y, z \mid \exists x \ (r(x, y, z) \land (x \neq c \lor x = z)) > .$$

The evaluation algorithm described above leads to

$$|Q| = \pi_{23}(\sigma_{1 \neq c \lor 1 = 3}(r))$$
$$= \{(b, c)\}.$$

The same answer is obtained with the model-theoretic approach as can be seen by building the set Ξ_B and then evaluating the query in every model in Ξ_B.

The following example illustrates the loss of completeness when the algebraic correspondent of a query contains joins in addition to projections and selections.

Example 6.48. Consider the database:
$$B = \{r(a, \omega_o, c)\}$$
and the query
$$Q = <x, z \mid \exists y \ (r(x,y,c) \wedge r(a,y,z))> .$$
By the model-theoretic approach (i.e., evaluation in all models), we have
$$\|Q\| = \{(a,c)\},$$
while the evaluation by the relational algebra gives
$$\begin{aligned}|Q| &= \pi_{15}\left(\sigma_{3=c}(r) \underset{2=2}{\bowtie} \sigma_{1=a}(r)\right) \\ &= \pi_{15}\left(\{(a,\omega_o,c)\} \underset{2=2}{\bowtie} \{(a,\omega_o,c)\}\right) \\ &= \pi_{15}(\{\}) \\ &= \{\}.\end{aligned}$$

The fact that the evaluation method for Codd's null values is complete when negative selections are involved must not be understood as an advantage of Codd's null values over marked null values. Actually, the negative selections that entail a loss of completeness with marked null values (as in Example 6.46) are not expressible with Codd's null values, and the negative selections that are handled correctly with Codd's null values are also handled correctly with marked null values [Imieliński and Lipski 84a].

6.6.5 Proof-theoretic approach

In the database B with marked null values of Example 6.38, containing

father$(\omega_1, Nimloth)$,
father$(\omega_2, Finrod)$,
father$(\omega_2, Galadriel)$,

the new constants ω_1 and ω_2 behave differently from the ordinary constants *Nimloth*, *Finrod* and *Galadriel*. Indeed, the ω_i's represent objects currently

6.6. INCOMPLETE INFORMATION

unknown, distinct or not from the ordinary constants of B. More precisely, for example, ω_1 is not necessarily distinct from *Galadriel* and ω_1 is not necessarily distinct from ω_2.

From a proof-theoretic point of view, the difference in behaviour between ordinary constants and marked null values can be represented by retaining only the unique-name axioms: those that express the difference of two ordinary constants of B. In other words, there are no unique-name axioms involving marked null values.

The restricted set of unique-name axioms is denoted by $UNA_\omega(B)$.

A theory T_B^{CA} associated with the database with marked null values B is defined as follows:

$$T_B^{CA} = B \cup CA(B) \cup DCA(B) \cup UNA_\omega(B).$$

In Example 6.38,
- $CA(B)$ contains $\forall x \forall y (father(x,y) \equiv (x = \omega_1 \wedge y = Nimloth) \vee (x = \omega_2 \wedge y = Finrod) \vee (x = \omega_2 \wedge y = Galadriel))$,
- $DCA(B)$ contains $\forall x\, (x = \omega_1 \vee x = \omega_2 \vee x = Nimloth \vee x = Finrod \vee x = Galadriel)$,
- $UNA_\omega(B)$ contains $Nimloth \neq Finrod$, $Finrod \neq Galadriel$ and $Nimloth \neq Galadriel$.

Note that the theory

$$T_B^{CWA} = B \cup CWA(B) \cup DCA(B) \cup UNA_\omega(B)$$

does not satisfactorily represent the information content of B. Indeed, for Example 6.38, $CWA(B)$ contains, among others, $\neg father(Finrod, Nimloth)$. As B contains $father(\omega_1, Nimloth)$, so necessarily $T_B^{CWA} \vdash \omega_1 \neq Finrod$ which is too strong, as the equality between ω_1 and $Finrod$ should remain undetermined.

Consequently, T_B^{CA} is the only acceptable theory associated with a database B with marked null values. This theory admits several models. The following two theorems state that it is consistent but, in general, not complete.

Theorem 6.25. T_B^{CA} *is consistent.*
Proof. This is a trivial consequence of the monotonicity of first-order logic and of a corollary of Theorem 6.2 that states that the theory $B \cup CA(B) \cup DCA(B) \cup UNA(B)$, which is a superset of T_B^{CA}, is consistent. □

Theorem 6.26. T_B^{CA} *is (in general) not complete.*
Proof. If the database B contains incomplete information, then it is clear

that T_B^{CA} is not complete. □

This is illustrated by Example 6.38, where neither

$$T_B^{CA} \vdash father(Finrod, Nimloth)$$

nor

$$T_B^{CA} \vdash \neg father(Finrod, Nimloth)$$

holds.

Queries are expressed in the relational language of Subsection 6.2.4. In particular, as T_B^{CA} is not complete, answers to yes–no queries have the following meanings:

- $\|<|Q>\| = \{()\}$ means **yes**,
- $\|<|Q>\| = \{\}$ means that the answer is indefinite,
- $\|<|\neg Q>\| = \{()\}$ means **no** as the answer to the query $<|Q>$.

6.6.6 Query evaluation in the proof-theoretic approach

The extended relational theories T_B^{CA} are decidable (because there are no functions, and the sets of constants and predicates are finite). Therefore, there exist sound and complete query evaluation algorithms, but their computational complexity is equivalent to that of general theorem provers, with the usual unacceptable efficiency.

One way to regain some efficiency without losing soundness is to sacrifice completeness to some extent.

First, the initial query is rewritten, through a classical transformation (see § 1.2.7 in [Thayse 88]), as an equivalent formula where negations directly bear on atoms. Next, the algebraic correspondent of the resulting query is constructed as described in Subsection 6.2.6, except that negation is handled differently. Thanks to the preprocessing of negation, differences are limited to subqueries of the form $<x \mid \neg p(x)>$ where $p(x)$ is an atom. Such queries are called *primitive negative queries*. Their treatment is described as follows.

PRIMITIVE NEGATIVE QUERIES

1. Particular case of equality:

    (a) $|<|a \neq b>| = \{()\}$ if $a \neq b \in UNA_\omega$ or $b \neq a \in UNA_\omega$
    $\phantom{|<|a \neq b>|} = \{\}$ otherwise,

    (b) $|<x \mid x \neq x>| = \{\}$,

6.6. INCOMPLETE INFORMATION

(c) $|<x\ |\ x\neq a>|=|<x\ |\ a\neq x>|$
$= \{c\ |\ c\neq a\in UNA_\omega \text{ or } a\neq c\in UNA_\omega\}$,

(d) $|<x,y\ |\ x\neq y>|=|<x,y\ |\ y\neq x>|$
$= \{(a,b)\ |\ b\neq a\in UNA_\omega \text{ or } a\neq b\in UNA_\omega\}$.

2. In the general case, if p is a predicate different from equality, then

$$|<x\ |\ \neg p(r)>|=D_{x,r}(|<x\ |\ p(x)>|),$$

where $D_{x,r}$ is a new relational operator, called the *difference operator* (because it replaces the classical set-theoretic difference), defined as follows.

Given a tuple x of m distinct variables, given a tuple r of n (with $n\geq m$) constants and/or variables such that var(x)=var(r)[12], and given a set T of tuples of n constants, $D_{x,r}(T)$ is defined as

$$D_{x,r}(T)=\{a\ |\ UNA_\omega\vdash t\neq r[x/a],\ \text{for every } t \text{ in } T\}.$$

The set $D_{x,r}(T)$ contains the m-tuples a such that $r[x/a]$ *disagrees* with all n-tuples of T. Two tuples a and b disagree if $UNA_\omega\vdash a\neq b$.

This new operator is illustrated in the following example.

Example 6.49. Given the database

$$B=\{p(\omega_1,b,a),p(b,a,b)\},$$

we have

$$\begin{aligned}|<x,y\ |\ \neg p(x,y,a)>|&=D_{(x,y),(x,y,a)}(\{(\omega_1,b,a),(b,a,b)\})\\ &=\{(a,a),(b,a),(a,\omega_1),(\omega_1,a)\}.\end{aligned}$$

The tuple (a,ω_1) is a solution because (a,ω_1,a) disagrees with (ω_1,b,a) and (b,a,b). Otherwise, $T_B^{CA}\vdash(\omega_1=a)\wedge(\omega_1=b)$ would hold and also $T_B^{CA}\vdash a=b$, which is absurd. The tuple (b,ω_1) is not a solution because (b,ω_1,a) does not disagree with (ω_1,b,a) as $\omega_1=b$ is not inconsistent with T_B^{CA}.

The following results hold for marked null values [Reiter 86]:

1. The answer to a query Q is sound if it is computed from its algebraic correspondent $|Q|$ as described above with the new difference operator definition. In other words, $|Q|\subset \|Q\|$ holds in this context.

---

[12] var(w) denotes the set of variables in w.

2. This query evaluation technique is complete (i.e., $|Q| \supset \|Q\|$) for conjunctive queries (i.e., queries built with only the $\vee$ and $\neg$ connectives) or for positive queries (i.e., queries built with only the $\wedge$ and $\vee$ connectives along with quantifiers), among others.

The loss of completeness for unrestricted queries comes from the translation of disjunctions and existential quantifications into their algebraic correspondent.

DISJUNCTIONS

Instead of the equality of Subsection 6.2.6, only the following inclusion holds:

$$\begin{aligned} |<x \mid F_1(x_1) \vee F_2(x_2)>| &= |<x \mid F_1(x_1)>| \cup |<x \mid F_2(x_2)>| \\ &\subset \|<x \mid F_1(x_1) \vee F_2(x_2)>\|. \end{aligned}$$

This weakening entails a loss of completeness. This is illustrated in the next example.

Example 6.50. Consider the database

$$B = \{p(b,a), p(a,\omega_1), q(\omega_1,a), q(b,b)\}$$

and the query

$$Q = <x,y \mid p(x,y) \vee \neg q(x,y)>.$$

On the one hand,

$$|<x,y \mid p(x,y)>| = \{(b,a),(a,\omega_1)\},$$

$$|<x,y \mid \neg q(x,y)>| = \{(a,b)\},$$

and consequently

$$|Q| = \{(b,a),(a,\omega_1),(a,b)\}.$$

On the other hand, $(a,a) \in \|Q\|$. Indeed,

$$T_B^{CA} \cup \{\omega_1 = a\} \vdash p(a,a)$$

and

$$T_B^{CA} \cup \{\omega_1 \neq a\} \vdash \neg q(a,a),$$

whence, $T_B^{CA} \vdash p(a,a) \vee \neg q(a,a)$, and consequently $(a,a) \in \|Q\|$.

6.6. INCOMPLETE INFORMATION

EXISTENTIAL QUANTIFICATIONS

If x and y are the only variables of the formula $F(x,y)$ then:

$$\begin{aligned} |<x \mid \exists y\, F(x,y)>| &= \pi_x(|<x,y \mid F(x,y)>|) \\ &\subset \|<x \mid \exists y\, F(x,y)>\|. \end{aligned}$$

Here again, inclusion holds instead of equality. This is the second cause for the loss of completeness. It can be seen as a particular case of the previous one; indeed, the quantification $\exists y$ is equivalent to a disjunction on the values of the domain. This is illustrated in the following example.

Example 6.51. Consider the database

$$B = \{p(a, e_2), p(w_1, e_1), q(e_2, w_1)\}$$

and the query

$$Q = <x, y \mid \exists z\, (p(x,z) \wedge \neg q(z,y))>.$$

On the one hand,

$$\begin{aligned} (a,a) &\in \|Q\| \\ &\Leftrightarrow T_B^{CA} \vdash \exists z\,(p(a,z) \wedge \neg q(z,a)) \\ &\Leftrightarrow T_B^{CA} \vdash \quad p(a,a) \wedge \neg q(a,a) \\ &\qquad\qquad \vee\ p(a,e_1) \wedge \neg q(e_1,a) \\ &\qquad\qquad \vee\ p(a,e_2) \wedge \neg q(e_2,a) \\ &\qquad\qquad \vee\ p(a,w_1) \wedge \neg q(w_1,a) \\ &\Leftrightarrow T_B^{CA} \vdash p(a,e_1) \vee \neg q(e_2,a) \vee p(a,w_1). \end{aligned}$$

On the other hand,

$$T_B^{CA} \cup \{w_1 = a\} \vdash p(a, e_1)$$

and

$$T_B^{CA} \cup \{w_1 \neq a\} \vdash \neg q(e_2, a),$$

whence, $T_B^{CA} \vdash p(a, e_1) \vee \neg q(e_2, a) \vee p(a, w_1)$, and consequently $(a,a) \in \|Q\|$. However,

$$\begin{aligned} |Q| &= \pi_{12}(|<x,y,z \mid p(x,z) \wedge \neg q(z,y)>|) \\ &= \{(w_1, e_1), (w_1, e_2), (w_1, a), (w_1, w_1)\} \end{aligned}$$

and $|Q|$ does not contain the tuple (a, a).

6.6.7 Correspondence between approaches

The model- and proof-theoretic approaches to databases with marked null values (§ 6.6.3 and § 6.6.5) have been developed independently, respectively in [Imieliński and Lipski 84a] and in [Reiter 86]. They are basically equivalent as the answer to a query Q is, in general, the same in either approach.

But the approaches may lead to marginally different results. The difference operator is not redefined in [Imieliński and Lipski 84a]; this prevents the correct handling of negation unlike [Reiter 86]. The handling of equality in [Imieliński and Lipski 84a] in qualifications of selections allows the detection of tautologies unlike [Reiter 86]. It is relatively easy to convince oneself that the advantages of both methods can be compounded by computing the algebraic correspondent of a query expressed in the relational language with

- the evaluation of qualifications of selections described in Subsection 6.6.4,

- the new difference operator defined in Subsection 6.6.6.

The new query evaluation method thus obtained would be sound in both the model- and proof-theoretic approaches and would be complete for a broader class of queries.

6.6.8 Alternative method of query evaluation

A sound, complete, and moderately efficient query evaluation method can be defined in the following way [Imieliński 84].

As seen in the previous sections, sound, incomplete, and efficient algorithms are available. They compute an answer $\|Q\|_<$ to a query Q with the property $\|Q\|_< \subset \|Q\|$. If, in addition, a non-sound, complete, and efficient algorithm is defined to compute an answer $\|Q\|_>$ with the property $\|Q\|_> \supset \|Q\|$, then the answer $\|Q\|$ can be computed by

$$\|Q\| = \|Q\|_< \cup \mathit{filter}(\|Q\|_> \setminus \|Q\|_<),$$

where *filter* is a function equivalent to a theorem prover that detects which tuples of the set $\|Q\|_> \setminus \|Q\|_<$ are in the answer $\|Q\|$.

If the computation of both $\|Q\|_<$ and $\|Q\|_>$ can be done efficiently (by invoking as often as possible the operations of the relational algebra), the method suggested here is interesting as it limits the use of a costly algorithm to a hopefully small set of candidate tuples.

6.7 Conclusion

The model-theoretic approach, inherited from classical relational databases, associates with a database a set of formulas directly describing extensions of the relations, and it focuses on the models of this set of formulas. A formula is considered true in the model-theoretic approach if it evaluates to true in all the models.

The proof-theoretic approach represents a database as a logical theory expressing the extension of the database and it adds axioms necessary for producing negative information. A formula is considered true in the proof-theoretic approach if it is a theorem of the theory.

In the proof-theoretic approach, the theory exactly represents the available knowledge about the real world. The appeal of the proof-theoretic approach lies in its ability to enrich the expressive power of databases by enabling extensions to the basic theory. For that, the proof-theoretic approach has available all the modelling power of first-order logic. A priori, any collection of formulas represents a database.

It is easy to understand that this kind of extension is more difficult in the model-theoretic approach, as describing sets of models is often more complex than building a theory. This chapter has shown, however, that this advantage of the proof-theoretic approach is somewhat illusory.

On one hand, the richness of the proof-theoretic approach is counterbalanced by the complexity of the associated evaluation mechanisms. It suggests, in effect, that query evaluation can be performed as a theorem proving process with a general inference method like resolution. But this is of course not realistic with large sets of facts and rules. Realistic extensions of relational databases are those with query evaluation algorithms based on specialized inference mechanisms, more efficient than resolution, invoking operations similar to those of the relational algebra, available from database management systems.

On the other hand, all the realistic extensions, presented in this chapter and recently developed in the research world, can be described with the model-theoretic approach. This suggests that the approach is powerful enough to describe the interesting extensions, amenable to implementation with reasonable evaluation algorithms.

From the point of view of logic, classical relational databases are very simple and both approaches are clearly equivalent. There exists a least Herbrand model in the model-theoretic approach and it is precisely the single model of the proof-theoretic approach. The same correspondence holds for hierarchical deductive databases.

Definite deductive databases and, more generally, stratified databases can also be described in both approaches. The proof-theoretic approach however is faced with some difficulties. The theories that we have described are partly implicit, as they rely on an inference rule for determining the ground instances of relations that are false in the theory. It is still possible to formulate the theories as finite sets of formulas, but this is more delicate [Hulin 88]. This is no longer possible when functions are introduced in definite databases and in stratified databases. Such extensions cannot, in general, be described by a recursive set of axioms [Apt et al. 88].

The generalized closed-world assumption can also be described in both approaches (with some nuances, as we have seen). Here also, the theory is partly implicit.

Incomplete information follows the same rule, which also applies to an extension of definite deductive databases with marked null values, as in traditional relational databases [Hulin 88].

To summarize, the state of the art is that the model- and proof-theoretic approaches represent two different ways of exploring extensions to relational databases, but that neither has proved definitely more powerful than the other for reasonable extensions, that is, those extensions that can be implemented with an acceptable efficiency.

Bibliography

[Abadi 87] M. Abadi, Temporal-logic theorem proving, PhD thesis, Rep. STAN-CS-87-1151, Stanford University, 1987.

[Aho et al. 74] A. Aho, J. Hopcroft and J. Ullman, *The Design and Analysis of Computer Algorithms*, Addison-Wesley, Reading, 1974.

[Aggarwal et al. 87] S. Aggarwal, C. Courcoubetis and P. Wolper, *Adding Liveness Properties to Coupled Finite-State Machines*, Université de Liège, 1987.

[Ajdukiewicz 35] K. Ajdukiewicz, Syntactic connexion, 1935, in S. McCall (editor), *Polish Logic, 1920–1939*, Clarendon Press, Oxford, 1967.

[Alpern and Schneider 85] B. Alpern and F.D. Schneider, Defining liveness, *Information Processing Letters*, Vol. 21, pp. 181–185, 1985.

[Alpern and Schneider 87] B. Alpern and F.D. Schneider, Recognizing safety and liveness, *Distributed Computing*, Vol. 2, pp. 117–126, 1987.

[Appelt and Konolige 88] D. Appelt and K. Konolige, A practical nonmonotonic theory for reasoning about speech acts, *Proc. 26th Annual ACL Meeting*, New York, 1988.

[Apt et al. 88] K.R. Apt, H. Blair and A. Walker, Towards a theory of declarative knowledge, in J. Minker (editor), *Foundations of Deductive Databases and Logic Programming*, pp. 89–148, Morgan Kaufmann, Los Altos, 1988.

[Apt and van Emden 82] K. Apt and M. van Emden, Contributions to the theory of logic programming, *Journal of the ACM*, Vol. 29, no 3, pp. 841–862, 1982.

[Atlan 86] H. Atlan, *A Tort et à Raison, Intercritique de la Science et du Mythe*, Editions du Seuil, Paris, 1986.

[Baldwin 75] T. Baldwin, The philosophical significance of intensional logic, *Proc. of the Aristotelian Society*, Vol. XLIX, p. 51, 1975.

[Bancilhon 85] F. Bancilhon, *Naive Evaluation of Recursively Defined Relations*, Technical Report DB-004-85, MCC, 1985.

[Banieqbal and Barringer 86] B. Banieqbal and H. Barringer, *A Study of an Extended Temporal Language and a Temporal Fixed Point Calculus*, Technical Report UMCS-86-10-2, Department of Computer Science, University of Manchester, 1986.

[Barendregt 80] H. Barendregt, *The Lambda Calculus, its Syntax and Semantics*, North-Holland, Amsterdam, 1980.

[Barwise and Perry 83] J. Barwise and J. Perry, *Situations and Attitudes*, MIT Press, Cambridge, Mass., 1983.

[Beeri and Ramakrishnan 87] C. Beeri and R. Ramakrishnan, On the power of magic, in *Proceedings of the Sixth ACM Symposium on Principles of Database Systems*, pp. 269–283, 1987.

[Bennett 76] M. Bennett, A variation and extension of a Montague fragment of English, in B. Partee (editor), *Montague Grammar*, pp. 119–163, Academic Press, New York, 1976.

[Besnard et al. 87] P. Besnard, J. Houdebine and R. Rolland, *Circonscriptions: Cas de Complétude et d'Incomplétude*, IRISA, Université de Rennes I, Rapport Interne no 373, September 1987.

[Besnard and Siegel 88] P. Besnard and P. Siegel, The preferential-models approach to non-monotonic logics, in P. Smets et al. (editors), *Non-Standard Logics for Automated Reasoning*, pp. 137–161, Academic Press, London, 1988.

[Bidoit and Froidevaux 87] N. Bidoit and C. Froidevaux, Minimalism subsumes default logic and circumscription in stratified logic programming, *Proc. LICS-87*, pp. 89–97, 1987.

[Biskup 83] J. Biskup, A foundation of Codd's relational maybe-operations, *ACM Transactions on Database Systems*, Vol. 8, no 4, pp. 608–636, December 1983.

[Bossu and Siegel 85] G. Bossu and P. Siegel, Saturation, non monotonic reasoning and the closed world assumption, *Artificial Intelligence*, Vol. 25, no 1, pp. 13–65, 1985.

[Boyer et al. 86] R. Boyer, E. Lusk, W. McCune, R. Overbeek, M. Stickel and L. Wos, Set theory in first-order logic : clauses for Gödel's axioms, *J. Automated Reasoning*, Vol. 2, no 3, pp. 287–327, September 1986.

[Brown 87] F.M. Brown (editor), *The Frame Problem in Artificial Intelligence (Proc. of the 1987 Workshop)*, Morgan Kaufmann, Los Altos, 1987.

[Büchi 62] J. Büchi, On a decision method in restricted second order arithmetic, in *Proc. Internat. Congr. Logic, Method and Philos. Sci. 1960*, pp. 1–12, Stanford University Press, Stanford, 1962.

[Bundy 83] A. Bundy, *The Computer Modelling of Mathematical Reasoning*, Academic Press, New York, 1983.

[Carnap 36] R. Carnap, *The Logical Syntax of Language*, Routledge and Kegan Paul, London, 1936.

[Chang 80] C. L. Chang, On the evaluation of queries containing derived relations in a relational data base, in H. Gallaire and J. Minker (editors), *Advances in Data Base Theory*, Vol. 1, pp. 235–260, Plenum Press, New York, 1980.

[Chisholm et al. 87] P. Chisholm, G. Chen, D. Ferbrache, P. Thanisch and M.H. Williams, Coping with indefinite and negative data in deductive databases: A survey, *Data and Knowledge Engineering*, Vol. 2, pp. 259–284, 1987.

[Chomsky 55] N. Chomsky, *The Logical Structure of Linguistic Theory*, Mimeographed, M.I.T. Library, Cambridge, Mass., 1955.

[Chomsky 57] N. Chomsky, *Syntactic Structures*, Mouton, The Hague, 1957.

[Chomsky 65] N. Chomsky, *Aspects de la Théorie Syntaxique*, Le Seuil, Paris, 1965.

[Chou 87] S. Chou, A method for the mechanical derivation of formulas in elementary geometry, *J. Automated Reasoning*, Vol. 3, no 3, pp. 291–300, September 1987.

[Chou 88] S. Chou, An introduction to Wu's method for mechanical theorem proving in geometry, *J. Automated Reasoning*, Vol. 4, no 3, pp. 237–268, September 1988.

[Chou and Schelter 86] S. Chou and W. Schelter, Proving geometry theorems with rewrite rules, *J. Automated Reasoning*, Vol. 2, no 3, pp. 253–274, September 1986.

[Church 40] A. Church, A formulation of a simple theory of types, *Journal of Symbolic Logic*, Vol. 5, pp. 56–68, 1940.

[Church 41] A. Church, The calculi of lambda-conversion, *Annals of Mathematics Studies*, Vol. 6, Princeton, 1941.

[Clark 78] K. Clark, Negation as failure, in H. Gallaire and J. Minker (editors), *Logic and Data Bases*, pp. 293–322, Plenum Press, New York, 1978.

[Clarke et al. 86] E. Clarke, E. Emerson and A. Sistla, Automatic verification of finite-state concurrent systems using temporal logic specifications, *ACM Toplas*, Vol. 8, pp. 244–263, 1986.

[Clarke and Gabbay 88] M. Clarke and D. Gabbay, An intuitionistic basis for non-monotonic reasoning, in P. Smets et al. (editors), *Non-Standard Logics for Automated Reasoning*, pp. 163–178, Academic Press, London, 1988.

[Codd 70] E. F. Codd, A relational model for large shared data banks, *Communications of the Association for Computing Machinery*, Vol. 13, no 6, pp. 377–387, 1970.

[Codd 79] E. F. Codd, Extending the database relational model to capture more meaning, *ACM Transactions on Database Systems*, Vol. 4, no 4, pp. 397–434, December 1979.

[Coelho and Pereira 86] H. Coelho and L. Pereira, Automated reasoning in geometry theorem proving with Prolog, *J. Automated Reasoning*, Vol. 2, no 4, pp. 329–390, December 1986.

[Cooper and Parsons 76] R. Cooper and T. Parsons, Montague grammar, generative semantics and interpretative semantics, in B. Partee (editor), *Montague Grammar*, pp. 311–362, Academic Press, New York, 1976.

[Courtoy 87] C. Courtoy (editor), L'articulation du rationnel et du raisonnable dans les sciences, *Revue des Questions Scientifiques*, Vol. 158, no 1, Namur, Belgium, 1987.

[Cresswell 78] M. Cresswell, Semantic competence, in F. Guenthner and M. Guenthner-Reutter (editors), *Meaning and Translation (Philosophical and Linguistic Approaches)*, New York University Press, New York, 1978.

[Date 81] C. J. Date, *An Introduction to Database Systems*, Addison-Wesley, Reading, 1981.

[Davidson 69] D. Davidson, Truth and meaning, in Daws, Hockney and Wilson (editors), *Philosophical Logic*, Reidel, Dordrecht, 1969.

[Davio et al. 78] M. Davio, J.-P. Deschamps and A. Thayse, *Discrete and Switching Functions*, McGraw-Hill, New York, 1978.

[Davis 80] M. Davis, The mathematics of non-monotonic reasoning, *Artificial Intelligence*, Vol. 13, pp. 73–80, 1980.

[Demolombe 79] R. Demolombe, Semantic checking of questions expressed in predicate calculus language, in *Proceedings of the Fifth International Conference on Very Large Databases*, pp. 444–450, Rio de Janeiro, 1979.

[Demolombe and Royer 87] R. Demolombe and V. Royer, Evaluation strategies for recursive axioms: a uniform presentation, in *Actes des Journées FIRTECH Systèmes et Télématique: Bases de Données et Intelligence Artificielle*, pp. 227–248, Paris, April 1987.

[Dijkstra 81] E. Dijkstra, *An Assertional Proof of a Program by G.L. Peterson*, EWD 779, Burroughs Corp., 1981.

[Dowty 79] D. Dowty, *Word Meaning and Montague Grammar*, Reidel, Dordrecht, 1979.

[Dowty et al. 81] D. Dowty, R. Wall and S. Peters, *Introduction to Montague Semantics*, Reidel, Dordrecht, 1981.

[Emerson and Halpern 86] E. Emerson and J. Halpern, Sometimes and not never revisited: on branching versus linear time, *Journal of the ACM*, Vol. 33, pp. 151–178, 1986.

[Emerson and Lei 85] E. Emerson and C. Lei, Temporal model checking under generalized fairness constraints, *Proc. 18th Hawaii International Conference on System Sciences*, Hawaii, 1985.

[Emerson and Sistla 84] E. Emerson and A. Sistla, Deciding full branching time logic, *Information and Control*, Vol. 61, pp. 175–201, 1984.

[Etherington 87] D. Etherington, Formalizing nonmonotonic systems, *Artificial Intelligence*, Vol. 31, no 1, pp. 41–85, January 1987.

[Etherington 88] D. Etherington, *Reasoning with Incomplete Information*, Research Notes in Artificial Intelligence, Pitman, London, 1988.

[Etherington et al. 84] D. Etherington, R. Mercer and R. Reiter, On the adequacy of predicate circumscription for closed-world reasoning, *Proc. AAAI-Workshop on Non-Monotonic Reasoning*, New Paltz, New York, pp. 70–81, October 1984.

[Fitting and Ben-Jacob 88] M. Fitting and M. Ben-Jacob, Stratified and three-valued logic programming semantics, *Proc. Fifth Int. Conf. Symp. on Logic Programming*, pp. 1054–1069, 1988.

[Flores and Winograd 79] F. Flores and T. Winograd, *Understanding Cognition as Understanding*, Ablex, Norwood, 1979.

[Frege 52] G. Frege, Über Sinn und Bedeutung, *Zeitschrift für Philosophie und philosophische Kritik*, Vol. 100, pp. 25–50, 1893; translatred in : P. Geach and M. Black (editors), *Translation from the Philosophical Writings of Gottlob Frege*, pp. 56–78, Basil Blackwell, Oxford, 1952.

[Froidevaux 86] C. Froidevaux, Taxonomic default theory, *Proc. ECAI-86*, pp. 123–129, Brighton, July 1986.

[Gabbay et al. 80] D. Gabbay, A. Pnueli, S. Shelah and J. Stavi, The temporal analysis of fairness, in *Seventh ACM Symposium on Principles of Programming Languages*, pp. 163–173, Las Vegas, 1980.

[Gabbay and Guenther 83] D. Gabbay and F. Guenther (editors), *Handbook of Philosophical Logic*; Vol. 1: *Elements of Classical Logic*, Reidel, Dordrecht, 1983.

[Gabbay and Guenther 84] D. Gabbay and F. Guenther (editors), *Handbook of Philosophical Logic*; Vol. 2: *Extensions of Classical Logic*, Reidel, Dordrecht, 1984.

[Gabbay and Guenther 86] D. Gabbay and F. Guenther (editors), *Handbook of Philosophical Logic*; Vol. 3: *Alternatives to Classical Logic*, Reidel, Dordrecht, 1986.

[Gabbay and Guenther 88] D. Gabbay and F. Guenther (editors), *Handbook of Philosophical Logic*; Vol. 4: *Topics in the Philosophy of Language*, D. Reidel, Dordrecht, to appear.

[Gallaire et al. 84] H. Gallaire, J. Minker and J.-M. Nicolas, Logic and databases : a deductive approach, *ACM Computing Surveys*, Vol. 16, no 2, pp. 153–185, June 1984.

[Gardarin and Simon 87] G. Gardarin and E. Simon, Les systèmes de gestion de bases de données déductives, *Technique et Science Informatiques*, Vol. 6, no 5, pp. 347–382, 1987.

[Gazdar et al. 85] G. Gazdar, E. Klein, G. Pullum and I. Sag, *Generalized Phrase Structure Grammar*, Basil Blackwell, Oxford, 1985.

[Gelfond 87] M. Gelfond, On stratified autoepistemic theories, *Proc. AAAI-87*, pp. 207–211, 1987.

[Gelfond and Lifschitz 88a] M. Gelfond and V. Lifschitz, The stable model semantics for logic programming, *Proc. Fifth Int. Conf. Symp. on Logic Programming*, pp. 1070–1080, 1988.

[Gelfond and Lifschitz 88b] M. Gelfond and V. Lifschitz, Compiling circumscriptive theories into logic programs, *Proc. AAAI-88*, Morgan Kaufmann, pp. 455–459, 1988.

[Gelfond and Przymusinska 86] M. Gelfond and H. Przymusinska, Negation as failure: careful closure procedure, *Artificial Intelligence*, Vol. 30, no 3, pp. 273–287, 1986.

[Genesereth and Nilsson 87] M. Genesereth and N. Nilsson, *Logical Foundations of Artificial Intelligence*, Morgan Kaufmann, Los Altos, 1987.

[Georgeff and Lansky 86] M. Georgeff and A. Lansky (editors), *Reasoning about Actions and Plans (Proc. of the 1986 Workshop)*, Morgan Kaufmann, Los Altos, 1986.

[Ginsberg 87] M. Ginsberg (editor), *Readings in Nonmonotonic Reasoning*, Morgan Kaufmann, Los Altos, 1987.

[Gochet 80] P. Gochet, Théorie des modèles et compétence pragmatique, in H. Parrek et al (editors), *Le Langage en contexte*, pp. 319–388, Benjamins, Amsterdam, 1980.

[Gochet 82a] P. Gochet, La Sémantique récursive de Davidson et de Montague, in M. Loi (editor), *Penser les Mathématiques*, Le Seuil, Paris, 1982.

[Gochet 82b] P. Gochet, L'originalité de la sémantique de Montague, *Etudes Philosophiques*, Vol. 2, pp. 149–175, 1982.

[Gochet 83] P. Gochet, La sémantique des situations, in F. Nef (editor), *Histoire, Epistémologie, Langage*, Vol. 5, Part 2 : *La Sémantique Logique*, pp. 195–212, 1983.

[Gochet 86] P. Gochet, *Ascent to Truth*, Philosophia Verlag, Berlin, 1986.

[Gödel 40] K. Gödel, *The Consistency of the Axiom of Choice and of the Generalized Continuum-Hypothesis with the Axioms of Set Theory*, Princeton University Press, Princeton, 1940.

[Goldblatt 87] R. Goldblatt, *Logics of Time and Computation*, Lecture Notes no 7, Center for the Study of Language and Information, 1987.

[Grant 77] J. Grant, Null values in a relational data base, *Information Processing Letters*, Vol. 6, no 5, pp. 156–157, October 1977.

[Grant 79] J. Grant, Partial values in a tabular database model, *Information Processing Letters*, Vol. 9, no 2, pp. 97–99, August 1979.

[Grant and Minker 86] J. Grant and J. Minker, Answering queries in indefinite databases and the null value problem, in P. Kanellakis (editor), *Advances in Computing Research*, Vol. 3, pp. 247–267, 1986.

[Grégoire 86] E. Grégoire, Raisonnement plausible : inférence non monotone et logiques autoépistémiques, *Proc. Int. Conf. on Information Processing and Management of Uncertainty in Knowledge-Based Systems*, pp. 376–379, Paris, July 1986.

[Grégoire 88a] E. Grégoire, A note on Moore's autoepistemic logic, in P. Smets et al. (editors), *Non-Standard Logics for Automated Reasoning*, pp. 132–133, Academic Press, London, 1988.

[Grégoire 88b] E. Grégoire, *Non-monotonicity and Rationality*, Unité d'Informatique, Université Catholique de Louvain, Louvain-la-Neuve, Belgium, 1988.

[Grégoire 89] E. Grégoire, About the relationship netween a skeptical theory of inheritance in semantic networks and general nonmonotonic logics, *Proc. of the Workshop on Formal Aspects of Semantic Networks*, Santa Catalina Island, February 1989.

[Gribomont 85] E. Gribomont, Synthesis of parallel programs invariants, Lecture Notes in Computer Sciences, Vol. 186, pp. 325–338, Springer-Verlag, Berlin, 1985.

[Guenthner and Guenthner-Reutter 78] F. Guenthner and M. Guenthner-Reutter, *Meaning and Translation (Philosophical and Linguistic Approaches)*, New York University Press, New York, 1978.

[Hanks and McDermott 86] S. Hanks and D. McDermott, Default reasoning, nonmonotonic logics, and the frame problem, *Proc. AAAI-86*, pp. 328–333, 1986.

[Harel 79] D. Harel, *First-order Dynamic Logic*, Lecture Notes in Computer Science, Vol. 68, Springer-Verlag, Berlin, 1979.

[Heeffer 86] A. Heeffer (editor), Non-classical logics for expert systems, *CC-AI*, Vol. 3, nos 1–2, 1986.

[Henschen and Naqvi 84] L. Henschen and S. Naqvi, On compiling queries in recursive first order databases, *Journal of the ACM*, Vol. 31, no 1, pp. 137–147, 1984.

[Henschen and Park 86] L. Henschen and H. Park, Indefinite and GCWA inference in indefinite deductive databases, *Proc. AAAI-86*, pp. 191–197, 1986.

[Henschen and Park 88] L. Henschen and H. Park, Compiling the GCWA in indefinite deductive databases, in J. Minker (editor), *Foundations of Deductive Databases and Logic Programming*, Morgan Kaufmann, Los Altos, California, 1988.

[Hilbert and Ackermann 28] D. Hilbert and W. Ackermann, *Grundzüge der theoretischen Logik*, Springer, Berlin, 1928; translated into English : *Principles of Mathematical Logic*, Chelsea Pub. Co., New York, 1950.

[Hintikka 62] J. Hintikka, *Knowledge and Belief: an Introduction to the Logic of the two Notions*, Cornell University Press, Ithaca, 1962.

[Hintikka 88] J. Hintikka, Model minimization – an alternative to circumscription, *J. Automated Reasoning*, Vol. 4, no 1, pp. 1–13, 1988.

[Hulin 88] G. Hulin, *A proof-theoretic perspective of deductive databases with marked null values*, Manuscript M273, Philips Research Laboratory Brussels, 1988.

[Hulin 89] G. Hulin, Parallel processing of Recursive Queries in Distributed Architectures, in *Proceedings of the Fifteenth International Conference on Very Large Databases*, Amsterdam, 1989.

[Imieliński 84] T. Imieliński, On algebraic query processing in logical databases, in H. Gallaire, J. Minker and J. Nicolas (editors), *Advances in Data Base Theory*, pp. 285–318, Plenum Press, New York, 1984.

[Imieliński 86] T. Imieliński, Automated deduction in databases with incomplete information, in J. Minker (editor), *Preprints of Workshop: Foundations of Deductive Databases and Logic Programming*, pp. 242–283, Washington, August 1986, submitted to ACM TODS.

[Imieliński and Lipski 81] T. Imieliński and W. Lipski, Jr., *The Relational Model of Data and Cylindric Algebras*, ICS PAS Reports 446, Institute of Computer Science, Polish Academy of Sciences, August 1981.

[Imieliński and Lipski 84a] T. Imieliński and W. Lipski, Jr., Incomplete information in relational databases, *Journal of the ACM*, Vol. 31, no 4, pp. 761–791, October 1984.

[Imieliński and Lipski 84b] T. Imieliński and W. Lipski, Jr., The relational model of data and cylindric algebras, *Journal of Computer and System Sciences*, Vol. 28, pp. 80–102, 1984.

[Israel 80] D. Israel, What's wrong with non-monotonic logic?, *Proc. AAAI-80*, pp. 99–101, 1980.

[Jaeger 86] G. Jaeger, Some contributions to the logical analysis of circumscription, *Proc. 8th Int. Conf. on Automated Deduction*, Lecture Notes in Computer Science, Vol. 230, pp. 154–171, Springer-Verlag, Berlin, 1986.

[Janssen 86a] T. Janssen, *Foundations and Applications of Montague Grammar, Part I : Philosophy, Framework, Computer Science*, Centrum voor Wiskunde en Informatica, Tracte 19, Amsterdam, 1986.

[Janssen 86b] T. Janssen, *Foundations and Applications of Montague Grammar, Part II : Applications to Natural Language*, Centrum voor Wiskunde en Informatica, Tracte 28, Amsterdam, 1986.

[Jarke and Koch 84] M. Jarke and J. Koch, Query optimization in database systems, *ACM Computing Surveys*, Vol. 16, no 2, pp. 111–152, June 1984.

[Jayez 88] J. Jayez, *L'inférence en langue naturelle*, Hermes, Paris, 1988.

[John and Bennett 82] J. John and M. Bennett, Language as a self organizing system, *Cybernetics and Systems*, Vol. 2, pp. 201–212, 1982.

[Kamp 84] J. Kamp, A theory of truth and semantic representation, in Groenendijk et al. (editors), *Truth, Interpretation and Information*, Foris Publications, Dordrecht, 1984.

[Katz 72] J. Katz, *Semantic Theory*, Harper and Row, New York, 1972.

[Katz and Fodor 64] J. Katz and J. Fodor, The structure of a semantic theory, reprinted in : *The Structure of Language*, Prentice-Hall, Englewood Cliffs, 1964.

[Kautz 86] H. Kautz, The logic of persistence, *Proc. AAAI-86*, pp. 401–405, 1986.

[Keenan and Faltz 85] E. Keenan and L. Faltz, *Boolean Semantics for Natural Language*, Reidel, Dordrecht, 1985.

[Keller 86] A. M. Keller, Set-theoretic problems of null completion in relational databases, *Information Processing Letters*, Vol. 22, pp. 261–265, April 1986.

[Kemeny 57] J. Kemeny, Semantics as a branch of logic, *Encyclopaedia Britannica*, Vol. 20, p. 311, 1957.

[Kim et al. 85] W. Kim, D. Reiner and D. Batory (editors), *Query Processing in Database Systems*, Springer-Verlag, Berlin, 1985.

[Konolige 84] K. Konolige, A deduction model of belief and its logics, Stanford Comp. Science. Memo CS-84-1022, Stanford University, 1984.

[Konolige 86] K. Konolige, *A Deduction Model of Belief*, Research Notes in Artificial Intelligence, Pitman, London, 1986.

[Konolige 87] K. Konolige, On the relation between default theories and autoepistemic logic, *Proc. IJCAI-87*, pp. 394–401, 1987.

[Kowalski and Sergot 86] R. Kowalski and M. Sergot, A logic-based calculus of events, *New Generation Computing*, Vol. 4, pp. 67–95, 1986.

[Kripke 63] S. Kripke, Semantic analysis of modal logic I : Normal propositional calculi, *Zeit. Math. Logik Grund. Math.*, Vol. 9, pp. 67–96, 1963.

[Kripke 71] S. Kripke, Semantical considerations on modal logic, in L. Linsky (editor), *Reference and Modality*, pp. 63–72, Oxford University Press, London, 1971.

[Kröger 87] F. Kroger, *Temporal Logic of Programs*, Springer-Verlag, Berlin, 1987.

[Kunen 87] K. Kunen, Negation in logic programming, *J. Logic Programming*, Vol. 4, no 4, pp. 289–308, 1987.

[Lacroix and Pirotte 77] M. Lacroix and A. Pirotte, Domain-oriented relational languages, in *Proceedings of the Third International Conference on Very Large Databases*, pp. 370–378, Tokyo, October 1977.

[Lacroix and Pirotte 81] M. Lacroix and A. Pirotte, Associating types with domains of relational databases, *ACM SIGMOD Record*, Vol. 11, no 2, pp. 144–146, February 1981.

[Lakoff 72] G. Lakoff, Linguistics and natural logic, in D. Davidson and G. Hartman (editors), *Semantics for Natural Language*, Reidel, Dordrecht, 1972.

[Lepore 87] E. Lepore (editor), *New Directions in Semantics*, Academic Press, London, 1987.

[Levesque 81] H. Levesque, The interaction with incomplete knowledge bases : a formal treatment, *Proc. IJCAI-81*, pp. 240–245, 1981.

[Levesque 86] H. Levesque, Knowledge representation and reasoning, *Ann. Rev. Comput. Sci.*, Vol. 1, pp. 255–287, 1987.

[Levesque 87] H. Levesque (guest editor), Forum: a critique of pure reason, *Computational Intelligence*, Vol. 3, no 3, pp. 149–237, 1987.

[Lichtenstein and Pnueli 85] O. Lichtenstein and A. Pnueli, Checking that finite state concurrent programs satisfy their linear specifications, *Proc. 12th ACM Symposium on Principles of Programming Languages*, pp. 97–107, New Orleans, January 1985.

[Lifschitz 85a] V. Lifschitz, Computing circumscription, *Proc. IJCAI-85*, pp. 121–127, 1985.

[Lifschitz 85b] V. Lifschitz, Closed-world databases and circumscription, *Artificial Intelligence*, Vol. 27, pp. 229–235, 1986.

[Lifschitz 86a] V. Lifschitz, On the satisfiability of circumscription, *Artificial Intelligence*, Vol. 28, pp. 17–27, 1986.

[Lifschitz 86b] V. Lifschitz, Pointwise circumscription: preliminary report, *Proc. AAAI-86*, pp. 406–410, 1986.

[Lifschitz 87a] V. Lifschitz, Formal theories of action, in F.M. Brown (editor), *The Frame Problem in Artificial Intelligence (Proc. of the 1987 Workshop)*, pp. 35–57, Morgan Kaufmann, Los Altos, 1987.

[Lifschitz 87b] V. Lifschitz, Circumscriptive theories: a logic-based framework for knowledge representation, *Proc. AAAI-87*, pp. 364–368, 1987.

[Lifschitz 88] V. Lifschitz, On the declarative semantics of logic programs with negation, in J. Minker (editor), *Foundations of Deductive Databases and Logic Programming*, pp. 177–192, Morgan Kaufmann, Los Altos, 1988.

[Lloyd 87] J. Lloyd, *Foundations of Logic Programming*, Springer-Verlag, Berlin, second, extended edition, 1987.

[Loui 87] R. Loui, Defeat among arguments: a system of defeasible inference, *Computational Intelligence*, Vol. 3, no 2, pp. 100–106, 1987.

[Louis and Pirotte 82] G. Louis and A. Pirotte, A denotational definition of the semantics of DRC, the domain relational calculus, in *Proceedings of the Eighth International Conference on Very Large Databases*, pp. 348–356, Mexico, September 1982.

[Lukaszewicz 84] W. Lukaszewicz, Considerations on default logic, *Proc. AAAI-Workshop on Non-Monotonic Reasoning*, New Paltz, New York, pp. 165–193, 1984.

[McCarthy 80] J. McCarthy, Circumscription: a form of non-monotonic reasoning, *Artificial Intelligence*, Vol. 13, nos 1–2, pp. 27–39, 1980.

[McCarthy 86] J. McCarthy, Applications of circumscription to formalizing common-sense knowledge, *Artificial Intelligence*, Vol. 28, pp. 89–116, 1986.

[McCarthy et al. 78] J. McCarthy, M. Sato, T. Hayashi and S. Igarashi, On the model theory of knowledge, Stanford Art. Int. Lab. Memo AIM-312, Stanford University, 1978.

[McCarthy and Hayes 69] J. McCarthy and P. Hayes, Some philosophical problems from the standpoint of artificial intelligence, in Meltzer and Richie (editors), *Machine Intelligence*, Vol. 4, Edinburgh University Press, Edinburgh, 1969.

[McDermott 82] D. McDermott, Non-monotonic logic II: non-monotonic modal theories, *Journal of the ACM*, Vol. 29, no 1, pp. 34–57, 1982.

[McDermott 87a] D. McDermott, A critique of pure reason, *Computational Intelligence*, Vol. 3, no 3, pp. 151–160, 1987.

[McDermott 87b] D. McDermott, Logic, problem-solving and deduction, *Ann. Rev. Comput. Sci.*, Vol. 2, pp. 187–229, 1987.

[McDermott and Doyle 80] D. McDermott and J. Doyle, Non-monotonic logic I, *Artificial Intelligence*, Vol. 13, no 1–2, pp. 41–72, 1980.

[Maier 83] D. Maier, *The Theory of Relational Databases*, Pitman, London, 1983.

[Manna and Pnueli 79] Z. Manna and A. Pnueli, The modal logic of programs, Lecture Notes in Computer Sciences, Vol. 71, pp. 385–409, Springer-Verlag, Berlin, 1979.

[Manna and Pnueli 84] Z. Manna and A. Pnueli, Adequate proof principles for invariance and liveness properties of concurrent programs, *Science of Computer Programming*, Vol. 4, pp. 257–289, 1984.

[Mendelson 79] E. Mendelson, *Introduction to Mathematical Logic*, Van Nostrand, Princeton, second edition, 1979.

[Miller and Perlis 87] M. Miller and D. Perlis, Proving self-utterances, *J. Automated Reasoning*, Vol. 3, no 3, pp. 329–338, September 1987.

[Minker 82] J. Minker, On indefinite databases and the closed world assumption, *Proc. 6th Conf. on Automated Deduction*, Lecture Notes in Computer Science , Vol. 138, pp. 292–308, Springer-Verlag, Berlin, 1982.

[Minker 88a] J. Minker, Perspectives in deductive databases, *J. Logic Programming*, Vol. 5, no 1, pp. 33–60, 1988.

[Minker 88b] J. Minker (editor), *Foundations of Deductive Databases and Logic Programming*, Morgan Kaufmann, Los Altos, 1988.

[Minsky 74] M. Minsky, A framework for representing knowledge, in P. Winston (editor), *The Psychology of Computer Vision*, pp. 34–57, McGraw-Hill, New York, 1974.

[Moinard 87] Y. Moinard, Donner la préférence au défaut le plus spécifique, *Proc. Congrès RFIA*, pp. 1123–1132, Antibes, 1987.

[Moinard 88] Y. Moinard, Raisonnement non monotone : Contribution à l'étude de la circonscription, Thèse d'Université, Université de Rennes I, 1988.

[Montague 73] R. Montague, The proper treatment of quantification in ordinary English, in J. Hintikka et al. (editors), *Approaches to Natural Language*, pp. 221–242, Reidel, Dordrecht, 1973.

[Montague 74a] R. Montague, On the nature of certain philosophical entities, *The Monist*, Vol. 53, pp. 159–194, 1960; reprinted in R. Thomason (editor), *Formal Philosophy : Selected Papers of Richard Montague*, pp. 148–187, Yale University Press, New Haven, 1974.

[Montague 74b] R. Montague, Pragmatics, in R. Klibansky (editor), *Contemporary Philosophy : A Survey*, pp. 102–122, La Nuova Italia Editrice, 1968; reprinted in R. Thomason (editor), *Formal Philosophy : Selected Papers of Richard Montague*, pp. 95–118, Yale University Press, New Haven, 1974.

[Montague 74c] R. Montague, English as a formal language, in B. Visentini et al. (editors), *Linguaggi nella e nella Tecnica*, pp. 189–224, Edizioni di Commutà, 1970; reprinted in R. Thomason (editor), *Formal Philosophy : Selected Papers of Richard Montague*, pp. 108–221, Yale University Press, New Haven, 1974.

[Montague 74d] R. Montague, Universal grammars, *Theoria*, Vol. 36, pp. 373–398, reprinted in R. Thomason (editor), *Formal Philosophy : Selected Papers of Richard Montague*, pp. 222–246, Yale University Press, New Haven, Conn., 1974.

[Moore 86] J. Moore, Editor's preface to 'Basic principles of mechanical theorem proving in elementary geometrics' by Wu Wen-tsun, *J. Automated Reasoning*, Vol. 2, no 3, pp. 219–220, September 1986.

[Moore 80] R. C. Moore, Semantical considerations on nonmonotonic logic, *SRI Artificial Intelligence Center Technical Note 284*, SRI International, Menlo Park, 1980.

[Moore 85] R. C. Moore, Semantical considerations on non-monotonic logic, *Artificial Intelligence*, Vol. 25, no 1, pp. 75–94, 1985.

[Moore 88] R.C. Moore, Autoepistemic logic, in P. Smets et al. (editors), *Non-Standard Logics for Automated Reasoning*, pp. 105–136, Academic Press, London, 1988.

[Mott 87] P. Mott, A theorem on the consistency of circumscription, *Artificial Intelligence*, Vol. 31, no 1, pp. 87–98, 1987.

[Naish 85] L. Naish, *Negation and Control in Prolog*, Lecture Notes in Computer Science, Vol. 238, Springer-Verlag, Berlin, 1985.

[Nef 84] F. Nef (editor), *L'analyse Logique des Langues Naturelles*, Editions du Centre National de la Recherche Scientifique, Paris, 1984.

[Owicki and Gries 76] S. Owicki and D. Gries, An axiomatic proof technique for parallel programs, *Acta Informatica*, Vol. 6, pp. 319–340, 1976.

[Papalaskari and Weinstein 87] M. Papalaskari and S. Weinstein, *Minimal Consequence in Sentential Logic*, Technical Report MS-CIS-97-43, University of Pennsylvania, 1987.

[Partee 75] B. Partee, Montague grammar and transformational grammar, *Linguistic Inquiry*, Vol. 6, no 2, pp. 203–300, Spring 1975.

[Perlis and Minker 86] D. Perlis and J. Minker, Completeness results for circumscription, *Artificial Intelligence*, Vol. 28, pp. 29–42, 1986.

[Perrault 87] C. Perrault, *An Application of Default Logic to Speech Act Theory*, SRI International, Technical Report, Menlo Park, 1987.

[Peterson 81] G. Peterson, Myths about the mutual exclusion problem, *Information Processing Letter*, Vol. 12, pp. 115–116, 1981.

[Pirotte 78] A. Pirotte, High level data base query languages, in H. Gallaire and J. Minker (editors), *Logic and Databases*, pp. 409–436, Plenum Press, New York, 1978.

[Pirotte 82] A. Pirotte, A precise definition of basic relational notions and of the relational algebra, *ACM SIGMOD Record*, Vol. 13, no 1, pp. 30–45, September 1982.

[Poole 85] D. Poole, On the comparison of theories: preferring the most specific explanation, *Proc. IJCAI-85*, pp. 144–147, 1985.

[Pratt 76] V. Pratt, Semantical considerations on Floyd-Hoare logic, *Proc. 17th IEEE Symp. on Foundations of Computer Science*, pp. 109–121, 1976.

[Przymusinski 86] T. Przymusinski, Query answering in circumscriptive and closed-world theories, *Proc. AAAI-86*, pp. 186–190, 1986.

[Przymusinski 88a] T. Przymusinski, On the semantics of stratified deductive databases, in J. Minker (editor), *Foundations of Deductive Databases and Logic Programming*, pp. 193–216, Morgan Kaufmann, Los Altos, 1988.

[Przymusinski 88b] T. Przymusinski, Perfect model semantics, *Proc. Fifth Int. Conf. Symp. on Logic Programming*, pp. 1081–1096, 1988.

[Przymusinski 88c] T. Przymusinski, On the relationship between logic programming and non-monotonic reasoning, *Proc. AAAI-88*, pp. 444–448, 1988.

[Przymusinski and Przymusinska 88] T. Przymusinski and C. Przymusinska, Weakly perfect model semantics for logic programs, *Proc. Fifth Int. Conf. Symp. on Logic Programming*, pp. 1106–1120, 1988.

[Quine 60] W. Quine, *Word and Object*, MIT Press, Cambridge, Mass., 1960.

[Quine 72a] W. Quine, *Logique élémentaire*, Colin, Paris, 1972.

[Quine 72b] W. Quine, Methodological reflexions on current linguistic theory, in D. Davidson and G. Harman (editors), *Semantics of Natural Language*, Reidel, Dordrecht, 1972.

[Reiter 78] R. Reiter, On closed-world data base, in H. Gallaire and J. Minker (editors), *Logic and Data Bases*, pp. 55–76, Plenum Press, New York, 1978.

[Reiter 80] R. Reiter, A logic for default reasoning, *Artificial Intelligence*, Vol. 13, nos 1–2, pp. 81–131, 1980.

[Reiter 84a] R. Reiter, Circumscription implies predicate completion (sometimes), *Proc. AAAI-Workshop on Non-Monotonic Reasoning*, New Paltz, New York, pp. 418–420, October 1984.

[Reiter 84b] R. Reiter, Towards a logical reconstruction of relational database theory, in M. Brodie, J. Mylopoulos and J. Schmidt (editors), *On Conceptual Modelling*, pp. 191–238, Springer-Verlag, Berlin, 1984.

[Reiter 86] R. Reiter, A sound and sometimes complete query evaluation algorithm for relational databases with null values, *Journal of the ACM*, Vol. 33, no 2, pp. 349–370, April 1986.

[Reiter 87a] R. Reiter, A theory of diagnosis from first principles, *Artificial Intelligence*, Vol. 32, no 1, pp. 57–95, 1987.

[Reiter 87b] R. Reiter, Nonmonotonic reasoning, *Ann. Rev. Comput. Sci.*, Vol. 2, pp. 147–186, 1987.

[Reiter and Criscuolo 81] R. Reiter and G. Criscuolo, On interacting defaults, *Proc. IJCAI-81*, pp. 270–276, 1981.

[Rescher 69] N. Rescher, *Many-Valued Logic*, McGraw-Hill, New York, 1969.

[Rescher and Urquhart 71] N. Rescher and A. Urquhart, *Temporal Logic*, Springer-Verlag, Berlin, 1971.

[Riche 89] J. Riche, Logique et théorie de la signification, in *Encyclopédie Philosophique Universelle*, Presses Universitaires de France, Paris, 1989.

[Roelants 87] D. Roelants, Recursive rules in logic databases, Manuscript, Philips Research Laboratory Brussels, March 1987.

[Rohmer et al. 86] J. Rohmer, R. Lescoeur and J. Kerisit, The Alexander method – a technique for the processing of recursive axioms in deductive databases, *New Generation Computing*, Vol. 4, no 3, pp. 273–285, 1986.

[Russell 05] B. Russell, On denoting, *Mind*, Vol. 14, pp. 479–493, 1905; reprinted in I. Coppi and J. Gould (editors), *Contemporary Readings in Logical Theory*, pp. 93–105, Macmillan, New York, 1967.

[Saccà and Zaniolo 86] D. Saccà and C. Zaniolo, The generalized counting method for recursive logic queries, in *Proceedings of the International Conference on Database Theory*, Rome, September 1986.

[Schilpp 44] P. Schilpp, *The Philosophy of Bertrand Russell*, Open Court, La Salle, 1944.

[Shepherdson 84] J. Shepherdson, Negation as failure: a comparison of Clark's completed data base and Reiter's closed world assumption, *J. Logic Programming*, Vol. 1, no 1, pp. 51–79, 1984.

[Shepherdson 85] J. Shepherdson, Negation as failure II, *J. Logic Programming*, Vol. 2, no 3, pp. 185–202, 1985.

BIBLIOGRAPHY

[Shepherdson 88a] J. Shepherdson, Introduction to the theory of logic programming, in F. Drake and J. Truss (editors), *Logic Colloquium '86*, Studies in Logic and the Foundations of Mathematics, Vol. 124, pp. 277–318, North-Holland, Amsterdam, 1988.

[Shepherdson 88b] J. Shepherdson, Negation in logic programming, in J. Minker (editor), *Foundations of Deductive Databases and Logic Programming*, pp. 19–88, Morgan Kaufmann, Los Altos, 1988.

[Shoham 86] Y. Shoham, Chronological ignorance, *Proc. AAAI-86*, pp. 389–393, 1986.

[Shoham 87] Y. Shoham, A semantical approach to nonmonotonic logics, *Proc. 2nd Int. Symposium on Logic in Computer Science (LICS 87)*, pp. 275–279, 1987.

[Shoham 88] Y. Shoham, *Reasoning about Change*, MIT Press, Cambridge, Mass., 1988.

[Sistla and Clarke 85] A. Sistla and E. Clarke, Complexity of propositional linear temporal logics, *Journal of the ACM*, Vol. 32, pp. 733–749, 1985.

[Sistla et al. 87] A. Sistla, M. Vardi and P. Wolper, The complementation problem for Büchi automata with applications to temporal logic, *Theoretical Computer Science*, Vol. 49, pp. 217–237, 1987.

[Smets et al. 88] P. Smets, E.H. Mamdani, D. Dubois and H. Prade (editors), *Non-Standard Logics for Automated Reasoning*, Academic Press, London, 1988.

[Smullyan 78] R. Smullyan, *What is the Name of this Book ?*, Prentice-Hall, Englewood Cliffs, 1978.

[Suppes 51] J. Suppes, A set of independent axioms for extensive quantities, *Portugaliae Mathematica*, Vol. 10, pp. 163–172, 1951.

[Tarski 44] A. Tarski, The semantic conception of truth, *Philosophy and Phenomenological Research*, Vol. 4, pp. 341–375, 1944; reprinted in L. Linsky (editor), *Semantics and the Philosophy of Language*, pp. 13–47, University of Illinois Press, Urbana, 1972.

[Tarski 72] A. Tarski, *Logique, Sémantique, Métamathématique (1923-1944)*, Colin, Paris, 1972.

[Thayse 84] A. Thayse, *P-functions and Boolean Matrix Factorization*, Lecture Notes in Computer Science, Vol. 175, Springer-Verlag, Berlin, 1984.

[Thayse 88] A. Thayse (editor), *From Standard Logic to Logic Programming, Introducing a Logic Based Approach to Artificial Intelligence*, Wiley, Chichester, 1988.

[Thistlewaite et al. 1988] P. Thistlewaite, M. McRobbie and R. Meyer, *Automated Theorem-Proving in Non-Classical Logics*, Research Notes in Theoretical Computer Science, Pitman, London, 1988.

[Thomas 81] W. Thomas, A combinatorial approach to the theory of ω-automata, *Information and Control*, Vol. 48, pp. 261–283, 1981.

[Tolkien 77] J. R. R. Tolkien, *The Silmarillion*, Unwin Paperback, 1977.

[Touretzky 84] D. Touretzky, Implicit ordering of defaults in inheritance systems, *Proc. AAAI-84*, pp. 322–325, 1984.

[Touretzky 86] D. Touretzky, *The Mathematics of Inheritance Systems*, Research Notes in Artificial Intelligence, Pitman, London, 1986.

[Turing 36] A. Turing, On computable numbers with an application to the Entscheidungsproblem, *Proc. London Math. Soc.*, Ser. 2-42, pp. 230–265, 1936.

[Turner 84] R. Turner, *Logics for Artificial Intelligence*, Ellis Horwood, Chichester, 1984.

[Ullman 82] J. Ullman, *Principles of Database Systems*, Pitman, London, second edition, 1982.

[Ullman 86] J. Ullman, Implementation of logical query languages for databases, *ACM Transactions on Database Systems*, Vol. 10, no 3, pp. 289–321, 1986.

[Van Emden and Kowalski 76] M. H. Van Emden and R. Kowalski, The semantics of predicate logic as a programming language, *Journal of the ACM*, Vol. 23, no 4, pp. 733–742, October 1976.

[van Gelder 88] A. van Gelder, Negation as failure using tight derivations for general logic programs, in J. Minker (editor), *Foundations of Deductive Databases and Logic Programming*, pp. 149–176, Morgan Kaufmann, Los Altos, 1988.

[Vardi 88] M. Vardi, A temporal fixpoint calculus, in *Proc. 13th ACM Symp. on Principles of Programming Languages*, San Diego, pp. 250–259, 1988.

[Vardi and Wolper 86] M. Y. Vardi and P. Wolper, An automata-theoretic approach to automatic program verification, *Proc. Symp. on Logic in Computer Science*, pp. 322–331, Cambridge, June 1986.

[Vardi and Wolper 88] M. Vardi and P. Wolper, *Reasoning about Infinite Computation Paths*, IBM Research Report RJ 6209 (61199), 1988.

[Vieille 86] L. Vieille, Recursive axioms in deductive databases: the query/subquery approach, in *Proceedings of the First International Conference on Expert Database Systems*, pp. 253–267, Columbia, 1986.

[Weyrauch 80] R. Weyrauch, Prolegomena to a theory of mechanized formal reasoning, *Artificial Intelligence*, Vol. 13, no 1–2, pp. 133–169, 1980.

[Wolper 82] P. Wolper, Synthesis of communicating processes from temporal logic specifications, PhD thesis, Stanford University, 1982.

[Wolper 83] P. Wolper, Temporal logic can be more expressive, *Information and Computation*, Vol. 56, nos 1–2, pp. 72–99, 1983.

[Wolper et al. 83] P. Wolper, M. Vardi and A. Sistla, Reasoning about infinite computation paths, in *Proc. 24th IEEE Symposium on Foundations of Computer Science*, pp. 185–194, Tucson, 1983.

[Wu Wen-tsun 86] Wu Wen-tsun, Basic principles of mechanical theorem proving in elementary geometries, *J. Automated Reasoning*, Vol. 2, no 3, pp. 221–252, September 1986.

[Yahya and Henschen 85] A. Yahya and L.J. Henschen, Deduction in non-Horn databases, *J. Automated Reasoning*, Vol. 1, pp. 141–160, 1985.

[Zadeh 81] L. Zadeh, PROF – a meaning representation language for natural languages, in E. Mamdani and B. Gaines (editors), *Fuzzy Reasoning and its Applications*, Academic Press, New York, 1981.

[Zadrożny 87] W. Zadrożny, A theory of default reasoning, *Proc. AAAI-87*, pp. 385–390, 1987.

Index

A

Abstraction 83
    lambda ∼ 83
Accepted
    language 173
    word 173
Accepting
    execution 173
    state 173
Accessibility relation 14, 17
Accessible 178
Adjective
    attributive ∼ 110
    predicative ∼ 110
Alethic
    logic 17
    modality 17
Algebra
    relational ∼ 285, 286, 292
Alphabet 173
Answer
    complete ∼ 340, 341, 346
    sound ∼ 339, 341, 345
Antecedent of a deduction rule 302
Arity 197
Assignment 207
    function 35, 69, 105
Assumption
    closed-world ∼ 247, 286, 295, 305, 306, 343
    domain-closure ∼ 250, 286, 295, 305
    extended generalized closed-world ∼ 261
    generalized closed-world ∼ 260
    restricted closed-world ∼ 251
    unique-name ∼ 286, 295, 305
Atom 10, 32, 198, 240
Atomic formula 10, 240
Attribute 282
Attributive adjective 110
Autoepistemic logic 238
Automaton
    Büchi ∼ 172
    eventuality ∼ 183
    local ∼ 183
    looping ∼ 195
    model ∼ 187
Axiom 12, 202
    circumscription ∼ 262, 263
    completion ∼ 296, 309
    domain-closure ∼ 297
    schema 12
    unique-name ∼ 297, 343
Axiomatic system 12, 23

B

Barcan formula 39
Base
    Herbrand ∼ 290
    predicate 303
Belief logic 18
Body
    of a clause 243
    of a deduction rule 302
Boolean semantics 160
Branching-time 27
    temporal frame 227, 228
    temporal interpretation 228
    temporal logic 227
Büchi automaton 172
    generalized ∼ 176

C

CA 296
Calculus
    domain ∼ 285, 288
    lambda ∼ 82, 84

INDEX

tuple ∼ 285
Canonical
    example 235
    Herbrand model 246
Categorial
    grammar 61, 122
    rule 126
Category
    syntactic ∼ 126
Characteristic function 72
Circumscription 262
    axiom 262, 263
    pointwise ∼ 268
    policy 276
    prioritized ∼ 268
Clausal form 242
Clause 243
    about a predicate 254
    body of a ∼ 243
    definite Horn ∼ 244
    empty ∼ 245
    extended ∼ 246
    ground ∼ 243
    head of a ∼ 243
    Horn ∼ 244
    negative Horn ∼ 245
    normal form of a ∼ 254
    solitary ∼ 256
    stratified set of ∼s 246
    unit ∼ 258
Closed-world assumption 247, 286, 295, 305, 306, 343
    axioms 296
    deductive database 319
    extended generalized ∼ 261
    generalized ∼ 260, 322, 350
        model-theoretic approach 322, 325
        proof-theoretic approach 322, 325
        properties 324
    semantic definition 322
    syntactic definition 323
    restricted ∼ 251
Closure of a formula 182
Codd's null values 335
    syntax 335
Complete
    answer 340, 341, 346
    logic 22
Completeness 12, 22
Completion
    axiom 296, 309
    formula 253, 254
    hierarchical database ∼ 308
    predicate ∼ 252, 309
    recursive rules ∼ 312
Compositionality principle 60, 65, 109
Compositional semantics 11, 119
Compound operator 231
Computation 208
Computation tree logic 230
Condition 207
    output ∼ 208
Connective
    logical ∼ 10
Consequence
    immediate ∼ mapping 313, 329
    logical ∼ 12, 67, 167
    model-theoretic ∼ relation 242
Consequent of a deduction rule 302
Consistent formula 11, 230
Constant
    function ∼ 32, 240
    global function ∼ 197
    global predicate ∼ 197
    individual ∼ 32, 240
    interpretation function for ∼ 198
    local individual ∼ 197
    logical ∼ 10, 33, 106
    non-logical ∼ 32, 64, 68, 106

372   INDEX

    predicate ~ 32, 240
    Skolem ~ 334
Constraint
    integrity ~ 284, 289, 298, 301
Construction
    referentially opaque ~ 95
Context
    non-referential ~ 95
Contingent formula 34
Control
    point 207
    state 208
Conversion
    lambda ~ 83
CWA 248, 286

D Database 283
    deductive ~ 301, 302, 303
        closed-world assumption 319
        negation 317
    definite ~ 304, 312
        query evaluation 314
    definite deductive ~ 301, 304, 349
        model-theoretic approach 305
        proof-theoretic approach 306
    extensional ~ 303
    extension of a ~ 284
    hierarchical ~
        completion 308
        query evaluation 311
    hierarchical definite deductive ~ 305, 308
    intensional ~ 303
    partial ~ 328
    relational ~ 334
        attribute 282
        domain 282
        logical approach 294
        model-theoretic approach 295, 298

        proof-theoretic approach 296, 298
        relation 282
        view 305
    schema 284
    stratified ~ 326, 327, 349
        model-theoretic approach 329
        proof-theoretic approach 328
        query evaluation 332
    with Codd's null values 336
        model-theoretic approach 338
        query evaluation 341
    with marked null values 335
        model-theoretic approach 336
        proof-theoretic approach 342
        query 337, 338
        query evaluation 339, 344
    with null values
        model-theoretic approach 336
Datalog 304
Data model 281
DCA 286, 297
Deadlock 215
De dicto reading 131
Deduction rule 302
    linear ~ 303
    recursive ~ 303
Deductive
    database 301, 302, 303, 312
        closed-world assumption 319
        negation 317
    definite ~ database 301
Default logic 238
Definite
    database 304
        query evaluation 314
    deduction rule 304
    deductive database 301, 304, 349
        model-theoretic approach 305
        proof-theoretic approach 306
Horn clause 244

INDEX

Denotation 59, 81, 102, 106, 107
Dependence 303
    direct ∼ 303
    strict ∼ 303
De re reading 132
Derivation 12
Determined logic 22
Direct dependence 303
Disagreement 345
Discourse
    representation structure 158
    representation theory 156
Domain 282
    interpretation ∼ 33, 105, 198, 241
    semantic ∼ 10
Domain-closure
    assumption 250, 286, 295, 305
    axiom 297
Domain calculus 285, 288
Dynamic
    first-order ∼ logic 42
    logic 18, 30

E

Eventualities
    realization of ∼ 182
Eventuality 182
    automaton 183
Execution
    accepting ∼ 173
Existential
    modal operator 13
    quantifier 33
Expression
    lambda ∼ 83
Extended temporal logic 192
Extension 59, 93, 96, 98, 107, 241
    of a database 284
    of a non-monotonic system 237
    of a relation 282
    operator 99, 107

Extensional
    database 303
    logic 59
    sentence 141, 146
    type-theoretic logic 61, 106

F

Fairness 213
Final state 208
Finitary
    infinite ∼ tree 227
Finite-state system 218
First-order
    dynamic logic 42
    formula 240, 241
    predicate logic 32
    temporal ∼ theory 202
Fixpoint 313
Flowchart 207
Form
    clausal ∼ 242
    functional ∼ 32, 197
    normal ∼ of a
        clause 254
        deduction rule 308
    predicate ∼ 198
Formal language 1, 4
Formula 10, 67, 198, 230
    almost universal ∼ 267
    atomic ∼ 10, 240
    Barcan ∼ 39
    closure of a ∼ 182
    completion ∼ 253, 254
    consistent ∼ 11, 230
    contingent ∼ 34
    first-order ∼ 240, 241
    inconsistent ∼ 11, 34, 230
    length of a ∼ 183
    path ∼ 227, 228, 230
    quasi-atomic ∼ 20
    separable ∼ 266
    solitary ∼ 266

state $\sim$ 227, 228, 230
true in a
  frame 15
  model 15, 41, 69, 70
universal $\sim$ 267
valid $\sim$ 11, 15, 34, 70, 167, 200, 230
Frame 14, 25
  branching-time temporal $\sim$ 227, 228
  formula true in a $\sim$ 15
  linear temporal $\sim$ 198
  temporal $\sim$ 166
Function
  assignment $\sim$ 35, 69, 105
  characteristic $\sim$ 72
  constant 32, 240
  global $\sim$ constant 197
  interpretation $\sim$ 105
    for constants 198
    for variables 199
  name 32
  temporal interpretation $\sim$ 166
  transition $\sim$ 173
Functional form 32, 197

G

GCWA 322
Generalization 34
Generalized
  Büchi automaton 176
  phrase structure grammar 162
Global function constant 197
Global predicate constant 197
Global variable 197
Grammar
  categorial $\sim$ 61, 122
  generalized phrase structure $\sim$ 162
Ground
  clause 243
  literal 241
  term 240
Guard 207

H

Head
  of a clause 243
  of a deduction rule 302
Herbrand
  base 290
  canonical $\sim$ model 246
  interpretation 246, 290
  least $\sim$ model 295, 299, 306, 313, 321, 349
  model 246
  universe 290
Hierarchical
  database
    completion 308
    query evaluation 311
  definite deductive database 305, 308
Horn clause 244
  definite $\sim$ 244
  negative $\sim$ 245
Hypothesis 12

I

Immediate consequence mapping 313, 329
Incomplete information 332, 350
Inconsistent formula 11, 34, 230
Individual
  constant 32, 240
  local $\sim$ constant 197
Inference rule 12
Infinite finitary tree 227
Information
  incomplete $\sim$ 332, 350
Initial
  node 207
  state 173
Integrity constraint 284, 289, 298, 301
Intension 59, 93, 96, 98, 107

INDEX   375

    operator 99, 107
Intensional
    database 303
    logic 59, 100
    sentence 141, 148
    type-theoretic logic 61, 93, 107,
        108
Interpretation 10, 11, 33, 241
    branching-time temporal $\sim$ 228
    domain 33, 105, 198, 241
    function 105
        for constants 198
        for variables 199
    Herbrand $\sim$ 246, 290
    relational language $\sim$ 290
    temporal $\sim$ 166, 198
        function 166
Invariance property 214
Invariant 214

K Knowledge logic 18

L Label 207
Lambda 82
    abstraction 83
    calculus 82, 84
    conversion 83
    expression 83
Language 10
    ω-regular $\sim$ 175
    accepted $\sim$ 173
    formal $\sim$ 1, 4
    multimodal $\sim$ 24
    multimodal type-theoretic $\sim$ 88
    natural $\sim$ 1, 3
    relational $\sim$ 288, 290, 292
    type-theoretic $\sim$ 75, 79
Least Herbrand model 295, 299, 306,
        313, 321, 349
Least model 260
Length of a formula 183
Linear

    deduction rule 303
    temporal frame 198
Literal 241
    ground $\sim$ 241
Liveness property 214
Local
    automaton 183
    individual constant 197
    proposition 197
Logic 21
    alethic $\sim$ 17
    autoepistemic $\sim$ 238
    belief $\sim$ 18
    branching-time temporal $\sim$ 227
    complete $\sim$ 22
    computation tree $\sim$ 230
    default $\sim$ 238
    determined $\sim$ 22
    dynamic $\sim$ 18, 30
    extended temporal $\sim$ 192
    extensional $\sim$ 59
    extensional type-theoretic $\sim$ 61,
        106
    first-order
        dynamic $\sim$ 42
        predicate $\sim$ 32
    intensional $\sim$ 59, 100
    intensional type-theoretic $\sim$ 61,
        93, 107, 108
    knowledge $\sim$ 18
    multimodal $\sim$ 24
    non-monotonic modal $\sim$ 238
    non-standard $\sim$ 9
    normal $\sim$ 22, 25
    normal temporal $\sim$ 26
    predicate modal $\sim$ 36, 38
    propositional $\sim$ 10
    propositional modal $\sim$ 13
    sound $\sim$ 22
    temporal $\sim$ 18, 25, 165
Logical

connective 10
consequence 12, 67, 167
constant 10, 33, 106
Looping automaton 195

M

Mapping
    immediate consequence ~ 313, 329
    monotonic ~ 314
Marked null values 334
    syntax 334
Meaning 57, 107
    postulate 112, 150, 152
Memory 207
    state 208
Minimal model 261, 270, 321
Mixed predicate 303
Modal
    existential ~ operator 13
    non-monotonic ~ logic 238
    predicate ~ logic 36, 38
    propositional ~ logic 13
    universal ~ operator 13
Modality
    alethic ~ 17
Model 14, 25, 35, 40, 82, 167, 241, 290, 336, 349
    automaton 187
    canonical Herbrand ~ 246
    data ~ 281
    formula true in a ~ 15, 41, 69, 70
    Herbrand ~ 246
        least ~ 295, 299, 306, 313, 321, 349
    intersection property 260, 305
    least ~ 260
    minimal ~ 261, 270, 321
    preferred ~ 268, 270
    relational ~ 281
Model-theoretic approach 349

database
    definite deductive ~ 305
    relational ~ 295, 298
    stratified ~ 329
    with Codd's null values 338
    with marked null values 336
    with null values 336
    generalized closed-world assumption 322, 325
Model-theoretic consequence relation 242
Modus Ponens 13, 170
Monotonic mapping 314
Montague's semantics 115
Multimodal
    language 24
    logic 24
    type-theoretic ~ language 88
Mutual
    exclusion 208
    recursion 303, 327

N

Name
    function ~ 32
    predicate ~ 32
Natural language 1, 3
Necessitation rule 22, 170
Negation
    as failure 258
    deductive database 317
Node
    initial ~ 207
Non-logical constant 32, 64, 68, 106
Non-monotonicity property 237
Non-monotonic modal logic 238
Non-referential context 95
Non-standard logic 9
Normal
    form of a clause 254
    logic 22, 25
    temporal logic 26

INDEX

Null values
    Codd's ~ 335
        model-theoretic approach 338
        query evaluation 341
        syntax of ~ 335
    comparison between the model- and proof-theoretic approaches 348
    marked ~ 334
        model-theoretic approach 336
        proof-theoretic approach 342
        query evaluation 339, 344
        syntax of ~ 334
    model-theoretic approach 336
    query evaluation 339, 348
    syntax 334

Omega-regular language 175
Operator
    compound ~ 231
    existential modal ~ 13
    extension ~ 99, 107
    intension ~ 99, 107
    temporal ~ 197
    universal modal ~ 13
Output condition 208

Partial
    database 328
    theory 328
Partly synchronized product 219
Path 229
    formula 227, 228, 230
Pluri-extensionality property 237
Policy
    circumscription ~ 276
Polysemy 4
Possible-world semantics 16
Postulate
    meaning ~ 112, 150, 152
Predicate 32
    base ~ 303
    completion 252, 309
    constant 32, 240
    first-order ~ logic 32
    form 198
    global ~ constant 197
    mixed ~ 303
    modal logic 36, 38
    mutually recursive ~ 303
    name 32
    recursive ~ 303
    relation ~ 289, 292
    virtual ~ 302, 303
Predicative adjective 110
Preferred model 268, 270
Preorder relation 269
Principle
    compositionality ~ 60, 65, 109
Product
    partly synchronized ~ 219
Program variable 207
Projection rule 57, 124
Proof 12
Proof-theoretic
    relation 242
    system 242
Proof-theoretic approach 349
    database
        definite deductive ~ 306
        relational ~ 296, 298
        stratified ~ 328
        with marked null values 342
    generalized closed-world assumption 322, 325
Property 97, 140
    invariance ~ 214
    liveness ~ 214
    model intersection ~ 260, 305
    non-monotonicity ~ 237
    pluri-extensionality ~ 237
    safety ~ 214

Q

Proposition 6, 10, 97, 103
    local ~ 197
Propositional
    logic 10
    modal logic 13
    vocabulary 10
Quantifier
    existential ~ 33
    universal ~ 33
Quasi-atomic formula 20
Query 285, 289, 291, 298, 311, 314, 332, 337, 338, 339, 341, 344, 348
    yes–no ~ 289, 294, 295, 298, 344
Query evaluation
    database
        definite ~ 314
        hierarchical ~ 311
        stratified ~ 332
        with Codd's null values 341
        with marked null values 339, 344
        with null values 348

R

Rationality
    logical component of the ~ 273
    subjective component of the ~ 275
Realization of eventualities 182
Reasoning
    ideally rational ~ 242
Recursion
    mutual ~ 303, 327
Recursive
    deduction rule 303
    predicate 303
    rule 312
        completion 312
    semantics 119
Reference 93, 96, 107

Referentially opaque construction 95
Relation 241, 282
    accessibility ~ 14, 17
    extension of a ~ 282
    model-theoretic consequence ~ 242
    predicate 289, 292
    preorder ~ 269
    proof-theoretic ~ 242
    schema of a ~ 283
    value of a ~ 282
Relational
    algebra 285, 286, 292
    database 334
        attribute 282
        domain 282
        logical approach 294
        model-theoretic approach 295, 298
        proof-theoretic approach 296, 298
        relation 282
        view 305
    language 288, 290, 292
    model 281
Representation structure
    discourse ~ 158
Representation theory
    discourse ~ 156
Rule
    categorial ~ 126
    deduction ~ 302
        antecedent of a ~ 302
        body of a ~ 302
        consequent of a ~ 302
        definite ~ 304
        head of a ~ 302
        linear ~ 303
        normal form of a ~ 308
        recursive ~ 303
    inference ~ 12

INDEX

projection ∼ 57, 124
recursive ∼ 312
  completion 312

S

Safety property 214
Satisfaction 58, 70
Scheduler 205
Schema
  axiom ∼ 12
  of a database 284
  of a relation 283
Semantic
  domain 10
  value 105, 107
Semantics 10
  Boolean ∼ 160
  compositional ∼ 11, 119
  Montague's ∼ 115
  possible-world ∼ 16
  recursive ∼ 119
Sense 93, 96, 107
Sentence 67, 126
  extensional ∼ 141, 146
  intensional ∼ 141, 148
Separable formula 266
Signature 202
Skolem constant 334
Skolemization 334
Solitary
  clause 256
  formula 266
Sound
  answer 339, 341, 345
  logic 22
Soundness 12, 22
State 166, 173
  accepting ∼ 173
  control ∼ 208
  final ∼ 208
  formula 227, 228, 230
  initial ∼ 173

memory ∼ 208
system ∼ 208
Statement 207
Stratification 327
  of a database 326
Stratified database 326, 327, 349
  model-theoretic approach 329
  proof-theoretic approach 328
  query evaluation 332
Stratum 327
Strict dependence 303
Subimplication 261
Substitutable term 200
Syntactic category 126
Syntax 10
System
  axiomatic ∼ 12, 23
  finite-state 218
  state 208

T

Tautology 11, 21
Temporal
  branching-time ∼
    frame 227, 228
    interpretation 228
    logic 227
  extended ∼ logic 192
  first-order theory 202
  frame 166
  interpretation 166, 198
  function 166
  linear ∼ frame 198
  logic 18, 25, 165
  normal ∼ logic 26
  operator 197
Term 32, 197, 240
  ground ∼ 240
  substitutable ∼ 200
Theorem 12, 21
Theory 21, 242
  complete ∼ 242
  first-order temporal ∼ 202

partial ~ 328
Time
    branching-~ 27
Totality 227
Transition 207
    function 173
Translation 137, 140
Tree
    infinite finitary ~ 227
True
    formula ~
        in a frame 15
        in a model 15, 41, 69, 70
Truth 57
Tuple calculus 285
Type 75, 76, 106
Type-theoretic
    extensional ~ logic 61, 106
    intensional ~ logic 61, 93, 107, 108
    language 75, 79
    multimodal language 88

U

UNA 286, 297
Unique-name
    assumption 286, 295, 305
    axiom 297, 343
Universal
    formula 267
    modal operator 13
    quantifier 33
Universe 17
    Herbrand ~ 290
Update 284

V

Valid formula 11, 15, 34, 41, 70, 167, 200, 230
Valuation 20, 241
Value
    of a relation 282
    semantic ~ 105, 107

Variable 32, 240
    global ~ 197
    interpretation function for ~ 199
    program ~ 207
View 305
Virtual predicate 302, 303
Vocabulary
    propositional ~ 10

W

Word
    accepted ~ 173

Y

Yes–no
    query 289, 294, 295, 298, 344